Calculus for
ENGINEERING TECHNOLOGY

Calculus for

ENGINEERING TECHNOLOGY

WALTER R. BLAKELEY

RYERSON POLYTECHNICAL INSTITUTE

TORONTO, CANADA

JOHN WILEY & SONS

New York • Chichester • Brisbane • Toronto • Singapore

Library of Congress Catalog Card Number: 67–29017
Printed in the United States of America
20   19   18   17   16   15   14   13   12

ISBN 0 471 07931 6

Preface

This book has been prompted by the scarcity of calculus texts specifically designed for the engineering technologist.

It is generally agreed that a knowledge of calculus is necessary for a full understanding of numerous aspects of engineering. Yet many texts in specific fields of engineering avoid its use, often going to great lengths in attempts at cumbersome methods of explaining the principles involved. The assumption is that calculus is beyond the grasp of the engineering technologist.

Equally ineffective is the method of presenting a traditional course in calculus, establishing rigorous and general proofs with no attempt to relate them to the reality of engineering practice. For the technologist, the natural development is from the particular and familiar problem to the general and basic concept.

It is my contention that a good foundation in calculus may readily be achieved by an intuitive approach in the initial stages. With this approach, many difficult facets of engineering become clear, and calculus may quickly be used as a tool in the engineering subjects. The power and consistency of mathematics gradually emerge as the reasoning is directed toward more general truths. The final sophisticated modes then acquire much more meaning than if they were thrust upon the student from the beginning.

This book is therefore presented with the aim of making calculus interesting and understandable, of effecting its rapid use in engineering subjects, and of finally reaching an appreciation of its broad scope.

With these aims in view, rigorous proofs have been avoided when possible. Techniques applying to the vast majority of cases are first established, and an exhaustive treatment of the exceptional case is omitted. Illustrative examples are used frequently as the mathematical reasoning progresses. In later chapters, these examples are drawn mainly from the electrical and the mechanical fields. However, the principles throughout should be general enough for any of the engineering branches.

I have tried to stress that there is seldom only one method of solving a problem. The student is encouraged to use two or three different methods. The fact that each path leads to the same inevitable point is an interesting demonstration of the consistency of mathematics. No dogmatic patterns can replace a certain amount of ingenuity.

Each chapter has several sets of exercises, and each of these progresses from simple to more involved problems. There should be enough variety in the exercises to keep the student alert to the individuality of each problem. Often problems that appear similar have entirely different solutions. Thus it is hoped that exercises will not be stereotyped repetitions of one technique. Later problems in the exercises will require considerable preliminary analysis. The solution of a mathematical problem provides a concise means of analyzing the main structure of a situation and only later carrying out the details within the substructure.

The order of chapters may be varied to some extent, especially in the separate branches of calculus following Chapter 13.

Prerequisites include basic algebra and trigonometry, as well as elementary plotting and complex number theory.

I wish to acknowledge the helpful assistance of Miss Sophie Sieczka, Mr. Geoffrey Boyes, and Mr. George Shiels in the preparation of the text.

Although I hope that this book will be of assistance to mathematics instructors, it is directed primarily toward the student. If the student finds it useful, it has accomplished its purpose.

Walter R. Blakeley
Toronto
November 1967.

Contents

CHAPTER 10

CHAPTER 11

CHAPTER 12

CHAPTER 13

Calculus for
ENGINEERING TECHNOLOGY

CHAPTER 1

Functions

1.1 VARIABLES AND CONSTANTS

The branch of mathematics known as calculus is concerned primarily with the phenomenon of change or variation in scientific or engineering quantities. The subject of variation must be first investigated before it is possible to see how calculus is applied to it.

Many examples of how one quantity varies with another may easily be brought to mind. One simple example is the variation of temperature with the time of day, where, during a typical day, the temperature may reach a peak value at about 3 P.M. and a low value at about 4 A.M. This variation can be seen more clearly when a graph of temperature against time is drawn as in Fig. 1-1. Here, hourly readings of temperature have been taken, and each point on the graph representing a recorded reading is joined to the next point by a straight line. Separate recorded points on a graph are often joined by a straight line in instances where the variation in the interval between the points cannot be predicted by a simple law.

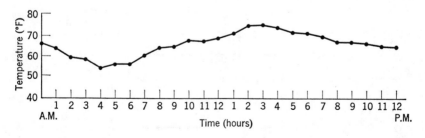

Fig. 1-1. Variation of temperature with time.

An example of a variation that can be predicted by a simple law is the variation of current with a varying resistance in a d-c circuit with a constant voltage (see Fig. 1-2).

Here, Ohm's law predicts the changing conditions, and thus $I = E/R$. According to the statement of the example, we see that the quantity that is originally changed from time to time is the resistance R. A rheostat could be used to control this variation. The voltage E remains constant. For each new setting of R, a new value of current I will flow. Thus, in general, I depends on R; R is originally varied at will and the value of I changes as a result of this original change in R. Mathematically, I is said to be a *function* of R. R, the quantity originally changed, is called an *independent variable*. I, the quantity changing as a result of the change in R, is called a *dependent variable*, since its value depends on the value of R. In the graph of this variation, the horizontal axis is used to scale off the independent variable and the vertical axis is used to scale off the dependent variable (Fig. 1-3).

In Fig. 1-3, the graph is a smooth curve since, if a continuous increase in R from a low value to a high value can be achieved, the current I will change smoothly and continuously in the manner indicated in the graph.

It is not necessary here to define rigidly the terms used in the previous discussion. However, the student should have a clear conception of their application. A quantity very seldom changes of its own accord; instead, its variation depends on a more independent variable. The independent variable of one problem may itself be controlled by changes in some more basic quantity; thus the terms *independent* and *dependent* are relative within the confines of the problem.

In any particular problem, then, it is usually possible to decide which of the variables is independent and which is dependent. The dependent variable is said to be a *function* of the independent variable. The word function covers many kinds of expression involving a variable; in this chapter the simpler types of functions will be studied.

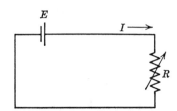

Fig. 1-2. A resistive circuit with constant voltage.

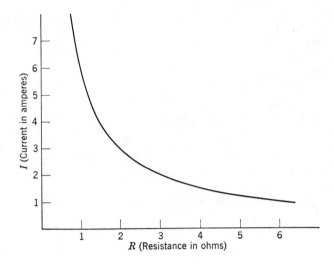

Fig. 1-3. Variation of current with resistance in a 6-volt circuit.

When one quantity is known to vary with another, but the law by which it varies is not specified, an abbreviated symbol is often used to represent the relationship. Thus, in our electrical example, I is a function of R, which may be written symbolically as $I = f(R)$. The symbol $f(R)$, however, may include many other expressions involving R in addition to the expression E/R. Using more general symbols to represent the variables, if y depends on x for its value, although we do not know or do not wish to specify the exact dependence, we may simply write $y = f(x)$. Other letters as well as f may be used for this symbol, the most common ones being ϕ, F, g, and G. If y is a function of x, and x is itself some different function of w, the relationships might be expressed symbolically as $y = f(x)$ and $x = \phi(w)$, the change of symbol indicating a different function. The same symbol, when used more than once in a particular problem, indicates the same function. Hence, if $f(x) = 4 - x^3$, $f(x - 1) = 4 - (x - 1)^3$; and $f(-2) = 4 - (-2)^3$. This symbolic method of indicating that one variable is a function of another is referred to as *functional notation.*

In most scientific laws representing the variation of one quantity with another, other quantities are involved that do not vary, which are called *constants.* Thus, in our stated electrical example, E is constant. Of course, the example might have been stated differently. If the voltage is varied and the resistance always remains at 2 ohms, I is a function of E, R being constant. This gives the simpler straight-line graph of Fig. 1-4.

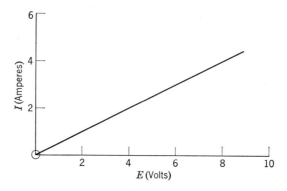

Fig. 1-4. Variation of I with E in a 2-ohm circuit.

If both E and R are varied at will, there would be two independent variables, and I would depend on both E and R. This type of variation could not be completely represented by a two-dimensional graph. At present, the discussion will be confined to functions of one independent variable.

In purely mathematical texts, the letter x is usually used to represent the independent variable, and the letter y to represent the dependent variable. This convention is used merely to generalize. The student of science or engineering must not become too stereotyped in this convention, however, since more specific symbols are usually used to represent the variables. This text will use x and y where generalization is necessary.

In the x-y system, unless the letters represent specific quantities, it may not be clear which letter represents the independent variable. Thus, if $y = x^2 + 2$, it is also true that $x = \sqrt{y - 2}$. Usually, in this system, y is considered to be the dependent variable, especially if y is shown as equal to a function of x.

EXAMPLE 1.   For the equation $y = 3 - 2/\sqrt{4 - x}$, express x as a function of y.

Solution: x must be isolated from the other parts of the equation. In preparation for this, the fraction involving x is first isolated. $2/\sqrt{4 - x} = 3 - y$. Squaring both sides,

$$\frac{4}{4 - x} = 9 - 6y + y^2.$$

$$(4 - x)(9 - 6y + y^2) = 4.$$

$$4 - x = \frac{4}{9 - 6y + y^2}.$$

$$x = 4 - \frac{4}{9 - 6y + y^2}.$$

This answer is satisfactory, although the right-hand side may be reduced to a common denominator, resulting in the equation

$$x = \frac{32 - 24y + 4y^2}{9 - 6y + y^2}.$$

EXERCISES

For the following statements, specify the independent variable, the dependent variable, and any constants involved. Put the statement into functional notation, using the first letter of the quantity involved or a standard letter symbol to represent that quantity.

1. The heat content of a quantity of water of constant volume depends on its temperature.

2. The resistance of a conductor of fixed length and cross-sectional area varies with its temperature.

3. The resistance of a conductor of fixed length and at constant temperature varies with its cross-sectional area.

4. When the amount of capacitance is changed in a circuit with a fixed a-c voltage source, the current in the circuit is affected.

5. If $f(x) = 2x - 1$ and $F(x) = 3x^2$, calculate the following:

(a) $f(x) - 2F(x)$.   (b) $f(x^2)$.

(c) $F(x - 2)$.   (d) $f(-3)$.

(e) $f[F(x)]$.   (f) $F[f(x)]$.

Express the following equations using the letter on the right-hand side as the dependent variable, and expressing it as a function of the letter originally on the left-hand side:

6. $y = 3x^2 - 5$.

7. $E = 4\sqrt{9 + X^2}$.

8. $I = 3/\sqrt{4 + x} - 2$.

9. $v = 3\sqrt{4 + (5 - X)^2}$.

10. $y = x^2 - 2x$.

1.2   THE POWER FUNCTION

There are many types of functions of an independent variable. In the example of Ohm's law, $I = E/R$, if R is kept constant and E varies, I may be said to *vary directly* with E, or I is said to be *directly proportional* to E. If E is doubled, so is I; or, in general, by whatever factor E is changed, I changes by the same factor. For *direct variation*, the dependent variable is always some constant times the independent variable. Here $I = (1/R)E$, $1/R$ being the constant multiplier. Direct variation will always produce a straight-line graph through the origin, as shown in Fig. 1-4. If y varies directly with x, $y = kx$, or the ratio $y/x = k$, where k is a constant.

If, on the other hand, R is varied and E remains constant, then, in the equation $I = E/R$, I *varies inversely* with R, or I is said to be *inversely proportional* to R. If R is doubled, I is halved, or, in general, by whatever factor R is changed, I changes by the reciprocal of that factor. For inverse variation, the dependent variable is always some constant times the reciprocal of the independent variable. Here, $I = E(1/R)$, E being the constant multiplier. *Inverse variation* will always produce a curved graph known as a rectangular hyperbola (see Fig. 1-3). If y varies inversely with x, $y = k/x$, or $xy = k$, where k is a constant.

EXAMPLE 2. y varies inversely with x, and, when $x = 2$, $y = 5$. Find y as an exact function of x, and sketch the graph for positive values of x. *Solution:* It may be stated that $y = k/x$, but this is not an exact equation, since the constant value of k is not yet known. k may be determined, however, by using the additional information in the problem. Since $y = k/x$ and $y = 5$ when $x = 2$, these specific simultaneous values of the variables satisfy the equation.

$$\therefore \ 5 = \frac{k}{2} \quad \text{and} \quad k = 10.$$

Thus the exact equation is $y = 10/x$. The graph is plotted by finding the values of y resulting from several values of x, independently chosen. The tabulation of several simultaneous values and the resulting graph are shown in Fig. 1-5.

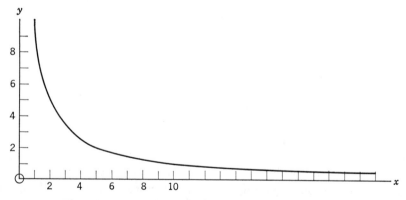

Fig. 1-5. Graph of $y = 10/x$ for positive values of x.

Both direct and inverse variation are particular examples of the power function. The *power function* may be defined as a function in which the independent variable is raised to any constant power. Thus $I = E/R$ is a power function of E of the first power or a power function of R of the minus first power, depending on whether E or R is cited as the independent variable. In general, the simple power function may be expressed as $y = Cx^k$, where C and k are positive or negative constants. k may also be zero; in this instance $y = Cx^0 = C$, a constant term.

A power function in a broader sense may have several terms in which the independent variable is raised to a constant power or simply a constant term. Thus $y = x^3 - 3x^2 + 2x - 4$ may be called a power function of x.

If powers of the independent variable are all positive whole numbers, some more explicit term may be used to describe the function. For example, $y = 3x^2 - 2x$ is a *quadratic function*; $y = 3x - 5x^3$ is a *cubic function*; $y = 5x^6 - 3$ could be called a power function of the sixth power, etc. Here the highest power occurring is taken to indicate the power of the function.

The graph of any power function may be sketched simply by choosing several arbitrary values of the independent variable and, by substitution in the equation, finding the value(s) of the dependent variable corresponding to each. When sufficient points on the graph are plotted, a smooth curve through these points can be sketched.

EXAMPLE 3.   Sketch the graph of $y = x^3 - 2x^2 - 1$. The solution is indicated by the tabulation and graph of Fig. 1-6.

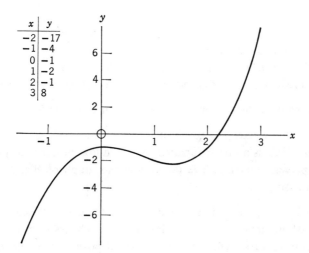

x	y
-2	-17
-1	-4
0	-1
1	-2
2	-1
3	8

Fig. 1-6. Graph of $y = x^3 - 2x^2 - 1$.

In Fig. 1-6, note that the scale of the *y*-axis has been contracted so that one unit on the *y*-axis covers much less distance than one unit on the *x*-axis. This is usual practice when space requirements make it necessary.

It will be found by sketching several of these graphs that the points arrange themselves to form a smooth curve, or, in some cases, a straight line. The straight line will always result from a case where both variables occur only to the first power with a possible additional constant term. Thus, $y = 2x$ (direct variation) has a graph which is a straight line through the origin; $y = 2x + 3$ is the same graph raised upwards three units, and thus it does not pass through the origin. Any function in which variables occur only to the first power, with the possibility of an extra constant term, is called a *linear function*, since its graph is a straight line; this type of variation is referred to as *linear variation*. Direct variation is thus a particular case of linear variation, where these is no extra constant term.

EXAMPLE 4.   *y* is a linear function of *x*, and the graph passes through the points $(-2,1)$ and $(2,3)$. (a) Find *y* as an exact function of *x*. (b) Find the value of *y* when $x = -4$.

Solution: Using letters to represent unknown constants, the function may be represented as $y = mx + b$, where *m* and *b* are to be determined. Substituting the pair of simultaneous values in this equation,

$$1 = -2m + b,$$

and

$$3 = 2m + b.$$

A simple algebraic solution of these equations gives $b = 2$; $m = \frac{1}{2}$.
(a) The function is therefore $y = \frac{1}{2}x + 2$.
(b) Substituting $x = -4$, $y = -2 + 2 = 0$.
The student should sketch this graph to be assured that it is linear.

A special case of some interest is examined in Example 5.

EXAMPLE 5.   Sketch the graph of $y = x + 1/x$.

Solution: This is a power function, with the first power and the "minus first" power occurring. The tabulation and graph of Fig. 1-7 indicate the solution.

In Fig. 1-7, the graph consists of two separate curves. When *x* is a very small negative number, with an absolute value much less than 1, *y* is a very large negative number. When *x* is a very small positive number, much less than 1, *y* is a very large positive number. For these extremely small values of *x*,

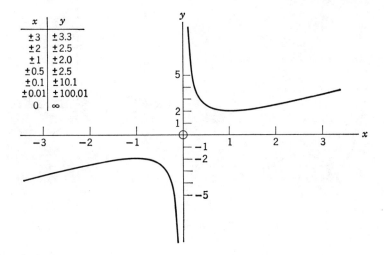

x	y
±3	±3.3
±2	±2.5
±1	±2.0
±0.5	±2.5
±0.1	±10.1
±0.01	±100.01
0	∞

Fig. 1-7. Graph of $y = x + 1/x$, a discontinuous function.

the corresponding values of y are too far down or up to plot on any reasonable scale. Using mathematical terminology, we say that, as x approaches zero for negative values of x, y approaches an infinitely negative number, whose value is less than any negative number one may choose; and when a positive x approaches zero, y approaches an infinitely large positive number, larger than any number one may choose. The symbol for an infinitely large number is ∞, called *infinity*. As we proceed from left to right on this graph, there is a sudden jump at $x = 0$, where y changes radically. This is an example of a *discontinuous function*, with a *discontinuity* at $x = 0$.

If y is a power function of x in which only even positive or negative powers occur (with the possible inclusion of a constant term), then any positive value of x will give the same value of y as the corresponding negative value of x. For example, if $y = x^4 - 3x^2 + 2$, at $x = 2$, $y = 6$; and at $x = -2$, $y = 6$ again. The tabulation of values of x can be abbreviated by taking $x = \pm 1$, $x = \pm 2$, etc., in pairs. This will result in a graph that is symmetrical about the y-axis, as shown in Fig. 1-8a. This type of function is called an *even power function* and the symmetry about the y-axis is referred to as *even symmetry*.

If only odd positive or negative powers occur in the function, then every substitution of a negative value for x will give a value of y opposite in sign and equal in absolute value to the value obtained by substituting the corresponding positive value. Thus, if $y = x^3 - 6x$, when $x = 2$, $y = -4$; and when $x = -2$, $y = +4$, etc. This type of function, called an *odd power*

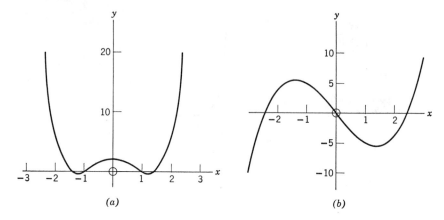

Fig. 1-8. Odd and even symmetry. (*a*) $y = x^4 - 3x^2 + 2$. (*b*) $y = x^3 - 6x$.

function, has *odd symmetry* (see Fig. 1-8*b*). A constant term such as $+2$ may be considered as $2x^0$, and a zero power is considered an even power.

The graphs of Fig. 1-9 illustrate commonly occurring power functions.

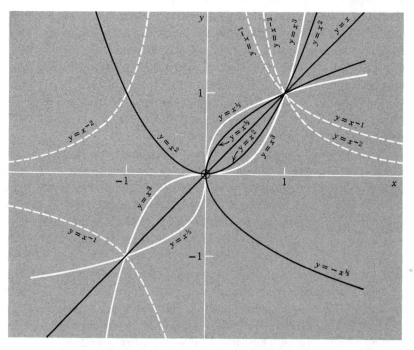

Fig. 1-9. Graphs of simple power functions.

The power function is only one of many types of functions. Examples of other types are $y = \sin x$; $y = 10^{2x}$; $y = \log x^2$. These other types of functions, known broadly as *transcendental functions*, are discussed in Chapter 9.

EXERCISES

1. Classify the following functions by as many of the following terms as apply: power function, direct variation, inverse variation, linear variation, quadratic function, cubic function, odd symmetry, and even symmetry.

(a) $y = x^4 - 3x^2$.   (b) $y = 7x$.   (c) $y = 3 - 5x$.

(d) $y = f(x) = f(-x)$.   (e) $v = 20t - 16t^2$.   (f) $y = 5/x^3 + 2x$.

(g) $f(-3) = -f(3)$.   (h) $I = E/X$, where E is constant.

(i) $i = 6(1 - 10^{5t})$.   (j) $i = 0.5t^2 - 1$.

Sketch graphs of the following functions, and classify them according to the terms in Exercise 1:

2. $y = 3x - 2$.

3. $y = (x^2 - 4)/2$, between $x = -4$ and $x = 4$.

4. $y = x^5 - 4x^3 + 2x$, between $x = -3$ and $x = 3$.

5. $i = 0.5t^2 - t$, between $t = -4$ and $t = 4$.

6. $v = t + 1/t^2$, between $t = -3$ and $t = 3$.

7. $v = t - 1/t$, between $t = -3$ and $t = 3$.

8. $I = 6/\sqrt{R^2 + 3}$ for positive values of both variables.

9. I varies directly with E, and when E is 20, I is 8. Find (a) I as an exact function of E and draw the graph and (b) the value of E to give an I of 3.

10. y varies directly with x. Complete the following tabulation of simultaneous values of x and y:

x	-1	0	1	2	3		
y				9		20	27

11. P varies inversely with V, and when $V = 50$, $P = 8$. Find (a) P when $V = 30$ and (b) V when $P = 30$.

12. y varies inversely with x. Complete the following tabulation of simultaneous values:

x	-1	0	1	2	3	
y				9		15

13. v is a linear function of t. When $t = 0$, $v = 20$ meters per second; and when $t = 3$, $v = 8$ meters per second. Find (a) v as an exact function of t and (b) v when $t = 5$.

14. y is a linear function of t. Complete the following tabulation of simultaneous values:

t	-1	0	1	2	3	4	
y				7		11	14

15. y is a quadratic function of x with even symmetry. Two points on the graph of the function are $(0, -3)$ and $(-2, 9)$. Find (a) y as an exact function of x, (b) y when $x = \pm 3$, and (c) x when $y = 6$.

16. q is a cubic function of t with odd symmetry.

(a) If $q = -10$ when $t = 2$, what is the value of q when $t = -2$?

(b) In addition to the simultaneous values of Part a, $q = 1$ when $t = 1$. Find q as an exact function of t.

(c) Find q when $t = 0.5$.

17. The weight of a body on or above the earth's surface varies inversely with the square of its distance from the earth's center. A man weighs 165 lb at the earth's surface. Assuming the earth's radius to be 3960 miles, (a) find a formula to express this man's weight w lb as a function of his height h *above the earth's surface*.

(b) What are the units of the constant in the numerator of this function of h?

18. If $I = E/R$, and $1/R = 1/r + \frac{1}{3}$, express I as a function of r.

19. If $P = i^2 r$ and i varies with r such that $4(i + ir/2) + ir = 6$, express P as (a) a function of i only and (b) a function of r only.

20. A man walks from one corner of a rectangular room of dimensions 18 ft by 24 ft on a diagonal toward the far corner at a steady rate of 3 feet per second. Express his distance from the 18 ft wall he is approaching (measured perpendicularly) as a function of time, measuring time from the start of his motion.

CHAPTER 2

Basic Analytic Geometry

2.1 SLOPE OF A LINE

Calculus is concerned with the change occurring in a dependent variable as a result of some variation in the independent variable for any type of function. These changes are always most clearly depicted by means of a graph of the function. Analytic geometry deals with the geometry of graphs and figures plotted in Cartesian coordinates, in which methods of proof are based mainly on principles of algebra. Since the fundamental ideas of calculus are derived from analytic geometry, it is necessary before proceeding farther to establish some essential principles of analytic geometry.

The *slope* of a straight line is concerned with the direction of the line. The *slope* of a straight line is defined as the ratio of rise to run between any two points on the line, which equals the tangent of the angle formed between the line and the positive horizontal direction. The slope of AB (Fig. 2-1) $= CB/AC = \tan \theta_1$. The slope of $DE = FE/DF = \tan \theta_2$; in this case, FE is a drop, or a negative rise, and DF is a positive run, so that the slope is negative —a fact that is confirmed if the slope is calculated as $\tan \theta_2$, a negative quantity since θ_2 is negative.

In proceeding from one point to another on a graph in the direction of increasing x (from left to right), the value of y may increase, as in AB, or decrease, as in DE. The slope in the second instance will always be negative.

The slope of the straight line between P_1, with coordinates (x_1, y_1), and P_2, with coordinates (x_2, y_2) is equal to $(y_2 - y_1)/(x_2 - x_1)$, which may be interpreted as the change in y divided by the change in x. This expression for the slope will always apply, even where some or all of the coordinates are negative, provided that correct algebraic procedure is followed in dealing with negative

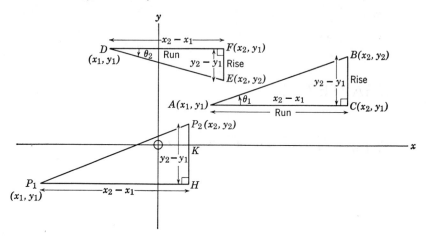

Fig. 2-1. Slope of a line.

numbers. In Fig. 2-1, both coordinates of the point P_1 are negative; for the line P_1P_2, the rise is $KP_2 + KH = y_2 - y_1$, since y_1 itself is negative; similarly, the run is $P_1H = x_2 - x_1$, and the slope is $(y_2 - y_1)/(x_2 - x_1)$.

EXAMPLE 1.   Find the slope of the line joining the following pairs of points:

(a) $(-5,2)$ and $(3,-1)$.
(b) $(2,-5)$ and $(6,-5)$.
(c) $(2,-5)$ and $(2,1)$.

Solution:

(a) Slope $= \dfrac{-1 - 2}{3 - (-5)} = -\dfrac{3}{8}.$

(b) Slope $= \dfrac{-5 - (-5)}{6 - 2} = \dfrac{0}{4} = 0$ (a horizontal line).

(c) Slope $= \dfrac{1 - (-5)}{2 - 2} = \dfrac{6}{0} =$ an infinite number.

This is a vertical line, where the run is zero.

2.2   THE ANGLE BETWEEN TWO STRAIGHT LINES

Two parallel lines will, of course, have the same slope; hence the condition for two lines to be parallel is that their slopes are equal.

Two lines perpendicular to one another also have slopes that are related by a definite law. Consider the two perpendicular lines AB and AC of Fig. 2-2. The relation between their slopes may easily be established if equal distances AB and AC are taken along the two lines and perpendiculars BD and CE constructed. The slope of $AB = DB/AD$, the rise and the run both being positive. The slope of $AC = CE/EA$, and taking EA as a positive run from left to right, CE is a negative rise (that is, a drop). It is a simple matter to establish that $\triangle ABD$ is congruent to $\triangle ACE$; therefore the slope of $AC = CE/EA = -AD/DB$, which is the negative reciprocal of the slope of AB.

If two lines are perpendicular, the slope of one line is the negative reciprocal of the slope of the other. $(m_1 = -1/m_2,$ and $m_2 = -1/m_1.)$

> EXAMPLE 2.   Find the slope of a straight line perpendicular to the straight line through the points $(3, -2)$ and $(-1, 0)$.
> *Solution:* The slope of the line through the two given points $= (0 + 2)/(-1 - 3) = -\frac{1}{2}$. Therefore the slope of the perpendicular line $= +1/\frac{1}{2} = +2$.

The angle between any two straight lines may be determined if the slopes of the lines are known. To determine the angle between two straight lines, we may either work from basic principles (see Example 3), or use a formula, which will be developed after we consider the example.

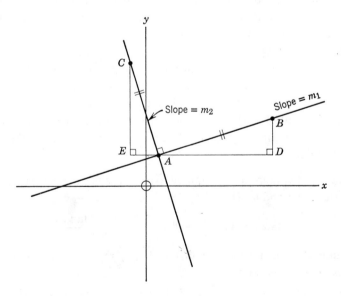

Fig. 2-2. Relation between the slopes of perpendicular lines.

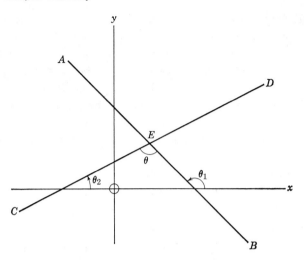

Fig. 2-3. The angle of intersection of two lines of slope -1 and 0.5, respectively.

EXAMPLE 3.   Find the angle of intersection of two straight lines with slopes of -1 and $+0.5$, respectively.

Solution: In Figure 2-3, line AB has a slope of -1 and line CD has a slope of $+0.5$, and it is required to find the angle of intersection θ. Note that either the angle CEB or the angle BED may be taken as the angle of intersection of the two lines, and that these two alternate answers to the question are supplementary angles. To find the angle θ, it is seen that $\theta_1 = \theta + \theta_2$, since an exterior angle of any triangle is equal to the sum of the two interior and opposite angles of the triangle. Hence $\theta = \theta_1 - \theta_2$. Since the slope of $AB = -1$, $\tan \theta_1 = -1$, and $\theta_1 = 135°$; similarly $\tan \theta_2 = 0.5$, and $\theta_2 = 26.6°$. Therefore $\theta = 135° - 26.6° = 108.4°$. The angle of intersection may be taken as either $108.4°$ or $71.6°$, its supplement, the first being angle CEB and the second, angle BED.

More generally, suppose it is required to find the angle of intersection of two lines of slopes m_1 and m_2 (Fig. 2-4). Here also $\theta = \theta_1 - \theta_2$.

$$\therefore \ \tan \theta = \tan (\theta_1 - \theta_2) = \frac{\tan \theta_1 - \tan \theta_2}{1 + \tan \theta_1 \tan \theta_2},$$

by a familiar formula. Since $\tan \theta_1$ and $\tan \theta_2$ represent the slopes of the two lines, given as m_1 and m_2,

$$\boldsymbol{\tan \theta} = \frac{m_1 - m_2}{1 + m_1 m_2},$$

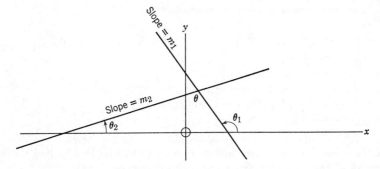

Fig. 2-4. The angle of intersection of two lines of slopes m_1 and m_2, respectively.

and θ may then be found from tangent tables or the tangent scale of the slide rule.

The condition for perpendicular lines, given previously, is a special case of this more general formula, since, for perpendicular lines, $\theta = 90°$ and $\tan \theta = \infty$; thus the denominator of the fraction $(m_1 - m_2)/(1 + m_1 m_2)$ is zero; $1 + m_1 m_2 = 0$, and $m_1 = -1/m_2$. The slopes of two perpendicular lines are therefore negative reciprocals, as previously stated.

EXAMPLE 4. By using the tangent formula developed above, find the angle of intersection (a) of the lines of Example 3 and (b) of two straight lines of slopes 1.5 and -1.5, respectively.
Solution: (a) $m_1 = -1$ and $m_2 = +0.5$;

$$\therefore \tan \theta = \frac{-1 - 0.5}{1 + (-1)(0.5)} = \frac{-1.5}{0.5} = -3.$$

θ must be less than $180°$. Thus $\theta = 108.4°$. If m_1 were taken as 0.5 and m_2 as -1, θ would be found to be $71.6°$, the supplement of $108.4°$, and one of two satisfactory answers, as discussed previously. Thus m_1 and m_2 are interchangeable.

$$\text{(b)} \tan \theta = \frac{1.5 - (-1.5)}{1 + (1.5)(-1.5)} = \frac{3}{-1.25} = -2.4.$$
$$\therefore \theta = 112.6°.$$

Hence the two lines intersect at angles of $112.6°$ and $67.4°$.

EXERCISES

1. Plot the following points and find the slope of the straight line joining them.
   (a) (1,2) and (3,5).   (b) (0,0) and (5, −2).   (c) (0, −8) and (0,5).
   (d) (−7,0) and (−2,0).   (e) (−1,1) and (−1, −3).

2. Find the slope of the line perpendicular to:

    (a) The y-axis.   (b) The x-axis.

    (c) A line of slope -4.   (d) A line of slope 0.75.

    (e) A line of slope -1.

3. Find the slope of a line which is

    (a) Parallel to the line joining $(3, -2)$ and $(7,0)$.

    (b) Perpendicular to the line joining $(0,0)$ and $(3,3)$.

    (c) Perpendicular to the line joining $(-2,5)$ and $(0,0)$.

    (d) Perpendicular to the line joining $(-3, -1)$ and $(-1,4)$.

4. For the triangle with vertices at $(2, -3)$, $(3,1)$, and $(-4,2)$, find the slopes of the three perpendiculars from a vertex to the opposite side.

5. Find the angles of intersection of two lines with the following slopes:

    (a) -3 and 2.   (b) -3 and -2.

    (c) -2 and -2.   (d) 0 and ∞.

    (e) 0 and -1.   (f) 2 and ∞.

    (g) -1 and 1.   (h) -1 and 3.

6. Find the three interior angles of the triangle with vertices A $(-3,2)$, B $(-4, -1)$, and C $(1, -1)$.

7. Find the angles of intersection of the two lines joining the following pairs of points:

    (a) $(0,0)$ and $(5,0)$; $(0,0)$ and $(3,2)$.

    (b) $(-2,2)$ and $(0,0)$; $(-1, -2)$ and $(2,1)$.

    (c) $(2, -3)$ and $(3, -2)$; $(1,2)$ and $(2,5)$.

    (d) $(-2, -1)$ and $(-3, -5)$; $(-2, -1)$ and $(-2,3)$.

8. Two straight lines intersect at an angle of $45°$, and one of them has a slope of -2. Find two possible values of the slope of the other line and illustrate.

9. In Fig. 2-5, AB is the diameter of the semicircle. Establish the coordinates of the point P and prove that angle $APB = 90°$.

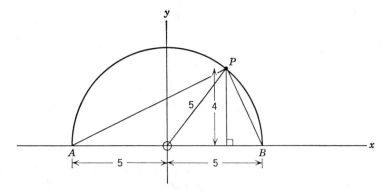

Fig. 2-5.

2.3   EQUATION OF A STRAIGHT LINE

Since the direction of a straight line never changes, the slope of a straight line will always be the same between any two points on it. On the other hand, the slope between two points on a curved path will vary, depending on where the two points are taken. For the purposes of analytic geometry, a straight line may be defined as a path whose slope is always constant; in general, a curved path is one whose slope varies.

The word *locus* is often used to designate the path traced by a point which moves according to some fixed law. The fixed law, which may be stated in words, indicates the restriction on the motion of the point; thus, for example, the locus of a point that moves so that it always remains on a straight edge is a straight line. The fixed law may also be stated in the form of an equation, which relates the dependent and independent variables and restricts the motion of the point to a smooth path. In most cases, if the law governing the motion of the point is given in words, the equation of the locus may be determined.

When the locus is a straight line, its equation may be determined, provided that two definite facts are known. One fact is not enough; for example, if it is known that a straight line passes through the point (1,2), an infinite number of lines with various slopes satisfy this condition; if, however, it is further stipulated that its slope have some definite value, the two known facts limit the path to one definite straight line.

For all problems in which the equation of a locus is to be determined, any indefinite position of the moving point is first taken with coordinates (x,y). Since these coordinates are the independent and dependent variable of the x-y system, their use as the coordinates of the moving point implies the fact that the point does move or vary. When the locus is a straight line, its equation is readily determined from the fact that the slope is constant; two independent expressions for the slope are therefore equated. No special formulas are necessary in setting up the equation of a straight line.

EXAMPLE 5.   (a) Find the equation of the straight line passing through (2,1) with a slope of -2.

(b) Find three other points on this line.

(c) Does the point $(-1,2)$ lie on this line?

(d) Find where this line crosses the two axes.

Solution: (see Fig. 2.6).

(a) Let $P\,(x,y)$ represent any point on the line. The slope between P

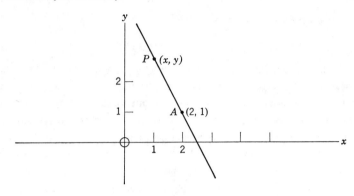

Fig. 2-6. The straight line through (2,1) with a slope of -2.

and A must equal the given slope of -2.

$$\therefore \frac{y-1}{x-2} = -2; \qquad y - 1 = -2x + 4; \qquad 2x + y = 5;$$

or, expressing the dependent variable as a function of x, $y = -2x + 5$ is the required equation.

(b) We may determine as many additional points on the line as required by substituting any value of the independent variable x, and finding the value of y corresponding to it in the established equation. When $x = -1$, $y = 7$; when $x = 0$, $y = 5$; when $x = 1$, $y = 3$. Therefore $(-1,7)$, $(0,5)$, and $(1,3)$ lie on the straight line.

(c) Substituting $x = -1$, $y = 7$, not 2. Thus the point $(-1,2)$ is not on the line.

(d) For the x-axis, $y = 0$ at all points, and therefore $y = 0$ is the equation of the x-axis; also $x = 0$ is the equation of the y-axis. Substituting $y = 0$, $x = \frac{5}{2}$; and substituting $x = 0$, $y = 5$. Therefore the line crosses the x-axis at $(\frac{5}{2},0)$ and the y-axis at $(0,5)$.

Another way of stating this is that the *x-intercept* is $\frac{5}{2}$ and the *y-intercept* is 5.

EXAMPLE 6.   (a) Find the equation of the straight line joining the points $(2,-1)$ and $(1,-5)$. (b) Find where this line crosses the line $2x + y = 5$. *Solution:* (a) Let $P(x,y)$ represent any variable point on the line. Referring to Fig. 2-7, we now have three points on the line, and the slope between any two points must equal the slope between any other two points:

$$\frac{y+1}{x-2} = \frac{-1+5}{2-1} \quad \text{or} \quad \frac{y+5}{x-1} = \frac{-1+5}{2-1} \quad \text{or} \quad \frac{y+1}{x-2} = \frac{y+5}{x-1}.$$

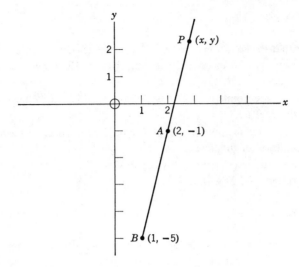

Fig. 2-7. The straight line through $(2, -1)$ and $(1, -5)$.

Any one of these forms, when cleared of fractions and simplified, results in the equation $4x - y = 9$, the required equation. (This statement should be verified.) The equation of the line may also be determined by expressing y as a linear function of x, of the form $y = mx + b$, and solving for m and b by using the fact that the line passes through $(2, -1)$ and $(1, -5)$, in a manner similar to that of Example 4 in Chapter 1.

(b) The one point where these two lines cross must satisfy both equations simultaneously. Therefore, the two equations are solved by algebra. Adding them directly gives $6x = 14$; $x = \frac{7}{3}$. By substitution, $y = \frac{1}{3}$. Therefore the point $(\frac{7}{3}, \frac{1}{3})$ is the point of intersection.

EXAMPLE 7.   Find the equation of the straight line with a y-intercept of -3 and a slope of 0.5.

Solution: Since the y-intercept is -3, the straight line passes through the point $(0, -3)$. Taking (x, y) as the variable point and equating two expressions for the slope, $(y + 3)/(x - 0) = 0.5$; $y = 0.5x - 3$, or $2y = x - 6$, either form being acceptable as the required equation.

By a method similar to that used in Example 7, it may readily be established that the equation of a straight line of slope m and y-intercept b is $y = mx + b$. While it is not necessary to learn any formulas for equations of straight lines, the equation $y = mx + b$, in which m is the slope and b the y-intercept, often may be used to quickly determine the slope and y-intercept of a particular line.

EXERCISES

1. Check that the equation $y = 2x - 3$ represents a straight line by finding at least three points on the locus and showing that the slope between any two of these points is always the same.

2. Show that the points $(-3,2)$, $(-1,1)$, and $(3,-1)$ are collinear (that is, lie on one straight line).

3. Find where the line joining $(-3,-2)$ and $(5,2)$ intersects the x-axis without finding the equation of the line.

4. Find the equations of the following straight lines:
   (a) Through $(-2,3)$ with a slope of 0.8.
   (b) Through $(1,-3)$ and $(-1,-2)$.
   (c) Through $(4,-1)$ and making an angle of $-45°$ with the positive horizontal.
   (d) Through the origin making an angle of $60°$ with the positive horizontal.
   (e) With x-intercept of -2 and y-intercept of 3.
   (f) A vertical line, 2 units to the left of the y-axis.
   (g) A straight line perpendicular to the y-axis through $(-3,-2)$.

5. A straight line has a slope of 0.5 and a y-intercept of -2. Find the value of y on the line where $x = 5$ by drawing a diagram and using first principles without finding the equation of the line.

6. The velocity v of an object is known to vary linearly with time t. When $t = 0$, $v = 5$, and when $t = 2$, $v = 11$. Find v as a function of t.

7. I is known to vary linearly with E in an electric circuit. When $E = 0$ volts, $I = 0$ amperes, and I increases at a rate of 2 amperes per volt. Find I as a function of E.

8. Find where the straight line with slope of -1 passing through the point $(2,-3)$ crosses the straight line through the origin and the point $(-2,-3)$.

9. (a) Find the slope of the straight line $y = mx + b$, where m and b are constants, by finding the slope between any two points on the line.
   (b) Using the result of (a), write down the slope of the line $y = -7x - 2$.
   (c) Using the result of (a), find the slope of the line $4x + 3y - 5 = 0$.

10. Find the equation of a line parallel to the line $y = 4x + 3$ and passing through the point $(1,-1)$.

11. Find the equation of the line perpendicular to the line $3x + 2y = 4$ and passing through the point $(-3,2)$.

12. A triangle has vertices A $(0,6)$, B $(-3,2)$, and C $(1,1)$.
   (a) Find the angles of the triangle.
   (b) Find the equations of all three perpendiculars from a vertex to its opposite side.
   (c) Prove that these three perpendiculars are concurrent.

2.4   THE CONICS

When a dependent variable is related to an independent variable by a power function of higher degree than the first, the graph of the function is a curve of some kind. The *conics* are particular examples of this type of function where the highest power occurring is a two-dimensional term of the form x^2, y^2, or xy. Conics include such curves as the circle, the parabola, the ellipse, and the hyperbola.

It is not within the scope of this text to examine the conics in detail. However, certain frequently occurring conics and the equations representing them will be presented very briefly.

The equation of a *circle* with center at the origin and radius r (Fig. 2-8) is $x^2 + y^2 = r^2$. This should be fairly easily established from the diagram.

The equation of a *parabola* with vertex at the origin and axis along one of the coordinate axes is either $y^2 = kx$, $x^2 = ky$, $y^2 = -kx$, or $x^2 = -ky$ as indicated in Fig. 2-9, in which k is positive.

The equation of an *ellipse* centered at the origin is of the form

$$\frac{x^2}{a^2} + \frac{y^2}{b^2} = 1,$$

where a and b are the constant lengths indicated in Fig. 2-10.

A *hyperbola* always has two symmetrical branches. The equation of a hyperbola in which the complete curve, including both branches, is symmetrical about both coordinate axes, is of the form either $x^2/a^2 - y^2/b^2 = 1$ or $y^2/b^2 - x^2/a^2 = 1$ (Fig. 2-11); the lines AB and CD are *asymptotes* to the hyperbolas.

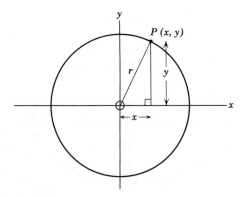

Fig. 2-8. Diagram of $x^2 + y^2 = r^2$, a circle with the center at the origin and radius r.

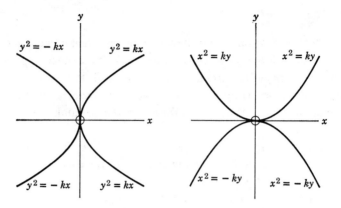

Fig. 2-9. The parabola in four standard positions.

If the asymptotes are at right angles, the hyperbola is said to be a rectangular hyperbola.

For the hyperbolas of Fig. 2-11, if $a = b$, the hyperbolas are rectangular. The graph of inverse variation, $y = k/x$ or $xy = k$ (Fig. 2-12) is a rectangular hyperbola for which the asymptotes are the coordinate axes.

If any of these conics is not situated in what has been called a standard position with reference to the coordinate axes, the form of its equation will be altered. In most instances it is possible to detect the type of conic represented by the form of its equation. All the conics have an equation with at least one two-dimensional term, but with no terms of higher dimensions. *If the equation contains terms in x^2, y^2 and xy, it may represent a parabola, a*

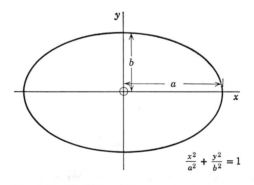

Fig. 2-10. The ellipse in a standard position.

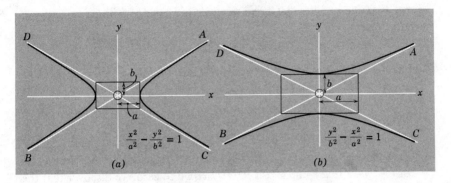

Fig. 2-11. The hyperbola in two standard positions.

hyperbola or an ellipse, and the detection of the conic will usually involve some plotting. Otherwise, the detection criteria are as follows:

For a Circle. There are terms in x^2 and y^2 with equal coefficients and the same sign. There is no xy term.

For a Parabola. One variable appears as a second-power term, but all other terms have no higher power than the first.

For an Ellipse. There are terms in x^2 and y^2 with unequal coefficients and the same sign, but no xy term.

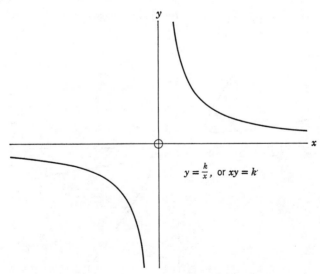

Fig. 2-12. The rectangular hyperbola of inverse variation. Equation: $y = k/x$ or $xy = k$.

For a Hyperbola. Either: There are terms in x^2 and y^2 with opposite signs, but no xy term,

Or: There is a term in xy with no x^2 or y^2 term.

EXAMPLE 8. Identify the following curves: (a) $x^2 - y^2 + 3x = 0$. (b) $x^2 + y^2 + 3x = 1$. (c) $x^2 + 2y^2 + 3x = 2$. (d) $x + 5y^2 + 3y = 7$. (e) $x^2 + y^2 + 3xy + 1 = 0$.

Solution: By following the detection criteria carefully, we see that (a) is a hyperbola, (b) is a circle, (c) is an ellipse, (d) is a parabola, and (e) requires a further test. When we attempt to find the x- and y-intercepts for (e) by putting y and x respectively equal to zero, we discover that the intercepts are imaginary; this means that the graph does not cross the axes. If $x = 1$, $y^2 + 3y + 2 = 0$; $(y + 2)(y + 1) = 0$; and $y = -2$ or -1. By a similar method, it may be determined that the following points lie on the graph:

x	1	1	−1	−1	2	2	−2	−2	5	5	−5	−5
y	−1	−2	1	2	−1	−5	1	5	−2	−13	2	13

When these points are plotted, it soon becomes apparent that the graph is a hyperbola with branches in the second and fourth quadrants, as shown in Fig. 2-13.

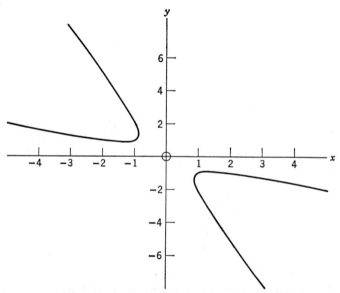

Fig. 2-13. Graph of $x^2 + y^2 + 3xy + 1 = 0$.

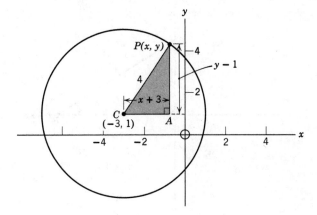

Fig. 2-14. A circle with the center at $(-3,1)$ and a radius of 4 units.

EXAMPLE 9.   Find the equation of the circle with center at $(-3,1)$ and radius of 4 units.

Solution: The circle is the locus of a point that moves so that it is always the same distance from the center. In Fig. 2-14, it is seen that the radius is the hypotenuse of a right-angled triangle, whose base and vertical side are easily determined. Using the theorem of Pythagoras,

$$CP = \sqrt{(x + 3)^2 + (y - 1)^2} = 4.$$

Squaring both sides,

$$(x + 3)^2 + (y - 1)^2 = 16.$$
$$x^2 + 6x + 9 + y^2 - 2y + 1 = 16.$$
$$x^2 + y^2 + 6x - 2y - 6 = 0, \quad \text{the required equation.}$$

EXERCISES

1. Identify the graphs of the following functions, and make a sketch of each by determining several points on the graph:

(a) $y = x^2 - 3$.   (b) $x^2 - 4x + y^2 = 0$.   (c) $16x^2 + 25y^2 = 400$.

(d) $y = 2/x$.   (e) $x - y^2 = 1$.   (f) $x^2 - y^2 = 1$.

2. Find the slope of the straight line joining the two points on the graph of $2y = x^2 + 2$ where $x = 2$ and $x = 4$.

3. Identify the following graphs:

(a) $xy + 4x = 40$.   (b) $x^2 + 4x + y^2 - 12 = 0$.

(c) $x^2 + 4x + y - 12 = 0$.   (d) $x^2 + 4x - y^2 - 12 = 0$.

(e) $x^2 + y^2 + xy = 7$.   (f) $x^2 + y^2 + 2xy - x + y = 2$.

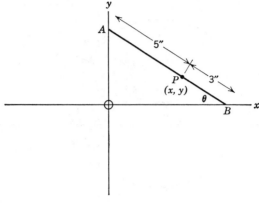

Fig. 2-15.

4. Find in simplest form the equation of the circle passing through $(4, -1)$ with center at $(0, -1)$.

5. Find in simplest form the equation of the locus of a point which moves so that the angle subtended at the point by the line joining $(-3,0)$ and $(3,0)$ is always a right angle. Identify the curve from its equation.

6. In a certain graph, x and y are related to one another through a parameter θ by the following pair of equations: $x = 4 \cos \theta$ and $y = 4 \sin \theta$. Find the direct relation between x and y by eliminating θ from the parametric equations, and identify the graph.

7. Find the equation relating x and y for the parametric equations $x = a \cos \theta$ and $y = b \sin \theta$, and identify the graph.

8. Find the points of intersection of the parabola $3y = x^2$ and the lines:
(a) $y = 2x$. (b) $y = 2x - 3$. (c) $y = 2x - 4$. Illustrate and interpret the geometrical significance of (b).

9. Prove that the line $x + 2y = 4$ is tangent to the ellipse $x^2 + 4y^2 = 8$, and find the point of contact.

10. Using the fact that any tangent to a circle is perpendicular to the radius drawn to the point of contact, find the equation of the tangent to $x^2 + y^2 = 169$ at the point in the first quadrant where $x = 12$.

11. In Fig. 2-15, AB is a rigid straight edge of length 8 in. which moves so that A always lies on the y-axis and B always lies on the x-axis. Find the equation of the locus of point P and identify the curve it traces.

2.5 SOME USEFUL FORMULAS

The following formulas may be established from first principles. However, if the principles involved form part of a more comprehensive problem, it is often convenient to apply these formulas directly.

The length of the straight line joining (x_1,y_1) and (x_2,y_2)

$$= \sqrt{(x_2 - x_1)^2 + (y_2 - y_1)^2}.$$

This has already been used in Example 9 and Fig. 2-14 for the length CP.

The coordinates of the midpoint of the line joining (x_1,y_1) and (x_2,y_2) are

$$\left(\frac{x_1 + x_2}{2}, \frac{y_1 + y_2}{2}\right).$$

The distance of the point (x_1,y_1) from the straight line $Ax + By + C = 0$, understood to be the perpendicular distance, is $(Ax_1 + By_1 + C)/\sqrt{A^2 + B^2}$.

The area of a triangle with vertices, in counterclockwise direction, of (x_1,y_1), (x_2,y_2), and (x_3,y_3) is $\frac{1}{2}[x_1(y_2 - y_3) + x_2(y_3 - y_1) + x_3(y_1 - y_2)]$.

EXAMPLE 10.   (a) Establish the equation of the line through $(2,-1)$ perpendicular to the line $x + y + 3 = 0$. (b) Find the coordinates of the point in (a) where the perpendicular meets the line. (c) Hence find the length of the perpendicular. (d) Check this length by formula. ·

Solution: (a) $x + y + 3 = 0$ may be put in the form $y = -1x - 3$, and, by comparison with the formula $y = mx + b$, the slope of the line is seen to be -1. Thus the slope of the perpendicular $= -1/-1 = 1$. The equation of the perpendicular through the point $(2,-1)$ is therefore $(y + 1)/(x - 2) = 1$, or $x - y - 3 = 0$.

(b) The coordinates of the point where the perpendicular meets the line will be the algebraic solution of the two equations $x + y + 3 = 0$ and $x - y - 3 = 0$: $x = 0$ and $y = -3$. Therefore the coordinates of the point are $(0,-3)$.

(c) The length of the perpendicular $= \sqrt{(2 - 0)^2 + (-1 + 3)^2} = \sqrt{8} = 2\sqrt{2}$.

(d) By the formula of Section 2.5, the distance from the point to the

$$\text{line} = \frac{1 \times 2 + 1(-1) + 3}{\sqrt{1^2 + 1^2}} = \frac{4}{\sqrt{2}} = \frac{4\sqrt{2}}{2} = 2\sqrt{2},$$

which checks with (c).

EXERCISES

1. Find the lengths of the three sides of the triangle with vertices at $(2,3)$, $(2,-2)$, and $(6,0)$.

2. Find the coordinates of the midpoints of the sides of the triangle with vertices at $(-1,-3)$, $(2,-1)$, and $(0,5)$.

3. One end of a line is at the point $(-3,2)$ and the coordinates of its midpoint are $(-1,-3)$. Find the coordinates of the other end of the line.

4. Find the equation of the median of the triangle of Exercise 2 from the vertex $(-1,-3)$ to the midpoint of the opposite side.

5. Show that the triangle with vertices at A $(1,6)$, B $(-2,1)$, and C $(6,3)$ is right-angled at A: (a) by comparing the slopes of AB and AC; (b) by showing that the midpoint of BC is equidistant from the three vertices, and hence that the circle with BC as diameter passes through A.

6. Find the length of the chord of the hyperbola $xy = 8$ between the points where $x = 1$ and $x = 4$.

7. Find the area of the triangle with vertices $(0,0)$, $(6,0)$, and $(4,3)$: (a) by sketching the triangle and using the formula area $= \frac{1}{2}$ base $\times$ height; (b) by using the formula for the area of a triangle given in Section 2.5.

8. Find the area of the triangle with vertices $(-1,-2)$, $(3,4)$, and $(5,-1)$.

9. Prove that the median of the triangle with vertices at A $(2,5)$, B $(-1,-2)$, and C $(5,-1)$ drawn from A to the midpoint of BC is a vertical line.

10. Prove that the three medians of the triangle of Exercise 9 are concurrent.

11. Show that the three right bisectors of the sides of the triangle with vertices at $(2,3)$, $(-2,-1)$, and $(4,-1)$ are concurrent.

12. Find the equation of the circle circumscribing the triangle of Exercise 11 by using the fact that the center of the circle is at the point of intersection of the right bisectors.

13. Find in simplest form the equation of the locus of a point which moves so that its distance from the point $(1,0)$ is always equal to its distance from the line $x = -1$. Identify the locus.

14. Find in simplest form the equation of a point which moves so that the sum of its distances from the two points $(3,0)$ and $(-3,0)$ is always equal to 10. Identify the locus.

15. Find in simplest form the locus of a point which moves so that the difference of its distances from the two points $(5,0)$ and $(-5,0)$, taken in either order, is always equal to 8. Identify and sketch the locus.

16. Derive in simplest form the equation of the curve for which the difference of the distances from any point (x,y) on the curve to the points $(-1,-1)$ and $(0,0)$, taken in either order, is always one unit. Identify the curve.

CHAPTER 3

Rate of Change and the Derivative

3.1 THE Δ NOTATION

In the study of the changes occurring in independent and dependent variables, the symbol Δ (a Greek capital letter, *delta*) is used to represent *the change in* or *an increment of*. For example, for $y = f(x)$, when x changes from -2 to 3 and y changes from 1 to -1 as a result, Δx represents the change in x, which is $3 - (-2) = 5$; and Δy represents the corresponding change in y, which is $-1 - 1 = -2$. Graphically, Δx represents the run and Δy represents the rise between two specific points on the graph of the function.

EXAMPLE 1. For the function $y = 3x^2 - 8$, find:
(a) Δx and Δy when x changes from 1 to 3.
(b) The average slope from $x = 1$ to $x = 3$.
(c) Δx and Δy when x changes from -1 to 3.
(d) The average slope from $x = -1$ to $x = 3$.
Solution: (a) $\Delta x = 3 - 1 = 2$.

$$\text{When } x = 1, y = 3(1)^2 - 8 = -5.$$
$$\text{When } x = 3, y = 3(3)^2 - 8 = 19.$$
$$\Delta y = 19 - (-5) = 24.$$

(b) The average slope over an interval is defined as the slope of the straight line joining the two points at the extreme ends of the interval; that is, average slope $= \Delta y / \Delta x$. For this example, the average slope $= \frac{24}{2} = 12$. For all functions except the linear function, the slope of the graph changes as we proceed along the graph; hence the necessity for using the qualification "average" to specify this particular slope.

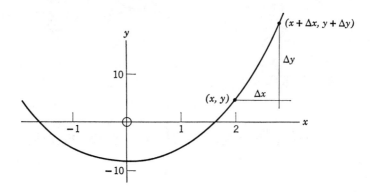

Fig. 3-1. Finding the average slope of the graph of $y = 3x^2 - 8$.

(c) When x changes from -1 to 3, $\Delta x = 3 - (-1) = 4$. Δy may be determined by the method of part a, but we will use a more general method here. (x,y) represents any point on the graph of the function (Fig. 3-1). If x is increased by an amount Δx, Δy means the corresponding increase in y, and thus the point $(x + \Delta x, y + \Delta y)$ also lies on the graph, and satisfies the equation.

$$\therefore \ y = 3x^2 - 8 \qquad (1)$$

and

$$y + \Delta y = 3(x + \Delta x)^2 - 8 \qquad (2)$$

Subtracting (1) from (2),

$$\Delta y = 3(x + \Delta x)^2 - 3x^2$$
$$= 3[x^2 + 2x\,\Delta x + (\Delta x)^2] - 3x^2$$
$$= 6x\,\Delta x + 3(\Delta x)^2$$

Thus Δy depends on two independent variables, x and Δx. For our example, $x = -1$ and $\Delta x = 4$.

$$\therefore \ \Delta y = 6(-1)(4) + 3(4)^2 = -24 + 48 = 24.$$

This method yields a value for Δy without finding the values of y at the two endpoints of the interval. It also provides a general expression for Δy, which may be used for any number of cases with varying values of x and Δx.

(d) The average slope in the interval from $x = -1$ to $x = 3$ is $\Delta y/\Delta x =$

$24/4 = 6$. This is not the same as the average slope between $x = 1$ and $x = 3$, calculated as 12 in part b. This indicates that the graph cannot be a straight line.

The graph of Fig. 3-1 has a contracted y-axis, resulting in a false picture of the steepness of the curve. Slope cannot be determined by any angle measurement.

EXAMPLE 2. Prove that an equation of the first degree in x and y represents a straight line.

Solution: In Chapter 1, it was discovered that all power functions of the first degree (linear functions) resulted in straight-line graphs. Now it is possible to prove this fact conclusively. Let the linear function be $y = mx + b$, where m and b are constants. To find the average slope, the method of Example 1(c) is used and Fig. 3-2 illustrates the method. Any point (x,y) is taken on the graph, and x takes on an increment Δx and y has a corresponding increment Δy, so that the point $(x + \Delta x, y + \Delta y)$ also lies on the graph.

$$\therefore \ y = mx + b \qquad (1)$$

$$y + \Delta y = m(x + \Delta x) + b \qquad (2)$$

Subtracting (1) from (2),

$$\Delta y = m(x + \Delta x) - mx = m\,\Delta x.$$

$$\therefore \ \frac{\Delta y}{\Delta x} = m, \text{ a constant.}$$

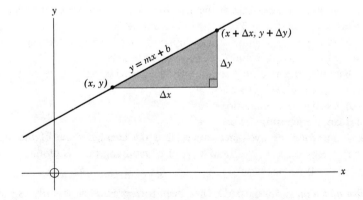

Fig. 3-2. The graph of the general linear function.

This means that the average slope over any interval is constant. Therefore the graph of a power function of the first degree is a straight line, and the function is called a linear function for this reason.

The form of linear equation $y = mx + b$ has been used several times, both here and in previous sections. It has just been proved that the constant m represents the constant slope. Moreover, if $x = 0$, $y = b$; therefore b represents the y-intercept.

EXERCISES

Using the Δ symbol throughout, find the average slope of the graphs of the following functions over the intervals indicated:

1. $y = 5 - 2x$ between $x = 2$ and $x = 4$.
2. $3x - 2y = -7$ between $x = -2$ and $x = 1$.
3. $i = t^2 - t$ between $t = 2$ and $t = 5$.
4. $y = 4/x + 3$ between $x = 1$ and $x = 4$.
5. $i = 4 \sin 100\pi t$ between $t = 1/600$ and $t = 1/200$, $100\pi t$ being in radians.
6. $v = 6(1 - 10^{-50t})$ between $t = 0.01$ and $t = 0.02$.
7. $i = 10^{-100t} \sin 100\pi t$ between $t = 0.0025$ and $t = 0.005$.

Find the average slope of the graphs of the following functions as a function of t and Δt; then use this general expression for the average slope to find the average slope between $t = 1$ and $t = 3$:

8. $y = 0.5t^2 + 1$.
9. $s = 20t - 16t^2$.
10. $v = t^2 - 5t - 2$.
11. $y = t^3 - 6t$.

12. Find the average slope of the graph of $Z^2 = 16 + R^2$ for positive Z's between (a) $R = 0$ and $R = 3$ and (b) $R = 3$ and $R = 5$.

13. Find the average slope of the ellipse $x^2/16 + y^2/9 = 1$ between $x = 3$ and $x = 4$ for (a) positive y's and (b) negative y's. Illustrate.

14. Establish the graphical conditions that give rise to:
    (a) a positive average slope.
    (b) a negative average slope.
    (c) an average slope of zero.

15. Using four- or five-place tables, find the average slope of the graph of $y = \cos \theta$ between $\theta = 29.9°$ and $\theta = 30.1°$ assuming (a) θ is plotted in degrees and (b) θ is plotted in radians.

Rate of Change. So far, $\Delta y/\Delta x$ has been interpreted graphically as the average slope over an interval of the graph. Graphs are simply a visual aid in interpreting a function, and the average slope has a more fundamental interpretation in the analysis of the function itself.

If the displacement s feet of a moving body from a fixed point, for example, varies with time t seconds according to the graph of Fig. 3-3a, the average slope $\Delta s/\Delta t$ is the ratio of a displacement measured in feet to a time in seconds, and its units will be feet per second; that is, $\Delta s/\Delta t$ represents average velocity over the interval of time Δt. In Fig. 3-3b, the velocity v feet per second of a moving body is plotted against time t in seconds, and here the slope $\Delta v/\Delta t$ is in units of feet per second per second; therefore $\Delta v/\Delta t$ represents average acceleration over the interval of Δt seconds. In Fig. 3-3c, current i amperes in a circuit is plotted against t seconds; here the slope $\Delta i/\Delta t$ is in units of amperes per second, and it represents the average rate of change of current with respect to time over the time interval Δt seconds. In each case, the change in a dependent variable is compared to the change in the independent variable that caused it, and the ratio of the changes represents the *rate* at which one thing is changing as compared to another (or *with respect to* the other).

EXAMPLE 3.   Find the average rate of change of i with respect to t between $t = 0$ and $t = 2$ for the function $i = 3t^2 - 5t$. Assume t is in seconds and i is in amperes, and attach units to the answer.
Solution: $\Delta t = 2$. At $t = 0$, $i = 0$; at $t = 2$, $i = 12 - 10 = 2$. $\therefore \Delta i = 2$. Therefore average rate of change of i with respect to $t = \Delta i/\Delta t = 1$ ampere per second.

EXAMPLE 4.   The velocity of a body in feet per second t seconds after

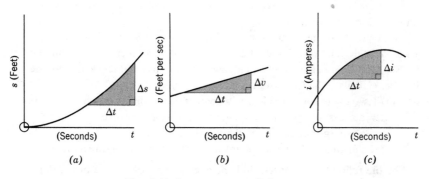

Fig. 3-3. Average rates of changes.

it is set in motion is known to be $v = 1.5t^2 + 2$. Find the average acceleration over an interval of time Δt as a function of t and Δt.

Solution: At time t,

$$v = 1.5t^2 + 2. \qquad (1)$$

At time $t + \Delta t$,

$$v + \Delta v = 1.5(t + \Delta t)^2 + 2. \qquad (2)$$

Subtracting (1) from (2),

$$\Delta v = 1.5(t + \Delta t)^2 - 1.5t^2$$

$$= 3.0(\Delta t)t + 1.5(\Delta t)^2.$$

Dividing by Δt,

$$\frac{\Delta v}{\Delta t} = 3.0t + 1.5(\Delta t).$$

Therefore average acceleration over an interval $\Delta t = 3.0t + 1.5(\Delta t)$ ft per second per second.

EXERCISES

1. A body moves in a straight line so that its displacement s feet from a fixed point varies with time t seconds according to the function $s = t^2 - 4t$. Calculate its average velocity between (a) $t = 4$ and $t = 6$, (b) $t = 0$ and $t = 2$, (c) $t = 2$ and $t = 4$, and (d) $t = 0$ and $t = 4$. Illustrate graphically.

2. A body moves in a straight line so that its velocity v feet per second varies with time t seconds according to the function $v = 8t - t^2$. Calculate its average acceleration from (a) $t = 1$ to $t = 4$ and (b) $t = 8$ to $t = 10$. Explain the significance of a negative answer.

3. The acceleration of a body is given by $a = 6t - t^3$, the basic units being feet and seconds. Find the average rate of change of acceleration between $t = 0$ and $t = 2$, attaching units to the answer.

4. A point on the circumference of a rotating machine moves counterclockwise through 150° in 0.2 sec, then a further 40° counterclockwise in the next 0.2 sec, and then begins moving clockwise and moves 90° clockwise in the next 0.2 sec. Calculate the following:

(a) Its average angular velocity in degrees per second for each of these 0.2-sec intervals, assuming counterclockwise velocity as positive.

(b) Its average angular velocity over the total 0.6 sec.

(c) Its angular velocity at the instant it starts changing direction.

For the following Exercises, find the average rate of change of the dependent variable (on the left) with respect to the independent variable for the given interval.

5. $v = 12 + 6t$ for any interval of t (t in seconds, v in feet per second).

6. $q = 3t^2$ between $t = 0.4$ and $t = 0.6$ (t in seconds, q in coulombs).

7. $P = 1000/V^{1.5}$ between $V = 9$ and $V = 16$ (P in pounds per square inch, V in cubic feet).

8. $v = 10^{-5t} - 10^{-10t}$ between $t = 0.1$ and $t = 0.2$ (t in seconds, v in volts).

9. $i = 3 \sin (100\pi t - \pi/6)$ between $t = 1/300$ and $t = 1/150$ (t in seconds, angles in radians, i in amperes).

Find an expression in two independent variables to represent the average rate of change of the variable on the left-hand side with respect to the variable on the right-hand side for the following functions.

10. $y = x^2$.

11. $s = 2t^2 - 1$.

12. $v = -3t + 2t^3$.

3.2 THE DERIVATIVE

The idea of an instantaneous rate of change is probably familiar to the student from the concept of the speed of a car at any instant of time; or from natural growth, as illustrated by the growth of a small tree, where the rate of growth with respect to time at any instant is established, even though there can be no actual growth at an instant of time. The study of calculus really begins with the concept of an instantaneous rate of change.

For example, consider the motion of a body in a straight line in which displacement $s = 0.5t^2$, as illustrated in the graph of Fig. 3-4.

The average velocity in any interval of time Δt, starting at $t = 2$, is $\Delta s/\Delta t = \tan \theta$. Now let us examine what happens when Δt is made smaller and smaller. The point B will move backward along the curve and come closer and closer to point A. There will still be a definite value for the average velocity over the small interval of time, no matter how small Δt and Δs become. Thus, if $\Delta t = 0.000000020$ and $\Delta s = 0.000000042$, $\Delta s/\Delta t = 42/20 = 2.1$; this can be calculated just as accurately as if Δt were equal to 20 and Δs equal to 42.

As B approaches A, Δt and Δs approach zero. Their values may be extremely small, but it is still possible to calculate $\Delta s/\Delta t$. However, it is not possible to calculate this ratio when B falls exactly on A and Δt and Δs are both exactly zero. $0/0$ cannot be determined in itself and is called an *indeterminate expression*. We can, however, evaluate what this ratio approaches if the exact way in which Δt and Δs approach zero is known.

Geometrically, in Fig. 3-4, as Δt approaches zero the triangle ABC approaches a point triangle with all sides equal to zero. However, the angle θ is determined by the direction of the line AB, extended in both directions

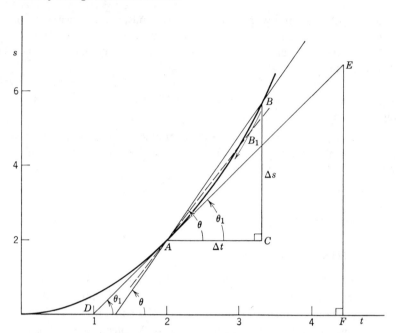

Fig. 3-4. Instantaneous rate of change developed from the average rate.

as far as we wish, and not by its length. When B falls on A, the extended line AB becomes a tangent to the curve, the point A being its point of contact. Thus, as Δt approaches zero, the value of $\Delta s/\Delta t$ approaches the slope of the tangent at A. In mathematical terminology

$$\lim_{\Delta t \to 0} \frac{\Delta s}{\Delta t} = \tan \theta_1, \text{ measured as a ratio of rise to run.}$$

This may be measured on the graph by extending the tangent to any length, such as DE, and dropping a perpendicular EF from E. Thus

$$\lim_{\Delta t \to 0} \frac{\Delta s}{\Delta t} = \frac{EF}{DF}.$$

In this way the point triangle has, as it were, been magnified to a large right-angled triangle, so that the *instantaneous velocity* at $t = 2$ may be determined. EF is measured in units of s and DF in units of t.

The numerical quantities involved in this approach of B toward A are shown in the following calculation and tabulation:

$$s = 0.5t^2,$$
$$s + \Delta s = 0.5(t + \Delta t)^2.$$

Subtracting,

$$\Delta s = 0.5(t + \Delta t)^2 - 0.5t^2 = t\,\Delta t + 0.5(\Delta t)^2.$$

When $t = 2$, $\Delta s = 2(\Delta t) + 0.5(\Delta t)^2$.

Δt	$\Delta s = 2(\Delta t) + 0.5(\Delta t)^2$	$\Delta s / \Delta t$
0.5	1.125	2.25
0.1	0.205	2.05
0.01	0.02005	2.005
0.0001	0.000200005	2.00005
1×10^{-10}	$2 \times 10^{-10} + 0.5 \times 10^{-20}$	2.00000000005

It is seen that, for the function $s = 0.5t^2$, at $t = 2$,

$$\lim_{\Delta t \to 0} \frac{\Delta s}{\Delta t} = 2,$$

since, by choosing a value of Δt as small as we wish, we arrive at a value of $\Delta s / \Delta t$ as close to 2 as we wish. Thus a ratio which is theoretically $0/0$ almost always approaches a definite value, called the *limiting value* or the *limit* as its numerator and denominator separately approach zero according to some fixed law.

The expression

$$\lim_{\Delta t \to 0} \frac{\Delta s}{\Delta t}$$

is shortened symbolically to ds/dt, and is called the *derivative* of s with respect to t. Therefore the *derivative* of one variable with respect to another is defined as the *instantaneous rate of change* of the first variable with respect to the second. Graphically, the derivative is the slope of the tangent at a point on the graph. In the particular example considered previously, ds/dt represents *instantaneous velocity*.

So far, there have been no actual calculations to determine the value of the derivative. The following examples will show how this is done.

EXAMPLE 5.   Find the derivative of the function $y = x^2 - 3$: (a) as a function of x and (b) at the point where $x = -3$.
Solution: (a) Let x take on an increment of Δx and y a corresponding increment of Δy.
Thus

$$y = x^2 - 3. \tag{1}$$

$$y + \Delta y = (x + \Delta x)^2 - 3. \tag{2}$$

Subtracting (1) from (2),

$$\Delta y = (x + \Delta x)^2 - x^2 = 2x(\Delta x) + (\Delta x)^2.$$

Dividing by Δx, $\Delta y/\Delta x = 2x + \Delta x$. This relationship is true no matter how small Δx may be.

$$\therefore \frac{dy}{dx} = \lim_{\Delta x \to 0} \frac{\Delta y}{\Delta x} = 2x + 0 = 2x.$$

(b) This gives dy/dx as a function of x; in this case the derivative is directly proportional to x, implying that the graph is not a straight line. At $x = -3$, the value of the derivative is the value of $2x$ when $x = -3$;

$$\therefore \text{ at } x = -3, \, dy/dx = 2(-3) = -6.$$

Graphically, the slope of the tangent at $x = -3$ is -6.

EXAMPLE 6.   Find the slope of the tangent to $y = x + 1/x$ at the point where $x = 2$.

Solution: Let x and y take on increments of Δx and Δy, respectively.

$$y = x + \frac{1}{x}. \qquad (1)$$

$$y + \Delta y = x + \Delta x + \frac{1}{x + \Delta x}. \qquad (2)$$

Subtracting (1) from (2),

$$\Delta y = \Delta x + \frac{1}{x + \Delta x} - \frac{1}{x}.$$

$$\frac{\Delta y}{\Delta x} = 1 + \frac{1}{\Delta x(x + \Delta x)} - \frac{1}{x(\Delta x)}.$$

As Δx approaches zero, the expression for $\Delta y/\Delta x$ approaches $1 + \infty - \infty$, which cannot be determined, since ∞ is undefined and $\infty - \infty$ cannot be assumed to be zero. This difficulty can be overcome if the two fractions are expressed as a single fraction over a common denominator. Thus,

$$\frac{\Delta y}{\Delta x} = 1 + \frac{x - (x + \Delta x)}{x(\Delta x)(x + \Delta x)} = 1 - \frac{\Delta x}{x(\Delta x)(x + \Delta x)}.$$

$$\therefore \frac{\Delta y}{\Delta x} = 1 - \frac{1}{x(x + \Delta x)},$$

no matter how small Δx may be.

$$\therefore \frac{dy}{dx} = \lim_{\Delta x \to 0} \frac{\Delta y}{\Delta x} = 1 - \frac{1}{x(x + 0)} = 1 - \frac{1}{x^2}.$$

This gives the slope of the tangent at any point (x,y) on the graph. Therefore the slope of the tangent at $x = 2$ is $1 - \frac{1}{4} = \frac{3}{4}$.

The slope of the tangent to a curve is taken to indicate the direction of the curve at the point of contact. This is often called the *slope of the curve* at this point.

EXERCISES

Find the derivatives of the following functions, from first principles, as in the preceding examples.

1. $y = 3x$.
2. $y = 3x + 2$.
3. $y = x^2 - 8$.
4. $y = x^2 - 8x$.

5. $y = \dfrac{2}{x}$.
6. $y = x^3 - 4$.

7. $y = \dfrac{1}{3 + x}$.
8. $y = \dfrac{1}{(3 + x)^2}$.

9. Find the slope of the tangent where $x = 2$ for Exercises 1 to 8.

Find the instantaneous velocity at $t = 3$ where the displacement s is given by the following functions. Assume basic units of feet and seconds:

10. $s = 20t - 16t^2$.
11. $s = t^3$.
12. $s = v_0 t + \frac{1}{2}at^2$, where v_0 and a are constants.

Find the acceleration at $t = 2$ where the velocity v is given by the following functions. Assume units of meters and seconds.

13. $v = 9.8t$.

14. $v = 6t + \dfrac{6}{t + 1}$.

15. The current in amperes in any circuit is defined as the instantaneous rate at which the charge q passes any point in the circuit (in coulombs per second). Find the current at $t = 0.5$ in a circuit for which the charge q coulombs is given as a function of t seconds by:

(a) $q = 6t - 0.5t^2$.
(b) $q = \dfrac{3}{(t + 0.5)^2}$.

16. Find the voltage at $t = 0.8$ sec across an inductance of $L = 2 \times 10^{-2}$ henry, where the current i in the inductance is given by $i = 50t^2 + 2$. [*Note:* Voltage across an inductance $= L(di/dt)$.]

3.3 DERIVATIVES OF POWER FUNCTIONS BY RULE

Finding the derivative of a function is referred to as *differentiating* the function, and this operation forms the basis of differential calculus. Differentiating power functions by the method illustrated in Section 3.2 takes considerable time and many of the steps are repetitive for each separate function. Thus far the general mathematical techniques involved in differentiating the power function have not been analyzed to find a pattern that can be used for all cases. This we now proceed to do.

If we examine the derivatives of several simple power functions, a general formula may be determined to apply to all particular cases.

$$\text{If } y = x, \qquad \frac{dy}{dx} = 1.$$

$$\text{If } y = x^2, \qquad \frac{dy}{dx} = 2x.$$

$$\text{If } y = x^3, \qquad \frac{dy}{dx} = 3x^2.$$

$$\text{If } y = x^4, \qquad \frac{dy}{dx} = 4x^3.$$

$$\text{If } y = \frac{1}{x} = x^{-1}, \quad \frac{dy}{dx} = -x^{-2}.$$

$$\text{If } y = \frac{1}{x^2} = x^{-2}, \quad \frac{dy}{dx} = -2x^{-3}.$$

In each case, the derivative is formed by taking the index as a multiplier and reducing the actual index of the independent variable by 1. This cannot be taken as a general rule, however, until some fractional indices are considered.

EXAMPLE 7.   Find the derivative of $y = x^{1/2}$.
Solution: If $y = x^{1/2}$,

$$y^2 = x, \qquad (1)$$

$$(y + \Delta y)^2 = x + \Delta x. \qquad (2)$$

Subtracting (1) from (2),

$$(y + \Delta y)^2 - y^2 = \Delta x.$$

$$2y\,\Delta y + (\Delta y)^2 = \Delta x.$$

Dividing by Δx,

$$2y\frac{\Delta y}{\Delta x} + \frac{\Delta y}{\Delta x}\Delta y = 1.$$

As Δx approaches zero, $\Delta y/\Delta x$ becomes dy/dx, and Δy approaches zero.

$$\therefore\ 2y\frac{dy}{dx} + 0 = 1.$$

$$\frac{dy}{dx} = \frac{1}{2y} = \frac{1}{2x^{\frac{1}{2}}} = \tfrac{1}{2}x^{-\frac{1}{2}}.$$

In Example 7, the derivative is formed again by taking the index as a multiplier and reducing the actual index of the independent variable by 1. We may now show that this will always be the case, no matter what index of the independent variable occurs in the function. If

$$y = x^n \qquad (1)$$

$$y + \Delta y = (x + \Delta x)^n,$$

which can be expanded by the binomial theorem to

$$x^n + nx^{n-1}\Delta x + \frac{n(n-1)}{2!}x^{n-2}(\Delta x)^2 + \cdots$$

$$\text{to an infinite number of terms.} \qquad (2)$$

Subtracting (1) from (2),

$$\Delta y = nx^{n-1}\Delta x + \frac{n(n-1)}{2!}x^{n-2}(\Delta x)^2 \cdots.$$

$$\therefore\ \frac{\Delta y}{\Delta x} = nx^{n-1} + \frac{n(n-1)}{2!}x^{n-2}(\Delta x) + \text{further terms in higher powers of }\Delta x.$$

$$\frac{dy}{dx} = \lim_{\Delta x \to 0}\frac{\Delta y}{\Delta x} = nx^{n-1},$$

for any constant value of n.

Comparing the function $y = kx^n$, where k is a constant, with the function $y = x^n$, it should be readily seen that for any Δx, the corresponding Δy for the first function will be k times the corresponding Δy for the second one; therefore its derivative will be k times the derivative of $y = x^n$. Thus, if $y = kx^n$, $dy/dx = knx^{n-1}$.

EXAMPLE 8.   Verify that, if $y = kx^3$, $dy/dx = 3kx^2$.
Solution:

$$y = kx^3. \tag{1}$$

$$y + \Delta y = k(x + \Delta x)^3 = kx^3 + 3kx^2(\Delta x) + 3kx(\Delta x)^2 + k(\Delta x)^3. \tag{2}$$

$$\therefore \Delta y = 3kx^2(\Delta x) + 3kx(\Delta x)^2 + k(\Delta x)^3.$$

$$\therefore \frac{\Delta y}{\Delta x} = 3kx^2 + 3kx(\Delta x) + k(\Delta x)^2.$$

$$\therefore \frac{dy}{dx} = \lim_{\Delta x \to 0} \frac{\Delta y}{\Delta x} = 3kx^2.$$

If $y = k$, where k is a constant, the graph will be a straight horizontal line for which the slope is zero; hence $dy/dx = 0$. This result may also be obtained by realizing that $k = kx^0$, and, using the preceding formula,

$$\frac{dy}{dx} = k0x^{-1} = 0.$$

If $y = $ a power function with two or more terms, its derivative is the sum of the separate derivatives of the terms. Suppose

$$y = k_1 x^{n_1} + k_2 x^{n_2};$$

$$y + \Delta y = k_1(x + \Delta x)^{n_1} + k_2(x + \Delta x)^{n_2}.$$

$$\Delta y = k_1[(x + \Delta x)^{n_1} - x^{n_1}] + k_2[(x + \Delta x)^{n_2} - x^{n_2}].$$

It is apparent by this time that each step in the solution will yield two separate terms, and the final derivative will consist of the sum of the two derivatives of the original separate terms.

In summary: **If $y = kx^n$, $dy/dx = knx^{n-1}$.**
    The derivative of a constant term is zero.
    The derivative of a function with several terms is the sum of the derivatives of the separate terms.

EXAMPLE 9.   Find the derivative of $y = 2x^3 - 5x^2 + 7x + 4$.
Solution: By successively using the power formula on each term,

$$\frac{dy}{dx} = 6x^2 - 10x + 7.$$

Note here that this function would have the same derivative if the constant term 4 were omitted. This stands to reason, since the graph of

$y = 2x^3 - 5x^2 + 7x + 4$ is the same as the graph of $y = 2x^3 - 5x^2 + 7x$, except that each point on the first graph is moved vertically upward by four units, as shown in Fig. 3-5. This will not affect the slope of the graph in any way; therefore the derivative is the same for both.

The symbol dy/dx means the derivative with respect to x of y. Taking its meaning in the same literal order as this interpretation, it may be symbolized as $(d/dx)(y)$. In this way, d/dx itself becomes a symbol which means "the derivative with respect to x" of what follows. Thus, if $y = 3x - 5$, for example, dy/dx means $(d/dx)(3x - 5)$, and the problem may be stated, "Find $(d/dx)(3x - 5)$." The dependent variable y does not appear in this statement of the problem, although it is always implied. This method of stating the problem is used simply for brevity.

Other symbols for the derivative are also frequently used. The symbol D_x, or simply D, to mean d/dx is coming into frequent usage; the subscript x may be omitted when it is obvious which letter represents the independent variable; $D(3x - 5)$ would need no clarification; however, $D(uv - v^2)$ would be ambiguous, since either u or v may be the independent variable,

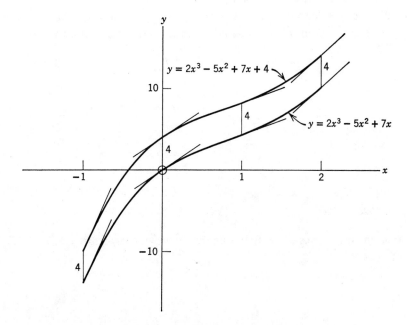

Fig. 3-5. Functions differing by a constant term have the same derivative.

the other letter being considered constant. $D_v(uv - v^2)$ is easily understood; we are asked to differentiate *with respect to v*, and the answer is $u - 2v$.

The symbol y' is also used to mean the derivative of y, but again this may be ambiguous unless the independent variable is clearly indicated. In functional notation, when a generalization is used to include all type of functions, the derivative of $f(x)$ is usually denoted by $f'(x)$; in this instance it is quite clear that x is the independent variable.

EXAMPLE 10. Find $f'(2)$, given $f(x) = 3x^2 - \sqrt{2x}$.

Solution: $f'(2)$ is a symbol indicating the value of $f'(x)$ when $x = 2$. $f(x) = 3x^2 - \sqrt{2} x^{\frac{1}{2}}$, the second term of the function being put in the form kx^n, a constant factor times a power of x, in preparation for differentiation.

$$\therefore f'(x) = 6x - \sqrt{2}\,\tfrac{1}{2}x^{-\frac{1}{2}} = 6x - \frac{\sqrt{2}}{2}\,x^{-\frac{1}{2}}.$$

$$\therefore f'(2) = 6(2) - \frac{\sqrt{2}}{2}\frac{1}{\sqrt{2}} = 12 - \tfrac{1}{2} = 11.5.$$

EXAMPLE 11. Find $(d/dt)(at^2 - 1/2bt + \sqrt{ct})$.

Solution: Since t is the only variable, it is important to separate constant multipliers from the variable factor in the second and third terms.

$$\therefore \frac{d}{dt}\left(at^2 - \frac{1}{2bt} + \sqrt{ct}\right) = \frac{d}{dt}\left(at^2 - \frac{1}{2b}t^{-1} + \sqrt{c}\,t^{\frac{1}{2}}\right)$$

$$= 2at + \frac{1}{2b}t^{-2} + \tfrac{1}{2}\sqrt{c}\,t^{-\frac{1}{2}}$$

$$= 2at + \frac{1}{2bt^2} + \frac{\sqrt{c}}{2\sqrt{t}}.$$

EXERCISES

Find the derivatives of the following functions with respect to x:

1. $y = 2x^3$.

2. $y = 2\sqrt{x}$.

3. $y = \dfrac{3}{x}$.

4. $y = 3x\sqrt{x}$.

5. $y = 4x^2 - 2$.

6. $y = 5x^3 - 7x + 3$.

7. $y = 3x - \dfrac{5}{x} - 7$.

8. $u = 4kx - \dfrac{3m}{x}$.

9. $i = \dfrac{5}{\sqrt{x}} + \dfrac{2}{x} - a$.

Differentiate the following functions with respect to t:

10. $v = 0.4t^2 - 2$.

11. $s = 3t + \dfrac{1}{t^2}$.

12. $i = at^2 - \sin 30° + \log 5$.

13. $y = \sqrt{2}\,t + \sqrt{2t}$.

Calculate the following:

14. $\dfrac{d}{dx}\left(\dfrac{6}{x} - 2\sqrt{x}\right)$.

15. $D\left(3t^{\frac14} + \dfrac{2}{\sqrt{t}} - 4\right)$.

16. $D_y(5y^2 - 2xy + x^2)$.

17. y', where $y = 3x - x^{-1} + \dfrac{2}{\sqrt{x}}$.

18. $f'(t)$, where $f(t) = v_0 t - 16t^2$.

19. $f'(3)$, where $f(x) = 2x - \dfrac{18}{\sqrt{3x}}$.

Find the slope of each of the following curves at the point indicated:

20. $y = x^5 - 3$ at $x = 2$.

21. $\sqrt{y} = 5x + 4$ at $x = -1$.

22. $y = (2x - 3)^3$ at $y = -27$.

Find the derivatives of the following functions:

23. $y = \dfrac{3t - 1}{t}$.

24. $y = (x^{\frac12} - 1)^2$.

25. $y = \sqrt{x^2 - 4x + 4}$.

26. $v = \dfrac{ax + b}{x^2}$, with respect to x.

27. $s = (t - 1)(t + 1)$.

28. $v = \dfrac{t^3 - t}{t + 1}.$

29. $y = \sqrt{x + \dfrac{1}{x} + 2}.$

30. Find the velocity and acceleration at time $t = 2$ sec of a body whose displacement s in feet is related to time t in seconds by the equation $s = \frac{1}{2}t^2 + \sqrt{t}$.

31. Find the slope of the normal to the curve $y = 4/x$ at the point where $x = 3$. (The normal is a line perpendicular to the tangent at the point of contact.)

32. Find the rate of change of q with respect to t, at $t = 5$, given

$$q = 10^{-2}t^2 + 10^{-1}t.$$

33. Find dy/dx at the point where $x = 0.25$, given

$$y = (3x)^2 + 4x/\sqrt{4x} - 1/2x^{1.5} + 4.$$

34. The reactance of an electrical circuit is given by $X = 2\pi f L - 1/2\pi f C$. Find, in terms of the constants, the rate of change of X with respect to:

  (a) A variable L, with all other quantities on the right-hand side constant.
  (b) A variable C, with all other quantities on the right-hand side constant.
  (c) A variable f, with all other quantities on the right-hand side constant.

3.4   PROBLEMS INVOLVING RATES OF CHANGE

The graphical interpretation of the derivative as the slope of the tangent to a curve, or as the slope of the curve itself, at any point on the graph, must constantly be kept in mind. The physical interpretation of the derivative as an instantaneous rate of change is even more important.

Many electrical quantities are directly proportional to the rate of change of a more basic quantity and are therefore evaluated by taking a derivative. Some of the more useful relationships involving derivatives are listed here.

Type of Relationship	Equation	Units
Linear velocity	$v = \dfrac{ds}{dt}$	Any consistent units
Linear acceleration	$a = \dfrac{dv}{dt}$	
Angular velocity	$\omega = \dfrac{d\theta}{dt}$	Any consistent units
Angular acceleration	$\alpha = \dfrac{d\omega}{dt}$	

Type of Relationship	Equation	Units
Power	$P = \dfrac{dW}{dt}$	Any consistent units

Type of Relationship	Equation	Units
Current in a circuit	$i = \dfrac{dq}{dt}$	t in seconds
Voltage across an inductor	$v = L\dfrac{di}{dt}$	q in coulombs i in amperes
Charge on a capacitor	$q = Cv$	v in volts
Current to a capacitor	$\therefore\ i = C\dfrac{dv}{dt}$	L in henrys C in farads
Induced voltage in a coil	$v = N\dfrac{d\phi}{dt}$	N in number of turns ϕ in webers
Voltage in one coil due to current in another	$v_2 = M\dfrac{di_1}{dt}$	M in henrys

The problem of rates may be examined in another way for several basic examples. If displacement s, for example, varies directly with time t, velocity is the constant rate s/t; if s does not vary directly with t, the instantaneous velocity is calculated as ds/dt. The ratio of the variables represents a rate only for direct variation. The derivative represents a rate for any type of variation.

A partial list of quantities to which this applies is given in the following tabulation.

Independent Variable	Dependent Variable	If Dependent Variable Varies Directly With Independent Variable	All Types of Function
Time t	Displacement s	Velocity $v = \dfrac{s}{t}$ (constant)	$v = \dfrac{ds}{dt}$
Time t	Velocity v	Acceleration $a = \dfrac{v}{t}$ (constant)	$a = \dfrac{dv}{dt}$
Time t (sec)	Charge q (coulombs)	Current $i = \dfrac{q}{t}$ (amp) (constant)	$i = \dfrac{dq}{dt}$
Time t	Work done W (or energy used)	Power $P = \dfrac{W}{t}$ (constant)	$P = \dfrac{dW}{dt}$

Each equation in the last column of this tabulation is general enough to cover the simpler instance in the third column. For example, if q is directly proportional to t, then $q = kt$ and $i = q/t = kt/t = k$; also $dq/dt = d/dt(kt) = k$. Thus the cases in the third column are particular cases of those in the last column.

EXAMPLE 12.   Find the equation of the tangent to the curve $y = x^2 - 2\sqrt{x}$ at the point where $x = 4$.

Solution: To find the equation of a straight line, two facts must be known about the line. In this problem, the two facts are not directly given, but the given data must be used to find them. With a study of the information, it should be seen that it is possible to find the slope of the line and a point on it. Within these two main steps are details to be carried out.

To find the slope of the tangent: The slope at any point is the value of the derivative at that point. $dy/dx = 2x - \frac{1}{2}(2)x^{-\frac{1}{2}} = 2x - x^{-\frac{1}{2}}$. Therefore at $x = 4$, the slope of the tangent is $2(4) - 4^{-\frac{1}{2}} = 8 - \frac{1}{2} = \frac{15}{2}$.

To find a point on the line: The point of contact is a point on the line, where $x = 4$. Since this point also lies on the curve, its coordinates satisfy the equation of the curve.

$$\therefore\ y = 4^2 - 2\sqrt{4} = 12.$$

Therefore one point on the line is (4,12).

Now, equating two expressions for the slope, the equation of the tangent is $(y - 12)/(x - 4) = 15/2$. This equation, when cleared of fractions, becomes $15x - 2y = 36$. The (x,y) of this equation is not the (x,y) of the given equation of the curve.

EXAMPLE 13.   A body moving in a straight line has a displacement of $s = 10t - 6t^2$ ft, t being in seconds. Find the following:
(a) Its velocity at time $t = 1.5$ sec.
(b) The time at which its velocity is 6 ft per second.
(c) The time it is at rest.
(d) Its acceleration at the time its velocity is zero.
Solution: (a) $v = ds/dt = 10 - 12t$.
At $t = 1.5$, $v = 10 - 12(1.5) = 10 - 18 = -8$ ft per second.
(b) For this condition, $10 - 12t = 6$; $12t = 4$; $t = \frac{1}{3}$ sec.
(c) For this condition, $v = 0$; $10 - 12t = 0$; $t = \frac{5}{6}$ sec.
(d) $a = dv/dt = (d/dt)(10 - 12t) = -12$.
Therefore its acceleration is -12 ft per second per second at all times including $t = \frac{5}{6}$.

EXAMPLE 14. The voltage across a capacitance of $C = 2.0 \times 10^{-2}$ farad in an *RLC* series circuit is given by $v = 80t^2 + 4$ volts for a short interval of time t seconds. If the inductor has an inductance $L = 8.0 \times 10^{-3}$ henry, find (a) an expression for current i in the circuit and (b) an expression for voltage across the inductor.

Solution:

(a) $i = C(dv/dt) = 2.0 \times 10^{-2} \times 160t = 3.2t$ amperes.

(b) Voltage across the inductor is

$$L(di/dt) = 8.0 \times 10^{-3}(d/dt)(3.2t)$$
$$= 8.0 \times 10^{-3}(3.2)$$
$$= 2.56 \times 10^{-2} \text{ volt, a constant.}$$

EXERCISES

1. A rocket travels a distance s feet from the launching site in t seconds, where $s = 20t^3$.

(a) Find its velocity at the end of 2 sec.

(b) Find its velocity when it has traveled 1280 ft.

2. Find the velocity at time $t = 2$ sec of a body for which displacement $s = 3t - t^2$ feet after t seconds of motion.

3. Find the slope of the tangent to the curve $y = 2x^3 - 5x$ at $x = 1$.

4. For the curve $y = 3x^2 - 6x$, find (a) the coordinates of the point of contact of the horizontal tangent and (b) the equation of the horizontal tangent.

5. Find the acceleration at $t = 1$ sec of a body whose displacement s, t seconds after it has been set in motion, is given by $s = 3t - 2t\sqrt{t}$ ft.

6. The work done in moving a body is expressed as $W = 2t - 3/t$ ft-lb over an interval of time t seconds. Find the power of the force creating the motion at $t = 2$ sec.

7. Find the angular acceleration at time $t = 3$ sec of a rotating machine for which the angular displacement at time t seconds is given by $\theta = t^3 - 2.5t^2$ radians.

8. For a certain enclosed gas, the pressure P in pounds per square inch and the volume V in cubic feet are related by the function $PV^{4/3} = 100$. Find the rate at which the pressure is changing with volume when the pressure is 6.25 lb per square inch.

9. Find the equation of the tangent to the curve $y = 4x^{3/2} + 2$ which has a slope of 6.

10. Find the equation of the tangent to the curve $y = 3x^2 + 1$ which makes an angle of $60°$ with the positive horizontal.

11. Find two points on the curve $y = 2x^3 - 3x^2 - 37x + 50$ where the tangent makes an angle of $135°$ with the positive horizontal.

12. The displacement of a body as a function of time t in seconds is given by $s = t^2(t^2 - 3)^2$ feet. Find (a) at what times its velocity is zero and (b) its acceleration at these times.

13. The distance traveled by an airplane along the runway from the time of touchdown is given by $s = 100t - 2t^2$ feet, t being in seconds after touchdown. Find (a) its velocity at touchdown, (b) its acceleration, and (c) how far it travels before coming to a stop.

14. A transmission line hangs in an approximate parabolic shape whose equation is $y = 2.0 \times 10^{-3}x^2$, where x and y are in feet and the origin is taken as the center point of the line. If the supporting poles are 200 ft apart, find the approximate angle of slope of the line at the point of support. Express the answer to the nearest degree.

15. Find the point of intersection of the curves $y = 1/x^2$ and $y^2 = x$ and the angle between the tangents to the curve at this point. Make a sketch to illustrate.

16. Find the coordinates of all points on the curve $y = 2x^3 - 3x^2 - 4x + 2$ where the tangent is perpendicular to the line $8y + x - 18 = 0$.

17. Find the equation of the normal to the curve $y = x^2 - 2x^{3/2}$ at the point where $x = 4$.

18. The angular displacement in radians of a rotating wheel t seconds after it has been set in motion is given by $\theta = -2t^{3/2} + \frac{1}{4}t^2$. Find (a) the angular velocity at $t = 4$, and (b) the angular acceleration at the moment it reverses direction.

19. In Fig. 3-6, the graph of the charge q coulombs flowing to a capacitor against time t in seconds is shown for two separate cases. For each case, sketch a graph of the current against time.

20. A differentiating circuit consists of an input voltage impressed on a capacitance C, which is in series with a resistance R, the output voltage being the voltage across the resistance (current in amperes times resistance in ohms). If $R = 12$ ohms and $C = 0.050$ farad, sketch the output voltage against time for the waves of *voltage across the capacitance* shown in Fig. 3-7.

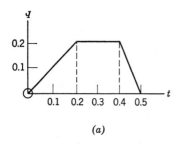

(a)

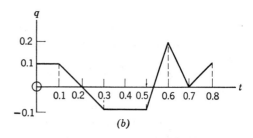

(b)

Fig. 3-6.

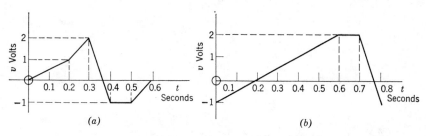

Fig. 3-7.

21. Find expressions for the following:

(a) The current in a circuit at t seconds in which the charge $q = 2.5t^{6/5} + 2t$ coulombs.

(b) The voltage across a 0.050-henry inductance at t seconds with current $i = 0.6t^2 + 2t$ amperes.

(c) The current flowing to a condenser of 3.0×10^{-2} farad over an interval t seconds if the voltage across the condenser is given by $v = 50\sqrt{t^3} + 10\sqrt{t}$ volts.

22. A coil of 20 turns is linked by a magnetic flux $\phi = 0.5t^3 - 0.8\sqrt{t}$ webers, t being in seconds. Find the induced voltage in the coil at time 0.8 sec.

23. The mutual inductance between two coils is 0.008 henry.

(a) The current in one is given by $i = 4t^{1/2} - 2t^{1/4}$ amperes, t being in seconds. Find the voltage induced in the other at time 0.0064 sec.

(b) If the current in one coil is constant, what voltage is induced in the other?

24. The charge q flowing to an inductance is given by $q = 0.050 \sin 100\pi t$, where q is in coulombs, t is in seconds, and the angle is in radians. Calculate the approximate current at $t = 0.004$ sec. (Since the derivative of a trigonometric function has not been developed, an approximate derivative may be calculated by taking a very small increase in t around $t = 0.004$ and, using tables, the corresponding increase in q may be determined; dq/dt will equal $\Delta q/\Delta t$ approximately. It is suggested that Δt be taken as the interval between $t = 0.0039$ sec and $t = 0.0041$ sec.)

CHAPTER 4

Maximum and Minimum

4.1 SECOND- AND HIGHER ORDER DERIVATIVES

We have seen that the derivative of a function represents the instantaneous rate of change of a dependent variable with respect to an independent variable. In analyzing the change in a function or in solving practical problems, it is often necessary to know at what rate the derivative itself is changing; interpreted graphically this is the rate at which the slope changes with respect to the independent variable.

It should be apparent that the rate of change of the derivative may be determined by taking a derivative of the function representing the derivative; this means that two successive derivatives of the original function must be taken, and the resulting function is called the *second derivative* of the original function. For $y = f(x)$, the second derivative means $(d/dx)(dy/dx)$, but any one of the contracted forms d^2y/dx^2, y'', $f''(x)$, D^2y may be used to represent the second derivative. The derivative of the original function [dy/dx, y', $f'(x)$ or Dy] may then be designated as the *first derivative* to avoid confusion.

Further derivatives beyond the second may also be taken, d^3y/dx^3, y''', or D^3y representing the *third derivative*, etc., as far as necessary. Each successive derivative represents the rate of change of the previous derivative with respect to the independent variable.

A second or higher order derivative cannot be determined directly in one step, since the calculation of each derivative depends on the function representing the derivative of one lower order. To calculate the fourth derivative, for example, four separate calculations are necessary.

The first derivative represents the rate of change of the dependent variable with respect to the independent variable at any value of the independent

54

variable; graphically, the first derivative represents the slope at any point. The second derivative represents the rate of change of the slope. If the slope increases with increasing x, then the rate of change of the slope with respect to x is positive. If the slope is increasing rapidly with respect to x, the rate of change of slope (that is, d^2y/dx^2) will be a large positive number. A positive second derivative thus indicates a positive rate of change of slope, hence an increasing slope; a negative second derivative indicates a decreasing slope.

A common example of successive derivatives is that of linear motion. Velocity at any instant is the rate of change of displacement with respect to time. Acceleration is the rate of change of velocity with respect to time.

$$v = \frac{ds}{dt}.$$

$$a = \frac{dv}{dt} = \frac{d}{dt}\left(\frac{ds}{dt}\right) = \frac{d^2s}{dt^2}.$$

and d^3s/dt^3 represents the rate of change of acceleration with respect to time.

EXAMPLE 1.   If $s = 6t^2 - t^3$ ft at t seconds, find the velocity, acceleration, and rate of change of acceleration at $t = 2$. Sketch graphs of s, v, a, and da/dt against time.

Solution:

$$v = \frac{ds}{dt} = 12t - 3t^2,$$

$$a = \frac{d^2s}{dt^2} = 12 - 6t,$$

$$\frac{da}{dt} = -6.$$

At $t = 2$,

velocity $= 12(2) - 3(2)^2 = 12$ ft per second,
acceleration $= 12 - 6(2) = 0$,
rate of change of acceleration $= -6$ ft per second per second
per second at all times.

The graphs are shown in Fig. 4-1.

EXAMPLE 2.   For the function $y = -4x^3 + 3x^2 + 5$, find for what values of x:
(a) The function is increasing with respect to x.
(b) The function is neither increasing nor decreasing.
(c) The slope is increasing with respect to x.

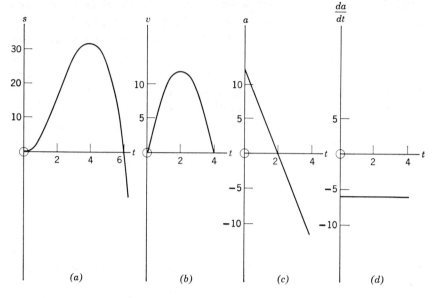

Fig. 4-1. Graphs of (a) s, (b) v, (c) a, and (d) $\dfrac{da}{dt}$ for $s = 6t^2 - t^3$.

(d) The slope is decreasing with respect to x.

(e) The rate of change of slope is increasing with respect to x.

Solution: (a) For an increasing function, the rate of change of the function is positive; in other words, dy/dx is positive.

$$\frac{dy}{dx} = -12x^2 + 6x = 6x(1 - 2x).$$

For this to be positive, x must have the same sign as $1 - 2x$. By trying various values of x, both positive and negative, it should finally be seen that $\frac{1}{2} > x > 0$. Thus for values of x between 0 and $\frac{1}{2}$, the function increases. For values of x less than 0 and greater than $\frac{1}{2}$ the function decreases.

(b) For a function which neither increases nor decreases, the rate of change of the function is zero; that is, $dy/dx = 0$. Since $dy/dx = 6x(1 - 2x)$, this can only be true if $x = 0$ or $\frac{1}{2}$.

(c) For an increasing slope, $(d/dx)(dy/dx)$ is positive; that is, d^2y/dx^2 is positive. $d^2y/dx^2 = -24x + 6$, and this is positive for all values of x less than $\frac{1}{4}$. Therefore, the slope increases with respect to x for all values of x less than $\frac{1}{4}$.

(d) From the reasoning of part c, the slope decreases with respect to x for all values of x greater than $\frac{1}{4}$.

(e) For the rate of change of slope to be increasing, $(d/dx)(d^2y/dx^2)$ must be positive; that is, $d^3y/dx^3 > 0$. $d^3y/dx^3 = -24$, a constant negative value. Therefore there are no values of x for which the rate of change of slope increases with respect to x.

EXERCISES

1. Find the second derivative of the following functions:

(a) $y = 3x^2 - 2x$.

(b) $s = 4t - 5t^2$.

(c) $i = 3t - 2\sqrt{t}$.

(d) $v = \dfrac{3}{t} - \dfrac{2}{\sqrt{t}}$.

2. Evaluate the second derivative of the following functions at the point where the independent variable equals 4.

(a) $y = \dfrac{16}{x^2} + 2x^{\frac{1}{2}}$.

(b) $P = \dfrac{300}{V^{1.5}}$.

(c) $q = 3t(t - \sqrt{t})$.

3. Find functions to represent (i) the acceleration and (ii) the rate of change of acceleration, at $t = 2$, for the following equations of motion, assuming units of feet and seconds.

(a) $s = 24t - 16t^2$.

(b) $s = \dfrac{4}{t} - \dfrac{3}{t^2}$.

(c) $v = 2t + \dfrac{4}{t}$.

4. For the function $y = 3x^2 - 12x$:

(a) Find for what values of x the function is increasing with increasing x.

(b) Find for what values of x the function is decreasing with increasing x.

(c) Find the coordinates of the point where the function is neither increasing nor decreasing.

(d) What is the slope at the point in (c)?

(e) Find the rate of change of the slope.

(f) Find the x-intercepts.

(g) Sketch the graph making use only of information calculated in the previous parts of this exercise.

5. For the function $y = x(x^2 - 3x - 9)$:

(a) Find the values of x for which the slope is zero.

(b) Is the slope positive or negative between the values of x in (a)?

(c) Find the rate of change of slope at $x = 2$.

(d) Find the limitations on the values of x such that the slope increases with increasing x.

(e) Find the values of x that produce a decreasing slope with increasing x.

(f) Find d^3y/dx^3.

6. Using the relationships $i = dq/dt$ and $v = L(di/dt)$, and given $q = t^2 - 4t^{3/2}$ and $L = 0.05$, find v at $t = 4$.

7. The charge on a capacitor t seconds after a given time is given by $q = 4t^3 - t^4$ coulombs. Find for what positive values of t (a) the current increases with respect to time and (b) the rate of change of current increases with respect to time.

4.2 GRAPHICAL ANALYSIS OF MAXIMUM AND MINIMUM

The graph of $y = x^3 - 3x^2 + 2$, Fig. 4-2, reaches a local highest point at A and a local lowest point at B. Variable y is said to have a *maximum* value at A and a *minimum* value at B, even though, in the outer regions of the graph, y reaches a higher value than at A and a lower value than at B. Maximum and minimum thus refer to a local high and low value of the dependent variable, higher or lower, respectively than at points immediately on either side of it.

In the usual type of continuous graph (for example, Fig. 4-2), it is noted that, as viewed from above, the curvature is convex at a maximum point and concave at a minimum point. Thus, between a successive maximum and minimum, there is a point where convexity changes to concavity. This point is called a *point of inflection*. At C, the point of inflection in Fig. 4-2, the tangent crosses the curve. This is not true at any point other than a point of inflection. Thus a point of inflection is often defined as a point at which the tangent crosses the curve.

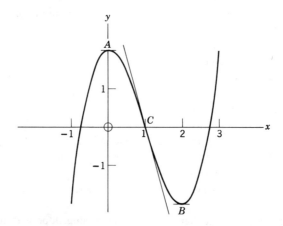

Fig. 4-2. Graph of $y = x^3 - 3x^2 + 2$.

The maximum and minimum points are designated, as a group, as *turning points*, and, as a group, all maxima, minima, and points of inflection are known as *critical points*.

Examination of the slope and how the slope changes as x increases from the extreme left to the extreme right in Fig. 4-2 results in the following observations:

At the extreme left, the slope has a very high positive value.
As x increases, the slope is positive but becomes less, up to point A.
At A, the slope is zero.
From A to C, the slope is increasingly negative; that is, the slope is still decreasing.
From C to B, the slope is negative, but it is becoming less negative, or increasing.
At B, the slope is zero.
To the right of B, the slope is positive and increasing.

These conclusions may be verified by the following tabulation:

x	-1.0	-0.5	0	0.5	1.0	1.5	2.0	2.5	3.0
Slope $= 3x^2 - 6x$	9	3.75	0	-2.25	-3.0	-2.25	0	3.75	9
$y'' = 6x - 6$	-12	-9	-6	-3	0	$+3$	$+6$	$+9$	$+12$
Analysis	Positive slope	Max.		Negative slope			Min.	Positive slope	
	—Decreasing slope—				Infl.	—Increasing slope—			

The function previously analyzed is typical enough to permit some general conclusions. For any continuous function, the slope (that is, the first derivative) is zero at maximum and minimum points. The basic feature that distinguishes a maximum from a minimum is that for a maximum the slope changes from positive to negative with increasing x through the critical point, whereas for a minimum the slope changes from negative to positive with increasing x through the critical point. In some cases it may be necessary to evaluate the first derivative at points on either side of the critical point and close to it in order to distinguish whether the point produces a maximum or a minimum y. This procedure may be lengthy, and there are usually simpler tests to distinguish the type of turning point involved.

Since, for a maximum, the slope changes from positive to negative continuously, the rate of change of the slope is negative on both sides of the maximum point. This usually means that the rate of change of the slope (that is, the second derivative) at the maximum point is negative, although in some special cases it may be zero at this one point. Thus, at a maximum point, the first derivative is zero and the second derivative is almost always negative. Certainly, if the first derivative is zero and the second derivative is negative at a point, that point will be a maximum point.

Similarly, for a minimum, the slope changes from negative to positive and the second derivative is positive on both sides of the minimum point and usually at the minimum point as well. Thus, if the first derivative is zero and the second derivative is positive at a point, that point will be a minimum point.

At a point of inflection, d^2y/dx^2 changes sign (in this function from negative to positive, but in other cases the change may be from positive to negative). Therefore at a point of inflection, d^2y/dx^2 must be zero. Another way of expressing this is that, at a point of inflection, the slope itself reaches a minimum value (here, -3) or, in some other instances, a maximum value; this means that the first derivative of the slope is zero, or $d^2y/dx^2 = 0$.

The following tabulation therefore indicates what usually are the quickest tests for critical points:

Value of first derivative	0	0	Any value including zero
Value of second derivative	Negative	Positive	0
Conclusion	Maximum	Minimum	Usually a point of inflection

Some fine points of logic are involved here, and we must be careful to interpret correctly. For example, it is false to say that, if the first derivative is zero, at this point there is a maximum or a minimum, since it should be apparent that the point may be a point of inflection. In Fig. 4-3, the point of inflection C occurs when the slope is zero.

In addition, a straight horizontal line has a slope (or first derivative) of zero, which does not indicate any critical point. Any linear function has a second derivative of zero, which does not indicate a point of inflection.

Thus setting the first derivative equal to zero could be a condition for many types of point, and to clarify whether the point is a critical point and, if so, of what type, further testing is required.

For all continuous functions, however, it is correct to state that for a maximum or a minimum the first derivative equals zero and for a point of

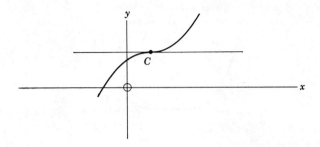

Fig. 4-3. A point of inflection where the slope is zero.

inflection, the second derivative is zero. The converses of these statements are not true; examples of false converses were given above.

In Fig. 4-4, y has its greatest value at A, and yet the slope at A is not equal to zero but has two distinct values. However, y is a discontinuous function, and the point A is not considered to be a maximum. Almost every rule has an exception; however, the rules given here, combined with an awareness that they may not always be interpreted blindly, should be sufficient for analysis.

In the preceding tabulation it is noted that when first and second derivatives are both zero, the point is *usually* a point of inflection. It has already been indicated that it may be a point on a horizontal line. It may also be a maximum or a minimum, where the second derivative is respectively negative or positive on both sides of the point and equal to zero at the point itself. This usually occurs where a linear function of y is proportional to a single power of a linear function of x.

Figure 4-5 illustrates the graphs of $y - 3 = x^4$, $y = 3 - x^4$, and $y - 3 = x^3$. A simple calculation shows that, for all three functions, the first and second derivatives are both zero at $x = 0$, and yet, at $x = 0$ one function reaches a

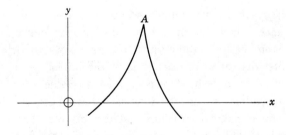

Fig. 4-4.

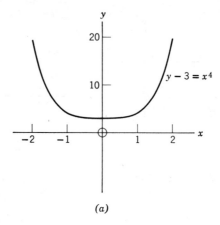

(a)

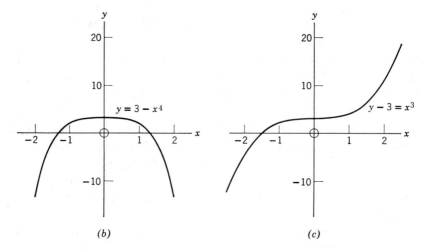

(b)   (c)

Fig. 4-5. (a) Plot of $y - 3 = x^4$. (b) Plot of $y = 3 - x^4$. (c) Plot of $y - 3 = x^3$.

minimum, one reaches a maximum, and the third has a point of inflection. In these instances, the value of the second derivative indicates no definite conclusion. More basic methods must be used, such as determining the sign of the first derivative on each side of the critical point. Thus where $y - 3 = x^4$, $y' = 4x^3$ and the slope is negative for values of x less than zero and positive for values of x greater than zero, the condition for a minimum point. In a similar manner, conclusive statements may be established for the other two functions. In this connection, note that the first derivative has the same sign on both sides of a point of inflection.

EXAMPLE 3.   For the function $y = x^3 - 6x^2 + 9x + 1$, find (a) all critical points and specify their type and, (b) any other values of x for which the function is equal to its maximum value.

Solution: The graph is shown in Fig. 4-6.

(a) $y' = 3x^2 - 12x + 9 = 3(x^2 - 4x + 3) = 3(x - 3)(x - 1)$.

$\quad y'' = 6x - 12$.

For maximum and minimum, $y' = 0$; hence $x = 3$ or $x = 1$.

At $x = 3$, $y'' = 6(3) - 12 = 6$, which is positive; therefore at $x = 3$, y is minimum. Minimum $y = 3^3 - 6(3)^2 + 9(3) + 1 = 1$. At $x = 1$, $y'' = 6(1) - 12 = -6$, which is negative. Therefore at $x = 1$, y has a maximum value of $1 - 6 + 9 + 1 = 5$.

For a point of inflection, $y'' = 0$; $6x - 12 = 0$; $x = 2$, and $y = 3$. Therefore at $(1,5)$, $(2,3)$, and $(3,1)$ there are a maximum, a point of inflection, and a minimum, respectively.

(b) The maximum value of y is 5. At any other point where the function is equal to its maximum value, $x^3 - 6x^2 + 9x + 1 = 5$. Therefore $x^3 - 6x^2 + 9x - 4 = 0$ at any point where $y = 5$.

This equation may present some difficulty; however, it should be realized from part a that $x = 1$ is a solution; in fact, since the line AD in Fig. 4-6, whose equation is $y = 5$, is a tangent at $x = 1$, then $x = 1$ is a double solution, since what would usually be two points of intersection have coincided at the one point A. This means that the function $x^3 - 6x^2 + 9x + 1$ has a factor $(x - 1)^2$. By long division the other factor is easily determined, and $x^3 - 6x^2 + 9x - 4 = (x - 1)^2(x - 4)$.

Therefore, at $y = 5$, $(x - 1)^2(x - 4) = 0$, and $x = 1$ or $x = 4$.

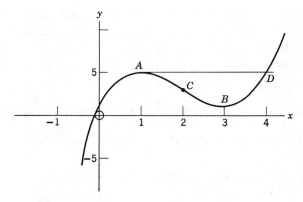

Fig. 4-6. Graph of $y = x^3 - 6x^2 + 9x + 1$.

Therefore, at $x = 4$ (point D in Fig. 4-6), the function repeats its maximum value.

EXAMPLE 4.   A body moves in a straight line so that, after t seconds, its displacement is given by $s = 3t^2 - t^3/2$ feet. Find (a) its velocity at $t = 2.5$, (b) at what time its displacement is a maximum, and (c) its maximum positive velocity.

Solution: (a) $v = ds/st = 6t - 3t^2/2$. At $t = 2.5$, $v = 15 - 18.75/2 = 5.63$ ft per second.

(b) For maximum s, $ds/dt = 0$ and d^2s/dt^2 is negative.

$$\therefore \ 6t - \frac{3t^2}{2} = 0; \qquad 3t\left(2t - \frac{t}{2}\right) = 0. \qquad t = 0 \text{ or } 4.$$

$d^2s/dt^2 = 6 - 3t$, and, at $t = 4$, d^2s/dt^2 is negative. Therefore at $t = 4$ sec, the displacement is maximum. (For maximum or minimum displacement, the velocity is zero.)

(c) For maximum positive velocity, $dv/dt = 0$ and d^2v/dt^2 is negative.
$d/dt(6t - 3t^2/2) = 0.$ $6 - 3t = 0.$ $t = 2.$
$d^2v/dt^2 = -3$, which is negative for all values of t.
Therefore at $t = 2$, the velocity reaches a maximum of $6(2) - 3(2)^2/2 = 6$ ft per second.

The qualification "positive" applied to a maximum is sometimes necessary to distinguish from a maximum negative, which would mean a negative quantity of greatest absolute value, that is, a mathematical minimum.

It is important to note that "maximum" and "minimum" refer to values of the dependent variable; graphically, they are vertical quantities, a highest and a lowest point. However, the application of calculus and algebra to the problem yields the value of the independent variable which produces a maximum or minimum dependent variable. In the "xy" system, we find the x which produces maximum or minimum y.

EXAMPLE 5.   Find all critical points on the graph (Fig. 4-7) of

$$y = 12x^5 - 15x^4 - 20x^3 + 30x^2 - 10.$$

Solution: $\dfrac{dy}{dx} = 60(x^4 - x^3 - x^2 + x).$

In this function, x is a common factor, and, since a substitution of $x = 1$

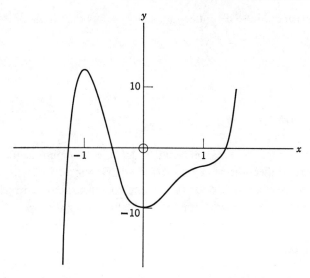

Fig. 4-7. Graph of $y = 12x^5 - 15x^4 - 20x^3 + 30x^2 - 10$.

makes the function equal to zero, $x - 1$ is also a factor. The remaining factor may be found by long division.

$$\frac{dy}{dx} = 60x(x - 1)(x^2 - 1) = 60x(x - 1)^2(x + 1).$$

$$\frac{d^2y}{dx^2} = 60(4x^3 - 3x^2 - 2x + 1) = 60(x - 1)(4x^2 + x - 1).$$

For maximum or minimum, $60x(x - 1)^2(x + 1) = 0$. Therefore $x = 0$, 1, or -1.

$x = 0$	$x = 1$	$x = -1$
$\dfrac{d^2y}{dx^2} = 60(-1)(-1) = 60$	$\dfrac{d^2y}{dx^2} = 0$	$\dfrac{d^2y}{dx^2} = -240$
(positive)	$\therefore x = 1$ gives a probable point of inflection. The graph verifies a point of inflection, with zero slope.	(negative)
$\therefore$ at $x = 0$, y has a minimum value of -10. (substitution)		$\therefore$ at $x = -1$, y has a maximum value of 13. (substitution)
	$y = -3$.	

For other points of inflection, $60(x - 1)(4x^2 + x - 1) = 0.$ $x = 1$ (already determined) or

$$4x^2 + x - 1 = 0.$$

$$x = \frac{-1 \pm \sqrt{1 + 16}}{8}$$

$$= -0.64 \text{ or } +0.39.$$

Therefore there is a maximum at $(-1, 13)$, a minimum at $(0, -10)$, and points of inflection at $x = -0.64$, $+0.39$ and $+1$. The graph may continuously be used to verify that the conclusions are correct and that the simple rules are valid for the problem.

EXERCISES

1. In Fig. 4-8, state whether the function is increasing or decreasing with respect to x for each of the four graphs.

2. In Fig. 4-8, state whether dy/dx is positive or negative for each case.

3. In Fig. 4-8, state whether the slope is increasing or decreasing with respect to x for each case.

4. In Fig. 4-8, state whether d^2y/dx^2 is positive or negative for each case.

5. Interpret the following characteristics of a function, using a characteristic of one of the derivatives in your answer.

(a) A decreasing function.

(b) An increasing slope.

(c) A function that is neither increasing nor decreasing.

(d) A slope that is neither increasing nor decreasing.

6. In Fig. 4-9, A, B, C, D, and E are the critical points. State between what two extreme critical points the following characteristics apply:

(a) An increasing function.

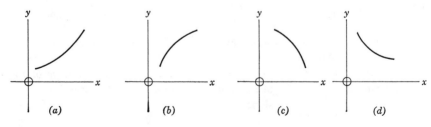

Fig. 4-8.

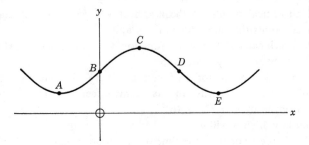

Fig. 4-9.

(b) A decreasing slope.

(c) A negative first derivative.

(d) A positive second derivative.

7. In Fig. 4-9, characterize the first and second derivatives as positive, negative, or zero for each of the five critical points.

8. Find, without graphing, whether the graph of $y = 2x^3 - 3x^2$ is convex or concave as seen from above at the following points:

(a) $x = 1$.   (b) $x = -2$.   (c) $x = 0.5$.   (d) $x = 0.4$.

9. State whether each of the following statements is true or false and analyze the reason in the case of false statements.

(a) If the derivative of a function of x is zero at a certain point, and the value of the function at that point is greater than its value for a greater value of x, the point is a maximum.

(b) If the derivative of a function of x is zero at a certain point, and the value of the function at that point is greater than its value for one greater value of x and for one lesser value of x, the point is a maximum.

(c) For a continuous function, any maximum has a greater value than any minimum.

(d) For a continuous function, any maximum has a greater value than its nearest minimum.

10. The following characteristics may indicate a maximum point. State in each case what other type(s) of point might be indicated.

(a) The first derivative at the point is zero.

(b) The first derivative at the point is zero, and the slope immediately to the left of that point is positive.

(c) The first derivative at the point is zero, the slope at an arbitrarily chosen point to the left is positive, and the slope at an arbitrarily chosen point to the right is negative.

(d) The second derivative at the point is negative.

11. Find any points of inflection on the graph of $y = x^2 - 3x + 2$.

12. Find all critical points on the graph of $y = x^2 - 3x + 2$, stating what type of critical points they are. Sketch the graph.

13. Find all critical points for the function $y = 0.5x^4$, stating what type of critical points they are. Sketch the graph.

14. Find all critical points for the function $y = x - 2x^4$, stating the type.

15. A ball is thrown upwards, and its height h feet at a time t seconds after it is thrown is given by $h = 32t - 16t^2$.

(a) Find how high it will go.

(b) What is the velocity at the time of maximum height?

(c) What is the acceleration at the time of maximum height?

(d) At what time does it return to its starting point?

(e) Sketch graphs of height, velocity, and acceleration against time.

16. Find all maxima, minima, and points of inflection for the function $y = 2x^3 - 3x^2 - 12x + 4$. Sketch the graph.

17. Find all critical points for the function $y = x^3 - 6x^2 + 12x$. Sketch the graph.

18. The equation $s = t(2t - 3)(t - 4)$ represents the displacement s feet of a moving body at time t seconds. Find (a) at what time it reaches a maximum displacement, (b) the time it reaches a maximum negative displacement, (c) the time of minimum velocity, and (d) the acceleration at the time of minimum velocity. Sketch s against t.

19. (a) Find all critical points for the function $y = x^4 - 6x^2 + 4$.

(b) Find other values of x for which the function of (a) has a value equal to its maximum.

(c) Sketch the graph from $x = -3$ to $x = 3$.

20. For the function $y = (x - 2)^2(x + 2)$, find (a) all critical points and (b) another value of x for which y equals its minimum value.

21. For the function $y = x^3 + 3x^2 - 9x + 5$, find (a) all critical points, (b) another value of x for which y equals its maximum value, and (c) another value of x for which y equals its minimum value.

22. Show that the function $y = x - 4/x$ has no critical points. Sketch the graph from $x = -4$ to $x = 4$ to illustrate.

23. (a) Find the maximum and minimum values of the function $y = x + 4/x$.

(b) Explain the apparent paradox of the answer to (a) with the aid of a graph.

24. For the curve $y = x^4 + 2x^3$:

(a) Find the slope at each point of inflection.

(b) Between what values of x is the slope decreasing as x increases?

(c) Sketch the curve to cover significant values of x.

25. Find the minimum value of the function $y = \dfrac{1 + 16t^3}{t^2}$, showing clearly that it is a minimum.

26. Find all critical points for the function $y = 6 + 4x + 3x^2 - x^3$.

27. Find all critical points for the function $y = 6 - 4x + 3x^2 - x^3$. Sketch.
28. Find all critical points for the function $y = 2\sqrt{x} + 27/x$. Sketch.

4.3 WORDED PROBLEMS

The application of the principles of maximum and minimum to worded problems involves a careful preliminary analysis of the way in which the quantities vary and an interpretation of the worded relationships into symbols and equations prior to the application of calculus and algebra.

Since the formation of the equation from the worded problem is often the most difficult part of the solution, some specific examples will be analyzed from which some hints on how to arrive at the proper equation will be developed.

The type of problem discussed is limited by the fact that only the simplest type of differentiation has been developed so far. As techniques for more advanced differentiation are developed, more comprehensive problems in maximum and minimum will be possible. The theory of maximum and minimum, however, as well as the method of translating worded problems into equations is always basically the same, whether the differentiation involved is simple or more difficult.

Certain electrical laws, which are probably familiar, will be used in some of the following problems. For convenience these laws are listed here.

SOME ELECTRICAL LAWS

The power in a circuit of resistance R ohms carrying current i amperes is equal to $i^2 R$ watts.

The average power in a circuit of resistance R ohms carrying an a-c effective current of I amperes is equal to $I^2 R$ watts.

The voltage drop across a resistance R ohms carrying a current I amperes is equal to IR volts.

The voltage drops across two or more circuits in parallel are equal to one another, and this is the voltage drop across the parallel branch.

The currents in two or more parallel resistive branches are inversely proportional to the resistances.

At a junction point, the sum of the incoming currents is equal to the sum of the outgoing currents.

Around any closed loop, the sum of the voltage drops is equal to the applied voltage in the loop.

EXAMPLE 6.   Figure 4-10 illustrates a constant voltage of 6.0 volts applied to a constant resistance of 3.0 ohms in series with a variable resistance. Find the value of the variable resistance to result in maximum power in this variable resistor, and find this maximum power.

Solution: The power in the variable resistor is equal to I^2R, and it is apparent that, as R is varied, I, the current in R, will also vary. In very general terms, when R is decreased, I will increase as a result. Thus the expression for power involves two variables, one of which is dependent on the other. Before calculus is applied, the expression for power must involve only one variable. It is therefore necessary to find how I depends on R, so that a substitution may be made for either I in terms of R or R in terms of I. This preliminary analysis is necessary to see clearly the structure of the problem before carrying out any details. To find the relationship between I and R, it is seen that, for any value of R, the sum of the voltage drops around the closed loop is equal to 6 volts, the applied voltage.

$$\therefore\ 3.0I + IR = 6.0.$$

This equation may be used to substitute either for I in terms of R or for R in terms of I in the power equation $P = I^2R$.

$$3I + IR = 6; \quad I = \frac{6}{3 + R} \quad \text{or} \quad R = \frac{6 \div 3I}{I}.$$

$$\therefore\ P = \frac{36R}{(3 + R)^2} \quad \text{or} \quad P = \frac{I^2(6 - 3I)}{I}.$$

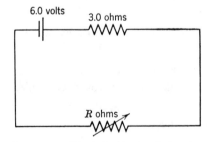

6.0 volts

3.0 ohms

R ohms

Fig. 4-10. To find the setting of R for the maximum power in R.

The first equation gives P as a function of R, and the second gives P as a function of I. Calculus may now be applied to either of these equations to find dP/dR in the first or dP/dI in the second. Both derivatives may be calculated; however, at this stage, the first function is difficult to differentiate, although it is easily differentiated after a study of quotients has been made. (*Caution:* The derivative of one function divided by another function is *not* the derivative of the first function divided by the derivative of the second function.)

The second of the two functions, however, may easily be expressed as the sum of two terms in powers of I, simply by dividing the numerator by I, as indicated. Thus $P = 6I - 3I^2$, and the equation is set up for applying calculus. For maximum P, $dP/dI = 0$.

$$\therefore\ 6 - 6I = 0. \qquad I = 1. \qquad \frac{d^2P}{dI^2} = -6, \quad \text{a negative value.}$$

Therefore P is a maximum when $I = 1$ ampere. Since $R = (6 - 3I)/I$ for all settings of R, the value of R to give maximum power is

$$\frac{6 - 3(1)}{1} = 3 \text{ ohms.}$$

The maximum power in the variable resistor $= I^2R = (1)^2(3.0) = 3.0$ watts. This result may be verified approximately by finding the power in the variable resistor for values of R above and below 3 ohms, as in the following tabulation:

R	1	2	3	4	5
$I = \dfrac{6}{3 + R}$	$\dfrac{3}{2}$	$\dfrac{6}{5}$	1	$\dfrac{6}{7}$	$\dfrac{3}{4}$
$P = I^2R$	2.25	2.88	3	2.94	2.81

EXAMPLE 7. A television manufacturer knows that if he produces 20,000 sets in a particular model the cost of manufacture will be $420 per set and the selling price will be $580 per set. He estimates that, for every additional 1000 sets up to a total of about 40,000 sets, the manufacturing cost for all sets will be reduced by $10 per set and the selling price for all sets will be reduced by $14 per set. Find from this estimate how many sets should be produced for a maximum profit.

Solution: Total profit equals the number of sets times the difference between cost price and selling price per set. The number of sets, the cost price per set and the selling price per set are all variables, and care must be taken not to have more than one variable in the expression for profit. It should, however, be apparent that both the cost and selling prices are dependent on the number of sets, so that it will be possible to represent the expression for profit as a function of one variable only.

This one independent variable may be designated as the total number of sets or as the number of sets in excess of 20,000; the latter choice will keep numerical portions of the equations to smaller values. The numerical coefficients may be made still smaller if the number of sets is designated in thousands. These ideas are not absolutely necessary for a correct solution; they will, however, simplify the details of calculation. Let n be the number of thousands of sets in excess of 20,000.

$$\text{Total number of sets} = (n + 20) \text{ thousand.}$$

$$\text{Manufacturing cost per set} = 420 - 10n \text{ dollars.}$$

$$\text{Selling price per set} = 580 - 14n \text{ dollars}$$

$\therefore$ Total profit

$$P = 1000(n + 20)[(580 - 14n) - (420 - 10n)].$$

$$P = 1000(n + 20)(-4n + 160) = 1000(-4n^2 + 80n + 3200).$$

For maximum P, $dP/dn = 0$ and d^2P/dn^2 is negative.

$$\therefore \quad 1000(-8n + 80) = 0; \qquad n = 10.$$

$d^2P/dn^2 = 1000(-8)$, a negative constant. Therefore, for $n = 10$, P is a maximum of

$$1000(10 + 20)(-40 + 160) = 3,600,000.$$

Therefore number of sets $= 20,000 + 10,000 = 30,000$, for a maximum profit of \$3,600,000.

The student should calculate the total profit for 29,000 and 31,000 sets, as an extra check on this result.

These problems in maximum and minimum may deal with a wide variety of subjects. In almost all situations, the greatest difficulties in solution seem to be the clear analysis of the variables involved and the setting up of the equation. The final steps of calculus and algebra are often very simple. The following general summary of steps to be taken in the solution of this type of problem may be of some assistance.

Summary of Steps in Solution of Worded Maximum and Minimum Problems

1. Form an expression for the quantity being tested for maximum or minimum, in terms of one independent variable. If more than one variable originally occurs in this expression, substitutions can be found to express them all in terms of one variable only. These substitutions result from either further information in the problem or basic laws or both.

2. Differentiate this expression with respect to the one variable remaining and equate the derivative to zero. Use algebra and calculus to finish the solution.

The next few examples show how the foregoing summary is constantly kept in mind in translating the words of a problem into an equation with one independent variable. These examples only take the solution as far as the formation of the equation, the completion of the problem being left to the student.

EXAMPLE 8.   Find two positive numbers whose product is 12 such that the sum of one of them and three times the other is a minimum.
Solution: An expression must be formed for the designated sum, since this is the quantity being tested for minimum. Each of the two required numbers may originally be treated as a separate variable, but sooner or later we see that they are not mutually independent. The further information in the problem that their product is 12 indicates that, if one of the numbers is x, the other is $12/x$.

Let the first number $= x$.

Then the second number $= \dfrac{12}{x}$.

The designated sum $S = x + \dfrac{3(12)}{x} = x + \dfrac{36}{x}$.

(The problem is completed by calculus and algebra.)

EXAMPLE 9.   A rectangular box with a square base is to be constructed from 36 ft² of wood. The base is to be of double thickness and a lid is to be made from the available material. Find the dimensions giving maximum volume.
Solution: An expression for volume must be formed. If the length and width of the square base are designated as x feet each and the height as h feet, volume $V = x^2h$. However, x and h are not mutually independent, due to the restriction of 36 ft² of wood. The relationship of x and h must be determined for a substitution.

Total material used (4 sides, 2 bottoms, and a top) $= 4hx + 3x^2$ ft^2.

$$\therefore \ 4hx + 3x^2 = 36, \quad \text{from which } h = \frac{36 - 3x^2}{4x}.$$

$$\therefore \ V = \frac{x^2(36 - 3x^2)}{4x} = 9x - \frac{3x^3}{4}.$$

The completion of this problem results in a length and width of 2 ft and a height of 3 ft.

EXAMPLE 10.   At a certain time two planes A and B are 50 miles apart, and A is directly west of B. A is tracking eastward at 400 miles per hour, and B is tracking northward at 200 miles per hour. Find how close they are at their nearest approach.

Solution: A diagram is necessary here (Fig. 4-11). The distance between planes is the quantity being tested for a minimum. Since this distance is variable, it is shown in the diagram as s, the distance apart after any time t has elapsed since their positions were at A and B. Let the units of t be hours, and t is fairly obviously the basic independent variable. s miles is to be expressed as a function of t.

Since A travels at 400 miles per hour, distance $AA_1 = 400t$ miles. Similarly, distance $BB_1 = 200t$ miles.

Distance $A_1B = 50 - 400t$ miles.

In the right-angled triangle A_1BB_1,

$$s^2 = (200t)^2 + (50 - 400t)^2$$

$$= 10^2(2000t^2 - 400t + 25).$$

$$s = 10\sqrt{2000t^2 - 400t + 25}.$$

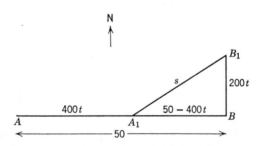

Fig. 4-11. Diagram for Example 10.

At this stage of differentiation, the technique of finding ds/dt for this function is not known. However, it may be reasoned that, since s must be positive, s^2 reaches its minimum at the same time that s reaches its minimum. Hence, for minimum s, $10^2(2000t^2 - 400t + 25)$ is a minimum; its first derivative must equal zero and its second derivative must be positive at the time calculated. In this way, the square root, which gives rise to the difficulty of differentiation, is avoided. The conclusion of the problem is again left for the student. The closest approach is slightly over 22 miles, after 6 min have elapsed.

EXAMPLE 11. A constant voltage of 12 volts is applied to a circuit with a 3-ohm resistor in series with three parallel-connected resistors, of 2 ohms, 4 ohms, and variable ohms, respectively. The circuit is diagrammed in Fig. 4-12. Find the setting of the variable resistor R that allows it to consume maximum power.

Solution: There are several approaches to this problem. One of the simplest is to let the current in the variable resistor be I amperes; the quantity being tested, the power in the variable resistor, P, equals $I^2 R$ watts. The problem now is that I and R both vary, but not independently. The next step is, therefore, to find the relationship between I and R in order to substitute for one in terms of the other in the equation $P = I^2 R$.

By applying simple electrical laws, the currents in all branches of the circuit may be expressed as functions of I and R; then the sum of the voltage drops around any closed loop may be equated to 12. The currents in the 4-ohm branch and the R-ohm branch are inversely proportional to the resistances; therefore the current in the 4-ohm branch $= (R/4)(I) = RI/4$. Similarly, the current in the 2-ohm branch $= RI/2$. These currents are easily verified by the fact that the voltage

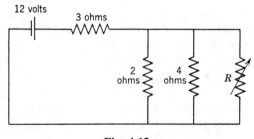

Fig. 4-12.

drop across each of the three parallel resistors, obtained as the product of current and resistance, is the same, equal to RI. Now the current in the 3-ohm branch is equal to the sum of the three currents in the branches of the parallel circuit. Therefore the current in the 3-ohm branch = $I + RI/4 + RI/2 = I + 3RI/4$. Taking the sum of the voltage drops around the extreme outside loop, or around any of the three loops which include the voltage source,

$$3\left(I + \frac{3RI}{4}\right) + RI = 12.$$

$$3I + \frac{13RI}{4} = 12.$$

$$12I + 13RI = 48. \qquad R = \frac{48 - 12I}{13I} \qquad \text{or} \quad I = \frac{48}{12 + 13R}$$

$$\therefore \; P = \frac{I^2(48 - 12I)}{13I} = \frac{48}{13}I - \frac{12}{13}I^2 \qquad \text{or} \quad P = \left(\frac{48}{12 + 13R}\right)^2 R.$$

The second function, in the form of a quotient, can be differentiated by a method not yet discussed. The first function, however, presents no problem. The value of I for a maximum value of this function is determined; from this the value of R may be calculated. This calculation shows that a setting of 0.92 ohms for R, when a current of 2 amperes is flowing in it, results in maximum power through the variable resistor.

EXERCISES

1. A beam of length L, uniformly loaded, has a bending moment, at any point at a distance x from its end, of $M = [wx(L - x)]/2$, where w is the load per unit length. Find by calculus at what point on the beam the bending moment is greatest.

2. When a thin 2-ft rod is suspended at one end and swung, the force tending to break it at any point is proportional to $h^2(2 - h)$, where h is the distance (in feet) of the point from the free end. Find where the most probable break will occur, and express the answer as a fraction of the length of the rod.

3. (a) Find how close the graph of $xy = 2$ comes to the origin. (Use the fact that, when a positive quantity is at its minimum, the square of that quantity is also at its minimum.)

   (b) Find how close the graph of $x^2y = 2$ comes to the origin.

4. The strength of a rectangular beam is proportional to the product of its width and the square of its depth. Find the dimensions of the strongest rectangular beam which may be cut from a cylindrical log 9 inches in diameter.

5. The total charge q flowing past a point in an electrical circuit over an interval of time t seconds is equal to $t^3 - 3t^4$ coulombs.

(a) Find the maximum charge q.

(b) Find the times at which maximum and minimum current occur.

6. During an interval of time t seconds, the current in a constant resistance of R ohms is $2t^2 - t$ amperes. Find the maximum power as a function of R and the time at which it occurs.

7. Two resistors of 2 ohms and 6 ohms are arranged in a more complex circuit such that the sum of their currents is always 4 amperes. (This does not imply that they are in parallel.) Find the current in each for minimum total power dissipation in the two resistors and find this minimum power.

8. A man wishes to fence off a 10,000-ft^2 rectangular plot of land, which already has a fence on one of its boundaries. Find the length and width of the plot in order that a minimum amount of new fencing is to be used.

9. A d-c source of constant voltage E with a constant internal resistance r ohms is applied to a variable resistance of R ohms, thus essentially placing r and R in series. Find the setting of R in terms of the other constants for greatest power dissipation in R.

10. A rectangular box with square ends and no lid is to have a volume of $10\frac{2}{3}$ ft^3. Find the dimensions so that a minimum area of material is used.

11. A rectangular box with square bottom and no lid is to have a volume of 8 ft^3. Each edge where two surfaces meet is to be reinforced with metal stripping. Find the dimensions so that a minimum length of stripping is used.

12. A rectangular sheet of tin has dimensions of 24 in. by 9 in. Equal squares are to be cut out of all corners and the resulting flaps folded up to form an open rectangular box. Find the length of the squares cut out in order that the box have greatest volume, and find the resulting volume.

13. Repeat Exercise 12 with dimensions of the tin sheet 12 in. by 18 in.

14. Repeat Example 10, reversing the speeds of the two planes, so that A's speed is 200 miles per hour and B's speed is 400 miles per hour.

15. A d-c voltage is applied to a circuit consisting of an 8-ohm resistor in series with a variable resistor. The applied voltage adjusts itself so that the variable resistor absorbs 25 watts of power for all its settings. Find the setting of the variable resistor resulting in minimum applied voltage.

16. A rectangular field of area 24,200 ft^2 is to be enclosed by a fence. The front fence costs 60 cents per foot and the fencing for the back and sides costs 20 cents per foot. Find the dimensions for minimum cost, and find the minimum cost.

17. An open tank to contain a definite quantity of water is to be rectangular with a square base. If it is to be completely lined, find the ratio of depth to width for the least amount of lining.

18. If the open tank of Exercise 17 is to be cylindrical, find the ratio of depth to diameter of cross section for minimum amount of lining.

19. At t seconds, a voltage of $t^2 - t^3$ volts is applied to a circuit consisting of a 3-ohm resistor and a $\frac{1}{9}$-farad capacitor in parallel. Using any of the electrical laws listed in Chapters 3 and 4, find (a) the current in the capacitor, as a function of t, (b) the total current from the voltage source, as a function of t, and (c) the maximum total current and when it occurs.

20. If a farmer harvests his crop on the first of August, he will have 600 bushels worth $2.00 per bushel. Every week he waits, the crop increases by 100 bushels, but the price drops by 20 cents per bushel. Find by calculus when he should harvest his crop in order to make the most money.

21. (a) Sketch on the same diagram the graphs of $y = x^2 + 1$ and $y = -x^2 + 7$, from $x = -2$ to $x = 2$.

(b) Find the maximum rectangular area which can be constructed with sides parallel to the axes and with its two upper corners on $y = -x^2 + 7$ and its two lower corners on $y = x^2 + 1$.

22. Find the vertical height and radius of a circular cone of maximum volume under the restriction that its slant height is to be 10 in.

23. A constant voltage E is applied to a circuit consisting of a 5-ohm resistor in series with a parallel combination of a 10-ohm and a variable resistor. Find the setting of the variable resistor enabling it to take maximum power.

24. A 6-volt, d-c source voltage is applied to a circuit consisting of a 2-ohm resistor in series with a three-branch parallel combination of a 3-ohm, a 4-ohm, and a variable resistor. Find the setting of the variable resistor allowing it to take maximum power.

25. The cost of a 3-phase, 115-kv transmission line will be $1,200,000 plus another sum which varies directly with the conductor size in circular mils. The power loss in the transmission line varies inversely with the conductor size. If a conductor size of 477,000 circular mils is to be used, the transmission line will cost $1,400,000 and the power loss at peak will be 3200 kw. Total transmission costs are considered as the sum of the line cost and the cost of the peak power loss, evaluated at $34 per kw.

(a) Find the most economical conductor size.

(b) If standard sizes are 795,000, 605,000, 477,000, 336,400, and 203,200 circular mils, which one will be most economical?

26. The fuel cost *per hour* to operate a ship is proportional to the square of its speed, and amounts to $40 per hour at a speed of 10 miles per hour. There is also a fixed expense of $120 per hour. Find the most economical speed for a trip of fixed length.

27. A circular cone is to have a volume of 24 in.³ and a minimum slant height. Find its height and the radius of its base.

28. A circular cylinder is to be cut from a circular cone of base diameter 6 in. and height 9 in., the base of the cylinder coinciding with the base of the

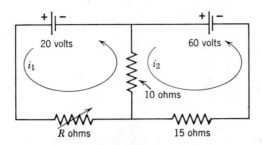

Fig. 4-13.

cone and the circumference of its upper surface lying on the surface of the cone. Find the base radius and height of the cylinder of maximum volume.

29. A gardener finds that, by planting 160 tomato plants in a certain land plot, each plant bears an average of 24 tomatoes. For every increase of 10 tomato plants above this, the average yield per plant for the entire crop decreases by one tomato. Find how many plants he should grow for the largest crop.

30. The *T*-circuit of Fig. 4-13 has resistances and emf's as shown, with a variable resistor *R*. Find the value of *R* for maximum power dissipation in *R*. (*Hint:* It is possible to determine one equation containing i_1 and i_2 and another containing i_1, i_2, and *R*; by eliminating i_2 from these two equations, a single equation containing i_1 and *R* may be found. This equation may be used for substituting in the expression for power in *R*.)

31. A cylindrical piece of iron of height $\frac{1}{2}$ in. and radius $1\frac{5}{8}$ in. is to have an inner concentric cylinder removed. The remaining portion is to be used as the core of a solenoid around which is to be wound several turns of wire, with 50 turns per inch of inner circumference. Assuming each turn is an approximate rectangle, find the radius of the inner cylinder to be removed so that the core will take the greatest possible length of wire.

CHAPTER 5

The Differential and Small Changes

5.1 MEANING OF THE DIFFERENTIAL

For any value of x in a continuous function $y = f(x)$, dy/dx has a value representing an instantaneous rate. When the symbol Dy or y' is used for the derivative, recognition is given to the fact that it has a definite value corresponding to each value of x and does not have to be expressed as the ratio of two values.

However, it is possible to split dy/dx into two components dy and dx, neither of which needs to be equal to zero; and for many purposes it is of advantage to split them.

In Fig. 5-1, the graph of $y = f(x)$ is shown, the function not being specified. The derivative at any point P on the graph means the slope of the tangent PL. To represent the derivative as a ratio with a numerical numerator and denominator, any length PN, in a horizontal direction, may be taken and NM drawn perpendicular to PN meeting the tangent at M. The slope of the tangent $= NM/PN$, and, since the symbol dy/dx is used for this slope, dx may arbitrarily be given a value equal to PN and the value of dy, dependent on dx, is therefore equal to NM. If any other point S is taken on the tangent, dx could arbitrarily be assigned a value equal to PR, and the dy corresponding to it equals RS. There is no limit to the number of right-angled triangles which can be constructed with hypotenuse along the tangent, all triangles being similar with the ratio of rise to run equal to dy/dx. dx may therefore be considered as an independent variable, with dy a dependent variable having a distinct value for each separate value of dx.

If $dy/dx = 1.5x$, then $dy = 1.5x\,dx$.

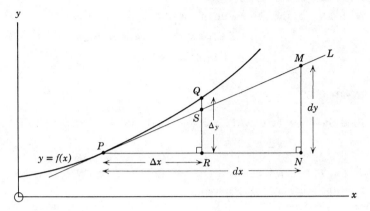

Fig. 5-1. The meaning of the differentials *dy* and *dx*.

In general $dy = (dy/dx)\, dx$; or, using other symbols, $dy = Dy\, dx$; or $dy = f'(x)\, dx$.

dx is called the *differential* of *x*, and *dy* is the *differential* of *y*.

Thus the symbol *d* may be interpreted as "the differential of."

EXAMPLE 1. Find the differential of *y* if $y = 3x^2 - 2 + 1/\sqrt{x}$.

Solution: $dy = (dy/dx)\, dx = (6x - \frac{1}{2}x^{-3/2})\, dx$, where *x* and *dx* are both independent variables.

EXAMPLE 2. Find $d(2t^3 - kt + m)$, assuming *t* as the independent variable.

Solution: In the same way that *dy* means $(dy/dx)\, dx$ if *x* is the independent variable, or $(dy/dt)\, dt$ with *t* as an independent variable, so

$$d(2t^3 - kt + m) = \frac{d}{dt}(2t^3 - kt + m)\, dt = (6t^2 - k)\, dt.$$

The differential of the dependent variable is always expressed in terms of the differential of the independent variable.

EXAMPLE 3. For the function $i = 4t^{3/2} - 2t^{1/2}$, find *di* and Δi at $t = 4$, given $dt = \Delta t = 5$.

Solution: $di = (6t^{1/2} - t^{-1/2})\, dt$. For the given values of the two independent variables *t* and *dt*,

$$di = (6 \times 2 - \tfrac{1}{2})(5) = \tfrac{23}{2}(5) = 57.5.$$

At $t = 4$,

$$i = 4 \times 4^{3/2} - 2 \times 4^{1/2} = 4 \times 8 - 2 \times 2 = 28.$$

With $\Delta t = 5$, the new value of t is $4 + 5 = 9$. At $t = 9$,

$$i = 4 \times 27 - 2 \times 3 = 102.$$
$$\therefore \ \Delta i = 102 - 28 = 74.$$

EXERCISES

1. Find the differential of the variable on the left-hand side of the following functions.

(a) $y = 5x^2 + 3\sqrt{x}$.   (b) $u = 7x - 5/x$.

(c) $I = E/R$, if R is constant.   (d) $I = E/R$, if E is constant.

2. Find the following differentials.

(a) $d(t^5 - 7t - t^{-3})$.   (b) $d(m^3 + 7m - p)$, where p is the only variable.

(c) $d\left(\dfrac{v^{3/2} - 8}{v + 2v^{1/2} + 4}\right)$.   (d) $d(x^2 - y^3 + x)$, where x and y both vary.

3. Evaluate the following expressions.

(a) $d\left(3 + \dfrac{2}{x^2}\right)$ at $x = 2$ and $dx = 0.5$.

(b) $d\left(2v - \dfrac{5}{\sqrt{v}}\right)$ if $v = 1$ and $dv = -0.5$.

4. For the function $y = x^2 - 2$, compute the values of dy and Δy when $x = 2$, $dx = 0.2$, and $\Delta x = 0.2$.

5. Find dy and Δy for the following functions with $x = 8$, $dx = 2$, and $\Delta x = 3$:

(a) $y = 3x - \dfrac{2}{x}$.   (b) $y = \sqrt[3]{x^2} + 2x$.   (c) $xy^2 = 3$.

6. When dx is always taken equal to Δx on a certain graph, it is found that dy always is equal to Δy for all values of x. What type of graph is it? Could $dy = \Delta y$ in certain isolated cases for any other type of graph? Illustrate.

5.2 APPROXIMATE EFFECT OF A SMALL CHANGE IN INDEPENDENT VARIABLE

In Fig. 5-1, if $dx = \Delta x = PR$, then $dy = RS$ and $\Delta y = RQ$. Then, provided Δx is quite a small increment and there is no sudden change or discontinuity in the function in this interval, Δy approximately equals dy, and thus

$$\frac{\Delta y}{\Delta x} \simeq \frac{dy}{dx} \quad \text{or} \quad \Delta y \simeq \frac{dy}{dx} \Delta x.$$

Therefore, to find the approximate change in the dependent variable resulting from a small change Δx in the independent variable, Δx is multiplied by the value of the derivative at the original value of x.

EXAMPLE 4. Find the approximate increase in displacement s when time t changes from 3.0 to 3.1 sec, if $s = 4t^2 - 2t$ ft.
Solution:

$$\frac{ds}{dt} = 8t - 2. \qquad \therefore \; \Delta s \simeq (8t - 2)\,\Delta t. \qquad \Delta t = 0.1 \qquad \text{and} \qquad t = 3.0.$$

$$\therefore \; \Delta s \simeq (24 - 2)(0.1) \simeq 2.2.$$

Therefore the increase in displacement is about 2.2 ft. If t had been taken as 3.05, its average value during the increment, the answer would have been slightly more accurate; however, this refinement is not usually necessary, since the answer is approximate in any case and it is easier to multiply by 3 than by 3.05.

Of course there are two other methods of getting a theoretically exact answer for Δs, the methods shown in Examples 1a and 1c of Chapter 3; the first of these methods involves finding the two values of the dependent variable corresponding to the values of the independent variable at the beginning and end of the increment; and the second method is to find an exact expression for the increment in the dependent variable. These methods are much more cumbersome than the approximate solution shown above, and care must be taken in using the first of these methods that portions of the calculation are done with considerable accuracy. This point of accuracy is further discussed in the following example.

EXAMPLE 5. In a resistive circuit with a constant d-c voltage source of 15 volts, find the change in current resulting from a change in resistance from 7.5 to 7.6 ohms.
Solution: Here Ohm's law applies, and $I = E/R$, E remaining constant.

$$\Delta I \simeq \frac{dI}{dR}\,\Delta R \simeq \frac{-E}{R^2}\,\Delta R \simeq \frac{-15}{7.5^2}\,(0.1) \simeq -0.0267 \;\text{(slide rule accuracy)}.$$

Therefore the change in current is a decrease of approximately 0.027 ampere.

If, in Example 5, ΔI is found by calculating the two separate currents, the following work is required:

$$I_1 = \frac{15}{7.5} = 2.0. \qquad I_2 = \frac{15}{7.6} = 1.9737, \quad \text{calculated by long division.}$$

$$\therefore \; \Delta I = 1.9737 - 2.0 = -0.0263.$$

This direct method will give a slightly more accurate answer, at the expense of time and inconvenience in performing the long division. In this method, if slide rule is used for the division, $I_2 = 1.97$ (slide rule accuracy). Now, when ΔI is calculated as $I_2 - I_1$, the result is -0.03, which is only accurate to one figure. The principle here is that when two numbers close in value are subtracted, the accuracy of the result is considerably lower than that of the two original numbers. In most cases, therefore, the method of using the derivative produces a fairly accurate answer in a minimum of time. The slide rule may be used throughout this calculation, since the small change itself is being calculated, and no subtraction is necessary. Since the method itself is approximate, the third figure of the answer is usually not justified, but the small change calculated is usually accurate to two figures for very small changes. Of course, this approximate method is only valid if the change in the independent variable is small.

EXAMPLE 6.   In the resistive circuit with any constant voltage source, find the percent change in current resulting from a 2% decrease in resistance.

Solution: A knowledge of percent change is necessary here, and it is recalled that percent change is the ratio of the actual change to the original quantity, multiplied by 100 to express it in percent. Thus a 2% decrease in resistance means a decrease of 2% of the original R.

$$\therefore \ \Delta R = -0.02R. \qquad I = \frac{E}{R}.$$

$$\therefore \ \Delta I \simeq \frac{dI}{dR} \Delta R \simeq \frac{-E}{R^2}(-0.02R) \simeq \frac{0.02E}{R} \simeq 0.02I.$$

The change in current is therefore 0.02 times the original current, which is a 2% increase in current (approx.).

EXERCISES

1. For what type of graph is Δy exactly equal to $(dy/dx)\,\Delta x$ for all values of x and Δx?

For the following functions, find the approximate change in the dependent variable when the independent variable changes by the indicated amount:

2. $y = 5x - 30/x^2$ when x changes from 10.0 to 10.1.

3. $s = t^3 - 4t^2 + 1$ when t changes from 6.0 to 6.05.

4. $R = (8.0 \times 10^{-2}L)/d^2$ if $L = 75$ and d is changed from 0.75 to 0.73.

5. If $PV^{1.4} = 1200$, find the approximate percent change in P if V is increased by 3%.

6. The power P in a circuit is calculated by measuring the current I and the resistance R in the circuit and using the law $P = I^2 R$. If R is measured accurately, but there is a 4% error in the measurement of I, find the percent error in the calculated P.

7. The period T of a simple pendulum may be found by using the law $T = 2\pi\sqrt{L/g}$, where L is its length and g is the gravitational constant. If L is measured 4% too low, find the percent error in the calculated T.

8. The resistance R of a conductor is calculated by measuring its length L and its diameter d and using a formula established as $R = 2.4 \times 10^{-1}L/d^2$. Find the percent error in the calculated resistance if (a) d is measured accurately, but L is measured 3% too high, and (b) L is measured accurately, but d is measured 3% too high.

9. The resonant frequency of an a-c circuit is given by $f = 1/2\pi\sqrt{LC}$. If C is held constant, find the approximate percent change in f resulting from a 5.0% decrease in L.

10. (a) Using calculus, find the approximate area of a flat circular ring with inner radius 8.0 in. and outer radius 8.05 in.

(b) From the calculations of (a), express the approximate area of a flat circular ring of small width in terms of the width and the radius of the inner boundary.

11. A spherical satellite of a planet, with a diameter of 960 miles, becomes coated with ice to a depth of 5.5 ft. Find the total volume of ice to two-figure accuracy.

12. The total current flowing to a parallel combination of a resistor and a capacitor is given by $i = v/R + C(dv/dt)$, where v is the voltage across the parallel circuit. If $v = 5t^2 - 2t^3$, find the approximate change in i from $t = 2.0$ to $t = 2.08$, assuming $R = 2.0$ and $C = 0.10$.

13. Two circular cones both have heights equal to twice the diameter of their bases. One cone is 3% higher than the other. Find the difference in their volume, expressed as a percentage.

5.3 IMPLICIT DIFFERENTIATION

The type of function where the dependent variable occurs by itself on the left-hand side is called an *explicit function*, *ex* indicating *outside*. y is isolated outside the function of x. If y is not isolated on one side, we have an *implicit relation*, *im* indicating *inside*. Thus $y = \sqrt{x^2 - 25}$ is explicit and $x^2 - y^2 = 25$ implicit, although they are both the same relation for $y \geq 0$. Many functions may be expressed either explicitly or implicitly.

So far, only explicit functions have been differentiated. The derivative may be found directly or a method using differentials may be used. To find the derivative of $y = 3x^2 - 2x$ by differentials, for example, it is seen that $dy = (6x - 2)\,dx$; dividing both sides by dx, which is not zero, $dy/dx = 6x - 2$.

The differentiation of implicit relations is accomplished by this method of differentials. A few examples should suffice to illustrate the procedure.

EXAMPLE 7.   If $y^2 - 2x = 0$, find y'.
Solution: If two expressions are equal, their differentials must be equal. Equating the differentials of the two sides,

$$2y\,dy - 2\,dx = 0.$$

Dividing by dx,

$$2y \frac{dy}{dx} - 2 = 0.$$

Solving for dy/dx by algebra, $dy/dx = 1/y$. In other words, $y' = 1/y$. If the answer is required to be a function of the independent variable, it is seen from the original equation that

$$y = \sqrt{2x}; \qquad \therefore\; y' = \frac{1}{y} = \frac{1}{\sqrt{2x}}.$$

This particular function may be expressed explicitly as $y = \sqrt{2}\, x^{\frac{1}{2}}$, and the derivative may be found directly as $(\sqrt{2}/2)x^{-\frac{1}{2}} = 1/\sqrt{2x}$, as above.

It is characteristic of implicit differentiation that the derivative itself first appears implicitly; algebraic methods must be used to express the derivative explicitly. The resulting answer may be a function of the dependent variable or of both the dependent and independent variables. Therefore there usually are two or more expressions for the derivative, all of which are perfectly correct. Explicit differentiation always produces an answer which is a function of the independent variable only.

EXAMPLE 8.   Find the rate of change of q with respect to t at the moment when $t = 0$ for the function $q^2 - t = t^3 + 4$.
Solution: Equating the differentials of the two sides,

$$2q\,dq - dt = 3t^2\,dt.$$

$$2q \frac{dq}{dt} - 1 = 3t^2.$$

$$\frac{dq}{dt} = \frac{3t^2 + 1}{2q}.$$

Values of both q and t at the specified time are required. From the original equation, at $t = 0$, $q^2 = 4$, and $q = \pm 2$. Therefore the rate of change of q at this time is

$$\frac{3(0) + 1}{\pm 4} = \pm \frac{1}{4}.$$

Often implicit relations cannot be expressed explicitly, and often, even if they can be, the explicit differentiation is difficult. In any case, an ability to differentiate implicitly adds flexibility to solutions and very often yields the simplest solution.

EXAMPLE 9.   If $i = \sqrt{t^2 - 3t + 6}$, find di/dt at $t = 1$.

Solution: In Chapter 6, a method of finding di/dt explicitly for this type of function will be developed. However, the implicit differentiation is the simplest solution. Squaring both sides in order to eliminate the square root,

$$i^2 = t^2 - 3t + 6.$$

$$2i \, di = (2t - 3) \, dt.$$

$$2i \frac{di}{dt} = 2t - 3.$$

$$\frac{di}{dt} = \frac{2t - 3}{2i}.$$

At $t = 1$, $i = \sqrt{1^2 - 3(1) + 6} = 2$ (assuming the positive square root only).

$$\therefore \frac{di}{dt} = \frac{2 - 3}{4} = -\frac{1}{4}.$$

EXERCISES

Find an expression for dy/dx for the following relations:

1. $x^2 + \dfrac{y^2}{4} = 1.$ 2. $2x^{3/2} - 3y^{3/2} = 5y + 7.$

3. $x - 2y = 3x^2 - y^2.$ 4. $\dfrac{y^2 - 2x}{3x^2 + 4y} = 6.$

Evaluate dy/dx at the specified point for the following relations:

5. $\dfrac{xy^3 - 3x^3}{x} = 14x^2 + 10$, at the point where $y = 3.$

6. $\dfrac{x^3 + 8y^3}{x^2 - 2xy + 4y^2} = x^2$, at the point where $x = 1$.

7. $\sqrt{3x} = 4y^2 - 5$, at the point where $y = -3/2$.

8. $y = \sqrt[3]{x^2 - y^3}$, at the point where $x = 4$.

9. Find the rate of change of Z with respect to R, assuming X is constant, if $Z = \sqrt{R^2 + X^2}$, expressing the answer as (a) a function of Z and R, (b) a function of R, and (c) a function of Z.

10. If $1/p^2 = 1/q^2 + 1/r^2$ and q is held constant at 2, find the rate of change of p with respect to r when r is equal to 2.

11. If $Z = \sqrt{R^2 + 4\pi^2 f^2 L^2}$ and $2\pi f L$ is held constant at 4 ohms, find the approximate change in Z resulting from a change in R from 3.0 to 2.9 ohms.

12. Find the maximum and minimum points on the ellipse

$$9x^2 + 4y^2 - 18x = 27.$$

13. If $1/y = 1/x_1 + 1/x_2$, and x_1 is constant, evaluate dy/dx_2 at the point where $x_2 = x_1$.

14. Find the maximum value of the function $y = 3x/(x^2 + 2)$. (*Hint:* Find $1/y$ as a two-termed function in powers of x and differentiate implicitly.)

15. If the power P in an a-c circuit is related to resistance R and source voltage E by the function $1/P = R/E^2 + 9/E^2 R$, find the value of resistance resulting in maximum power, assuming a constant E.

16. It may readily be established by the theorem of Pythagoras that a 26-ft ladder with its foot on the ground 10 ft from a vertical wall will reach 24 ft up the wall. Using this as a basis, and employing the method of approximate differentials, find how far from the foot of a wall to place the foot of a 27-ft ladder in order that it reach 24 ft up the wall.

17. A 26-ft ladder leans against a vertical wall with its foot on the ground. Find by approximate differentials how far the top of the ladder moves down the wall when (a) the foot of the ladder is 10 ft from the wall and is moved out $\frac{1}{2}$ foot, and (b) the foot of the ladder is 24 ft from the wall and is moved out $\frac{1}{2}$ foot.

18. The impedance of a circuit with R ohms resistance in series with X ohms reactance is given by $Z = \sqrt{R^2 + X^2}$ ohms. If X is constant and R is increased by a small amount ΔR, find the approximate increase in Z as a multiple of ΔR, (a) when $R = \frac{1}{2}X$, and (b) when $R = 2X$.

CHAPTER 6

Products, Quotients, and Composite Functions

6.1 PRODUCTS AND QUOTIENTS

Thus far in the study of calculus, only fairly simple power functions have been studied. The power function often occurs as one part of a more complex pattern, and this chapter is principally concerned with these combinations of simpler power functions. When differentiation of these more involved patterns has been established, a much wider variety of problems may be handled.

When two or more functions occur as separate terms of a larger function, differentiation is easily accomplished by taking derivatives term by term. If two or more functions occur in a product, however, the derivative cannot be found by simply taking the product of the derivatives of the factors. For example,

$$\frac{d}{dx} x(x-1) = \frac{d}{dx} (x^2 - x) = 2x - 1.$$

If the product of the derivatives of the separate factors is taken, the false answer of $1 \times 1 = 1$ is the result. It should therefore be clear that the derivative of a product is not the product of the derivatives.

However, products of two or more functions can be differentiated without multiplying them out. The following analysis goes back to first principles in dealing with the derivative of a product.

Let u and v each represent functions of x.

Let $\qquad\qquad\qquad\qquad y = u v \qquad (1)$

Let x take on an increment Δx, the corresponding increments in the dependent variables being Δu, Δv, and Δy.

Then
$$y + \Delta y = (u + \Delta u)(v + \Delta v)$$
$$= uv + u(\Delta v) + v(\Delta u) + (\Delta u)(\Delta v) \qquad (2)$$

Subtracting (1) from (2),

$$\Delta y = u(\Delta v) + v(\Delta u) + (\Delta u)(\Delta v).$$

Dividing by Δx,

$$\frac{\Delta y}{\Delta x} = u \frac{\Delta v}{\Delta x} + v \frac{\Delta u}{\Delta x} + \frac{\Delta u}{\Delta x} \Delta v.$$

As Δx approaches zero, the ratio of two increments becomes a derivative, and Δv approaches zero.

$$\therefore \frac{dy}{dx} = u \frac{dv}{dx} + v \frac{du}{dx}.$$

That is,

$$\frac{d}{dx}(uv) = u \frac{dv}{dx} + v \frac{du}{dx} \qquad \text{or} \qquad d(uv) = u\,dv + v\,du.$$

These are known as the *product formulas*, the first in terms of derivatives and the second in terms of differentials. It should be memorized; or, in a more basic way, its structure should become familiar. A structural analysis of this formula shows that, to find the derivative of a product, one factor at a time is differentiated and the result is multiplied by the remaining factor, resulting in two terms which are added.

The general pattern of differentiating only one factor at a time follows through to the instance where there are three or more factors in a product. Each successive derivative of one of the factors is multiplied by the remaining factors and these successive terms are added. Thus

$$\frac{d}{dx}(uvw) = uv \frac{dw}{dx} + uw \frac{dv}{dx} + vw \frac{du}{dx}.$$

Of course, if one of two factors in a function is a constant, it is not necessary to use the product formula, since the constant factor remains as a constant factor in the derivative. It would not be wrong to treat this as a product, however. Thus, as an example,

$$\frac{d}{dx} 3a(x^3 - 5x) = 3a(3x^2 - 5),$$

using the simplest solution; or, using the product formula,

$$\frac{d}{dx} 3a(x^3 - 5x) = 3a(3x^2 - 5) + (x^3 - 5x)0 = 3a(3x^2 - 5),$$

as before.

EXAMPLE 1.   Find

$$\frac{d}{dx} \sqrt{x^3} \left(7x + 3 - \frac{2}{x}\right).$$

Solution: Using the product formula, the derivative is

$$x^{3/2}(7 + 2x^{-2}) + \frac{3}{2} x^{1/2}\left(7x + 3 - \frac{2}{x}\right).$$

Removing parentheses and collecting terms, this simplifies to

$$\frac{35}{2} x^{3/2} + \frac{9}{2} x^{1/2} - x^{-1/2}.$$

The original expression could have been simplified by multiplication to a three-termed function of x with no products of variables occurring; the resulting function could then be differentiated term by term. The student should verify the answer by using this second method.

EXAMPLE 2.   Find the slope of the curve $y = (2x^{1/2} + 3)(x^2 - 5x + 3)$ at the point where $x = 4$.

Solution:

$$\frac{dy}{dx} = (2x^{1/2} + 3)(2x - 5) + x^{-1/2}(x^2 - 5x + 3)$$

Therefore at $x = 4$, the slope is

$$(4 + 3)(8 - 5) + \tfrac{1}{2}(16 - 20 + 3) = \tfrac{41}{2}.$$

If two functions are combined in a division, the derivative cannot be calculated by dividing the two separate derivatives. A division by any factor is equivalent to a multiplication by the negative first power of that factor. In this way the derivative of a quotient may be calculated as the derivative of a product. However, it is often more convenient to use a separate quotient formula, which is developed as follows.

Let u and v be functions of x.

$$d\left(\frac{u}{v}\right) = d(uv^{-1}) = u(-v^{-2})\, dv + v^{-1}\, du \quad \text{by product formula}$$

$$= -\frac{u}{v^2} dv + \frac{1}{v} du = \frac{v\, du - u\, dv}{v^2}.$$

The quotient formula, in its two forms, is therefore

$$\frac{d}{dx}\left(\frac{u}{v}\right) = \frac{v\,\dfrac{du}{dx} - u\,\dfrac{dv}{dx}}{v^2} \quad \text{or} \quad d\left(\frac{u}{v}\right) = \frac{v\,du - u\,dv}{v^2}$$

EXAMPLE 3.   Find the rate of change of i with respect to t at $t = 4$, given

$$i = \frac{t^{1.5} - t}{2t - 3}.$$

Solution: The techniques of implicit differentiation, the product formula, and the quotient formula often open up several ways of solving a problem. As more basic knowledge is acquired, solutions become more flexible. One method of solution verifies another and demonstrates the consistency of the mathematics involved.

For this problem, two methods of solution are presented:

Method 1.   Using the quotient formula,

$$\frac{di}{dt} = \frac{(2t - 3)(1.5t^{0.5} - 1) - 2(t^{1.5} - t)}{(2t - 3)^2} = \frac{t^{1.5} - 4.5t^{0.5} + 3}{(2t - 3)^2}.$$

At $t = 4$,

$$\frac{di}{dt} = \frac{8 - 4.5(2) + 3}{(8 - 3)^2} = \frac{2}{25}.$$

Method 2.   Multiplying by $2t - 3$,

$$i(2t - 3) = t^{1.5} - t.$$

Using implicit differentiation, and realizing that $i(2t - 3)$ is a product,

$$i(2\,dt) + (2t - 3)\,di = (1.5t^{0.5} - 1)\,dt.$$
$$(2t - 3)\,di = (1.5t^{0.5} - 1 - 2i)\,dt.$$

At $t = 4$,

$$i = \frac{8 - 4}{8 - 3} = \frac{4}{5}.$$

$$\therefore\ (8 - 3)\,di = (3 - 1 - \tfrac{8}{5})\,dt.$$

$$5\,di = \frac{2}{5}\,dt. \qquad \frac{di}{dt} = \frac{2}{25}.$$

A third method is to find $(d/dt)(t^{1.5} - t)(2t - 3)^{-1}$ by using the product formula. One portion of this calculation, the differentiation of $(2t - 3)^{-1}$, requires further basic techniques to calculate. These techniques are developed in Section 6.3.

EXAMPLE 4.   Locate all maxima and minima for the function

$$y = \frac{2x}{3 + 4x^2}.$$

Solution:

$$y' = \frac{2(3 + 4x^2) - 2x(8x)}{(3 + 4x^2)^2} = \frac{6 - 8x^2}{(3 + 4x^2)^2}.$$

For maximum or minimum y, $y' = 0$.

$$\therefore \ 6 - 8x^2 = 0. \qquad x = \pm\frac{\sqrt{3}}{2}.$$

The denominator of the fraction representing y' may have any value other than zero. Therefore, when $y' = 0$, the denominator may be ignored. It may therefore be assumed that, for maximum or minimum y, $x = \pm\sqrt{3}/2$. In applying the second derivative test to distinguish between maximum and minimum, the calculation of the second derivative may be shortened considerably. To calculate the actual second derivative, the quotient formula is again necessary, and the calculations become very cumbersome. Realizing that the denominator of the first derivative is a perfect square and therefore positive, and that the numerator of the first derivative is zero at the turning points, the sign of the second derivative may be determined without finding its actual value. Since the derivative of a quotient is of the structure "denominator times derivative of numerator minus numerator times derivative of denominator all over denominator squared," the value of y'' at the turning points will be "a positive quantity times derivative of numerator $(6 - 8x^2)$ minus zero times derivative of denominator $(3 + 4x^2)^2$ all over a positive quantity." A careful analysis of this involved expression shows that the second derivative has the same sign as the derivative of the numerator of the first derivative at the turning points. This means essentially that the denominator of the first derivative may once again be ignored in calculating the sign of the second derivative at the turning points.

Proceeding with a simplified calculation, the sign of y'' at the turning points is the same as the sign of $d/dx(6 - 8x^2) = -16x$ at these points. When $x = +\sqrt{3}/2$, the second derivative is therefore negative; and when $x = -\sqrt{3}/2$, the second derivative is positive. Therefore at $x = 0.866$, y reaches a maximum value of 0.29, by substitution; and at $x = -0.866$, y reaches a minimum value of -0.29.

This simplification of the second derivative test may always be used when the original function is a quotient.

EXERCISES

1. Find $(d/dx)x^2$ by expressing x^2 as $(x)(x)$ and using the product formula.

2. Find $(d/dx)x^2$ by expressing x^2 as x^3/x and using the quotient formula.

Find the following derivatives by product formula, and verify the answers by using a second method.

3. $\dfrac{d}{dx}(x^2 - 2)(2x + 1)$.

4. $\dfrac{d}{dt}(\sqrt{t} - 1)(t + \sqrt{t})$.

5. $\dfrac{d}{dv}v^2(3 - 2v)(v^2 + 4v)$.

6. $\dfrac{d}{dx}x^{3/4}(6x^{1/2} - 2x^{1/4})$.

7. $\dfrac{d}{dt}\dfrac{t^{-3}(t^2 - 5t - 6)}{t - 6}$.

8. $\dfrac{d}{du}3ku^{-1/4}\left(u^2 - \dfrac{1}{u}\right)$.

9. $\dfrac{d}{dx}(x - 2)(x^2 - 2)(x^3 - 2)$.

10. Find $(d/dx)uvw$, where u, v, and w are functions of x, by using the formula for the derivative of the product of two functions u and vw and then using the same formula in later details of the calculation.

11. Find $(d/dx)(u/v)$, where u and v are functions of x, from first principles, by taking small increments Δx, Δu, Δv, and Δy.

Find the following derivatives:

12. $\dfrac{d}{dw}\dfrac{2w + 3}{2w - 3}$.

13. $\dfrac{d}{d\theta}\dfrac{\theta^2 - 5\theta + 4}{\theta^2 - 3\theta - 4}$.

14. $\dfrac{d}{dy}\dfrac{y + 3y^{1/2} + 4}{y - 3y^{1/2} + 4}$.

15. $\dfrac{d}{dw}\dfrac{(2w + 3)^2}{2w - 3}$.

Use an implicit method to find the following:

16. $\dfrac{dZ}{dR}$, if $ZX = RX - RZ$, X being constant.

17. $\dfrac{dy}{dx}$ at $x = 1$, if $3x^2 - xy = 2x$.

18. $\dfrac{dy}{dx}$ at $x = 1$, if $3x^2 + \dfrac{x}{y} = 2x$.

19. $\dfrac{dy}{dx}$ at $x = 2$, if $y = \dfrac{3x^2 - 5x}{x + 2}$. (*Hint:* Cross-multiply and avoid quotient.)

20. $\dfrac{du}{dv}$, if $(u + 2v)(2u - v) = u - 2v$.

Find the equation of the tangent to each of the following curves at the point indicated.

21. $y = \sqrt{2x}(\sqrt{8x} - \sqrt{2}x)$ at $x = 4$.

22. $y = \dfrac{3x^2 - 5}{3 - x^2}$ at $x = -1$.

Find the second derivative of the following functions:

23. $y = \dfrac{3 - 2x}{3x + 1}$.

24. $v = \dfrac{t^2 - 0.5t}{4 - t}$.

25. $y = 3 - xy$. (*Hint:* Express the first derivative as a function of x.)

Test the following functions for maximum and minimum, specifying which type of turning point is involved.

26. $y = \dfrac{x^2}{4 - x}$.

27. $y = \dfrac{x - 2}{x^2 + 5}$.

28. $y = \dfrac{5 - x^2}{5 + x^2}$.

29. For the function of Exercise 26, the maximum y is less than the minimum y. Explain by the use of a graph.

30. The power in an electric circuit is given by $P = E^2R/(R^2 + X^2)$. If E and X are held constant, find the setting of R to give maximum P, and find maximum P.

31. The impedance of a circuit consisting of a resistor and an inductor in parallel is given by $Z = RX/(R + X)$. If X remains constant at 10 ohms, find by approximate differentials the change in Z resulting from a change in R: (a) from 0 to 0.4 ohm, and (b) from 10 to 10.4 ohms.

32. The current I in an a-c circuit is given by $I = E/\sqrt{R^2 + X^2}$. If E is held constant at 20 volts and X is held constant at 3.0 ohms, find the approximate change in I resulting from a change in R: (a) from 4.0 to 4.2 ohms, and (b) from 0 to 0.2 ohm. (*Hint:* Square both sides, cross-multiply, and differentiate implicitly.)

6.2   RELATED RATES

Frequently the independent variable in one equation is itself a dependent variable related to a third more basic variable. If, for example, $y = x^2 - 2$ and $x = 3t + 1$, the algebraic elimination of x from the two equations produces a third equation, $y = (3t + 1)^2 - 2$, simplifying to $y = 9t^2 + 6t - 1$. Two equations with a common variable always imply a third equation not involving the common variable.

From these equations,

$$\frac{dy}{dt} = 18t + 6, \qquad \frac{dx}{dt} = 3, \qquad \text{and} \qquad \frac{dy}{dx} = 2x = 2(3t + 1).$$

$$\frac{dy}{dx}\frac{dx}{dt} = 2(3t + 1)3 = 18t + 6 = \frac{dy}{dt}.$$

The fact that $(dy/dx)(dx/dt) = dy/dt$, established for the particular functions above, is readily established for $y = f(x)$ and $x = F(t)$, where the functional notation may designate any function whatever. The concept of

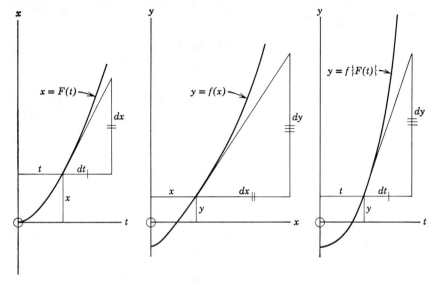

Fig. 6-1. $\dfrac{dy}{dx}\dfrac{dx}{dt} = \dfrac{dy}{dt}.$

differentials has shown that any arbitrary value may be assigned to the differential of an independent variable. Thus, by referring to Fig. 6-1, we see that any arbitrary value may be assigned to t and dt. From the equation $x = F(t)$, the corresponding values of x and dx may be found. If these values of x and dx are used in the equation $y = f(x)$, the corresponding value of dy may be found. This will be the value of dy corresponding to the original value of dt at the originally chosen value of t. This means that, in the expression $(dy/dx)(dx/dt)$, dx may be canceled out as in arithmetic, and the result is dy/dt, since, when the two dx's are equal, the resulting dy is the differential of y corresponding to the original dt.

By the same general reasoning, $dx/dy = 1/(dy/dx)$, since any derivative may be interpreted as the ratio of two numbers representing differentials.

Problems involving related rates may usually be solved by using these laws of related derivatives.

EXAMPLE 5. Two currents i_1 and i_2 are both functions of time t: $i_1 = t^2 - 2t$ and $i_2 = 6 - 3t^2$. Find (a) di_1/di_2 and (b) $d^2i_1/di_2{}^2$.

Solution: (a) $di_1/dt = 2t - 2.$ $di_2/dt = -6t.$

$$\therefore \; \frac{di_1}{di_2} = \frac{di_1}{dt}\frac{dt}{di_2} = (2t - 2)\frac{1}{-6t} = \frac{1 - t}{3t}.$$

(b) The laws of related derivatives only apply to first derivatives.

$\dfrac{d^2y}{dx^2}\dfrac{d^2x}{dt^2}$ is not equal to $\dfrac{d^2y}{dt^2}$, and $\dfrac{d^2x}{dy^2}$ is not equal to $\dfrac{1}{\dfrac{d^2y}{dx^2}}$.

In calculating second derivatives here, their meaning must be well analyzed. Thus, $d^2i_1/di_2{}^2$ means

$$\frac{d}{di_2}\left(\frac{di_1}{di_2}\right) = \frac{d}{di_2}\left(\frac{1-t}{3t}\right)$$

for this example. The derivative of a function of t must first be determined with respect to t. Using a law of related derivatives,

$$\frac{d}{di_2}(\ ) = \frac{d}{dt}(\ )\frac{dt}{di_2}.$$

$$\therefore\ \frac{d^2i_1}{di_2{}^2} = \frac{d}{dt}\left(\frac{1-t}{3t}\right)\frac{dt}{di_2},$$

and dt/di_2 is the reciprocal of di_2/dt determined in part a.

$$\therefore\ \frac{d^2i_1}{di_2{}^2} = \frac{d}{dt}\left(\frac{1}{3}t^{-1} - \frac{1}{3}\right)\frac{1}{-6t} = \frac{-1}{3t^2}\frac{-1}{6t} = \frac{1}{18t^3}.$$

A much neater solution to this whole example is achieved by using differentials. Thus

(a) $di_1 = (2t - 2)\,dt$ $di_2 = -6t\,dt.$

$$\therefore\ \frac{di_1}{di_2} = \frac{(2t - 2)\,dt}{-6t\,dt} = \frac{1-t}{3t}.$$

(b) $\dfrac{d^2i_1}{di_2{}^2} = \dfrac{d}{di_2}\dfrac{1}{3}(t^{-1} - 1) = d\left[\dfrac{1}{3}(t^{-1} - 1)\right]\dfrac{1}{di_2}$

$$= \frac{-1}{3t^2}\,dt\,\frac{1}{-6t\,dt} = \frac{1}{18t^3}.$$

EXAMPLE 6. A solid metallic sphere weighing 450 grams and with a 4.0-cm radius is heated and during expansion its radius increases at a rate of 0.20 cm per hour. Find: (a) the rate at which the volume is increasing at the end of 3.0 hr, (b) the rate of volume increase when the volume is 330 cm, (c) the rate of change of density after 2.0 hr of expansion.

Solution: (a) Given: $dr/dt = 0.20$; volume $V = \frac{4}{3}\pi r^3$. Required to find dV/dt.

$$dV = 4\pi r^2\, dr = 4\pi r^2(0.20\, dt). \qquad \frac{dV}{dt} = 4\pi r^2(0.20).$$

After 3.0 hr, $r = 4.0 + 0.60 = 4.60$. Therefore volume is increasing at $4\pi(4.6)^2(0.20) = 53$ cm³/hr.

(b) $V = \frac{4}{3}\pi r^3$. Thus $r = \sqrt[3]{3V/4\pi}$. Therefore, when $V = 330$, $r = \sqrt[3]{990/4\pi} = 4.3$.

$$\therefore \frac{dV}{dt} = 4\pi(4.3)^2(0.20) = 46 \text{ cm}^3/\text{hr}.$$

(c) Density ρ means mass per unit volume. Therefore $\rho = 450/V$.

$$\therefore \frac{d\rho}{dV} = \frac{-450}{V^2}.$$

$$\therefore \frac{d\rho}{dt} = \frac{d\rho}{dV}\frac{dV}{dt} = \frac{-450}{V^2}(4\pi r^2)(0.20) = \frac{-450}{(\frac{4}{3}\pi r^3)^2}(0.80\pi r^2)$$

$$= \frac{-450 \times 9 \times 0.80}{16\pi r^4}.$$

After 2 hr, $r = 4.4$. Therefore the rate of change of density is

$$\frac{-450 \times 9 \times 0.80}{16\pi(4.4)^4} = -0.17 \text{ gram/cm}^3/\text{hr}.$$

Notice in this problem the number of related variables: r is a function of t, V is a function of r, ρ is a function of V and depends on t through several links.

$$\frac{d\rho}{dt} = \frac{d\rho}{dV}\frac{dV}{dr}\frac{dr}{dt}.$$

EXERCISES

Find dy/dx, dx/dt, and dy/dt for the following functions without finding y as a function of t. Express all answers as functions of t.

1. $y = 3x^2 - 5$; $x = 4t + 3t^2$.

2. $y = \dfrac{2}{x} + 3x$; $x = t^2 + 1$.

3. $y = \dfrac{2}{x} + 3x$; $x^2 = t^2 + 1$.

4. $y = \sqrt{x}$; $xt^2 = 4$.

Find dy/dt, dx/dt, and dy/dx for the following functions:

5. $y = t^2 - 2t$; $x = 2 + 3t^2$.   **6.** $y = 2t + 1$; $x = \dfrac{t^2 - 3}{t^2 - 2}$.

7. $y = \dfrac{t + 1}{t - 1}$; $x = \dfrac{1}{t - 1}$.

Find d^2y/dx^2 for the following pairs of functions:

8. $y = 2t - t^2$; $x = 3t^2 - 2$.

9. $y = (t - 1)(t^2 + 1)$; $x = (t + 1)(t^2 - 1)$.

10. A current $i = t^{3/2} - t^{1/2}$ amperes flows through a resistor of 2 ohms at t seconds. Find the rate of change of power in the resistor, expressed in watts per second, at $t = 1$.

11. A 6-ft man is walking at 4 ft per second along level ground directly away from a light suspended 9 ft above the ground. Find the rate at which his shadow is lengthening in feet per second.

12. The horizontal component x ft and the vertical component y ft of the position of a rocket as measured from the point of firing are given by $x = (v_0 \cos \theta)t$ and $y = (v_0 \sin \theta)t - 16t^2$. The initial velocity is v_0 and θ is the acute angle of firing, both being constant. t is in seconds. Find expressions for (a) horizontal velocity, (b) vertical velocity, (c) the slope of the rocket path in space, (d) the time of maximum height, (e) the horizontal distance from the starting point at maximum height, and (f) the maximum height.

13. For a certain enclosed gas, the pressure P and volume V are related by the equation $PV^{1.2} = 960$, and the pressure is increasing at a rate of 2 units per minute. Find the rate at which the volume is changing in cubic units per minute at the moment when the volume is 32 cubic units.

14. A circular cylinder is increasing in volume at the rate of 2.0 ft³ per second, and the height of the cylinder is always one-third of the radius of its base. Find the rate at which the radius is changing with respect to time at the moment when the radius is 8.0 ft.

15. The impedance Z of a circuit with a resistor and inductor in parallel is given by $Z = RX/(R + X)$. If X remains constant at 4 ohms and R is increased at a rate of 2 ohms per second, find the rate at which Z is increasing in ohms per second when $R = 1$ ohm.

16. The effective resistance R of a parallel circuit of two resistances R_1 and R_2 is given by $1/R = 1/R_1 + 1/R_2$. If R_1 is constant at 8 ohms and R_2 is decreased at a rate of 2 ohms per second, find the rate at which R is changing in ohms per second at the moment when R_2 is 2 ohms.

17. The impedance Z of an RL series circuit is given by $Z^2 = R^2 + X^2$. If R remains constant at 4 ohms and X is increasing at a rate of 2 ohms per second, find the rate at which Z is changing in ohms per second when X is 8 ohms.

18. The velocity v and displacement s of a body are related by the equation $v = 2\sqrt{s}$, feet and seconds being the basic units. Find the acceleration by calculus.

19. The velocity v and displacement s of a body are related by the equation $v^2 = 1600 - 32s$, feet and seconds being the basic units. Find the acceleration by calculus.

20. In an *RLC* series circuit, the current i amperes is related to the capacitor charge q coulombs by the equation $i^2 = 16,000q$. The inductance $L = 3.0 \times 10^{-4}$ henry. Using electrical laws given in Section 3.4, find the voltage across this inductance.

21. A solid cube of metal is expanding with heat so that each of its linear dimensions increases at a rate of 4.0% per hour. Find the rate at which (a) the volume and (b) the area of one of its surfaces is increasing. Express the answer as a percent.

22. The pressure and volume of a gas are related by the equation $PV^{1.2} = 800$. If the volume is decreased at a rate of 15% per minute, find the percent rate of change of the pressure per minute.

23. Liquid is being poured into a hemispherical container of radius 14 in. whose open top is in the horizontal plane, at the rate of 75 in.^{3} per second. Find the rate at which the liquid is rising when its depth is 8.0 in. (The volume of a segment of a sphere of radius r, for which the perpendicular distance from the center of the flat surface to the curved surface is h, is given by $V = \pi(rh^2 - h^3/3)$.)

6.3   COMPOSITE FUNCTIONS

If u is a function of v, v is a function of w, and w is a function of x, then

$$\frac{du}{dx} = \frac{du}{dv}\frac{dv}{dw}\frac{dw}{dx}.$$

This rule is often referred to as the chain rule, since v and w form successive links between u and x. The rule may be extended, of course, to provide for as many related variables as necessary.

This rule provides a means of differentiating a much more complicated type of function than we have previously considered. The simple formula $(d/dx)kx^n = knx^{n-1}$ can now be broadened to provide not only for a constant power of a single variable but also for a constant power of any function of a single variable. If u is any function of x,

$$\frac{d}{dx}ku^n = \frac{d}{du}ku^n\frac{du}{dx} = knu^{n-1}\frac{du}{dx}.$$

EXAMPLE 7.   Find $(d/dx)[1/(x^3 - 4x)^9]$.

Solution: It is possible to work this out by letting $y = 1/(x^3 - 4x)^9$, expanding the denominator, cross-multiplying, and using an implicit method. The amount of calculation required may be realized when we discover that the expansion of the denominator contains ten terms. The simpler solution is as follows: Let $x^3 - 4x = u$. Then the derivative is

$$\frac{d}{dx}\frac{1}{u^9} = \frac{d}{dx}u^{-9} = \frac{d}{du}u^{-9}\frac{du}{dx} = -9u^{-10}\frac{du}{dx}.$$

$$\frac{du}{dx} = \frac{d}{dx}(x^3 - 4x) = 3x^2 - 4.$$

Therefore the required derivative is

$$-9u^{-10}(3x^2 - 4) = \frac{-9(3x^2 - 4)}{(x^3 - 4x)^{10}}.$$

The type of function considered here, where one function controls another function within itself, is often called a *composite function*. This type of function is not restricted to power functions. Other examples of a composite function are: $\sin(200t + \pi/6)$, $\log(3x^3 - 5)$, $10^{\cos x}$, etc.

EXAMPLE 8.   Find $(d/dt)\sqrt[3]{(t + 1/t)^2}$ at the point where $t = 1$.

Solution: The problem can be restated as $(d/dt)(t + 1/t)^{2/3}$. Let $t + 1/t = u$.

$$\frac{d}{dt}u^{2/3} = \frac{2u^{-1/3}}{3}\frac{du}{dt} = \frac{2}{3}\left(t + \frac{1}{t}\right)^{-1/3}\left(1 - \frac{1}{t^2}\right).$$

At $t = 1$, the derivative $= \frac{2}{3}(2)^{-1/3}(0) = 0$.

With a little practice, the student should eventually calculate this type of problem without actually substituting u for an inner function of the independent variable. The analysis is structural; it is necessary to see the main outer structure of the expression, ignoring the details at first, and to look into the details in a later step. In this way, a problem such as

$$\frac{d}{dx}\{[3(x^2 - 3)^{-2} + 4x^3]^{1/2} - 7\}^4$$

can be seen without detail as $d/dx\{\qquad\}^4$, which is $4\{\qquad\}^3$ times $d/dx\{\qquad\}$. The details inside the brace may be filled in afterward; they remain identical to the original complete expression inside this brace. Next we would have to deal with $d/dx\{[\quad]^{1\!/2} - 7\}$, which is $\frac{1}{2}[\quad]^{-1\!/2}$ times $d/dx[\quad]$. Later

$$\frac{d}{dx}[3(\quad)^{-2} + 4x^3] = -6(\quad)^{-3}\frac{d}{dx}(\quad) + 12x^2,$$

where

$$\frac{d}{dx}(x^2 - 3) = 2x.$$

The structure of our answer will then be

$$4\{\qquad\}^3\tfrac{1}{2}[\qquad]^{-1\!/2}[-6(\quad)^{-3}2x + 12x^2].$$

The brackets may now be filled in as they were originally, and the answer may require a good deal of algebraic simplification.

Before the student uses this structural analysis, the previous method of actually substituting a single letter, such as u, for an inner function should be used for several of the first examples in the following exercise.

EXAMPLE 9.   The impedance of an a-c circuit is given by

$$Z = \sqrt{R^2 + (\omega L - 1/\omega C)^2}.$$

Find (a) dZ/dR, with ω, L, and C constant.
   (b) dZ/dL, with R, ω, and C constant.
   (c) $dZ/d\omega$, with R, L, and C constant.
Solution:

$$\text{(a)}\quad \frac{dZ}{dR} = \frac{1}{2}\left\{R^2 + \left(\omega L - \frac{1}{\omega C}\right)^2\right\}^{-1\!/2}2R,$$

which simplifies to $R[R^2 + (\omega L - 1/\omega C)^2]^{-1\!/2}$.

$$\text{(b)}\quad \frac{dZ}{dL} = \frac{1}{2}\left\{R^2 + \left(\omega L - \frac{1}{\omega C}\right)^2\right\}^{-1\!/2}2\left(\omega L - \frac{1}{\omega C}\right)^1\omega$$

$$= \omega\left(\omega L - \frac{1}{\omega C}\right)\left\{R^2 + \left(\omega L - \frac{1}{\omega C}\right)^2\right\}^{-1\!/2}.$$

$$\text{(c)}\quad \frac{dZ}{d\omega} = \frac{1}{2}\left\{R^2 + \left(\omega L - \frac{1}{\omega C}\right)^2\right\}^{-1\!/2}2\left(\omega L - \frac{1}{\omega C}\right)^1\left(L + \frac{1}{\omega^2 C}\right)$$

$$= \frac{\omega^4 L^2 C^2 - 1}{\omega^3 C^2}\left\{R^2 + \left(\omega L - \frac{1}{\omega C}\right)^2\right\}^{-1\!/2}.$$

EXAMPLE 10. Find $(d/dt)[(t - 2/t^2)^{\frac{1}{2}}(3t^2 - 1)]$.

Solution: Here the structure of a product supersedes all inner structures. Hence, if the first expression in parentheses is replaced by u and the second by v, the form of the derivative is $u(dv/dx) + v(du/dx)$, the calculation of dv/dx and du/dx being details within this principal structure. Without actually using substitutions of u and v, but adhering to this structure, the derivative is

$$\left(t - \frac{2}{t^2}\right)^{\frac{1}{2}} 6t + (3t^2 - 1)\frac{1}{2}\left(t - \frac{2}{t^2}\right)^{-\frac{1}{2}}\left(1 + \frac{4}{t^3}\right)$$

$$= 6t\left(t - \frac{2}{t^2}\right)^{\frac{1}{2}} + \frac{(1 + 4/t^3)(3t^2 - 1)}{2(t - 2/t^2)^{\frac{1}{2}}}$$

$$= \frac{12t(t - 2/t^2) + (1 + 4/t^3)(3t^2 - 1)}{2(t - 2/t^2)^{\frac{1}{2}}}$$

$$= \frac{12t^5 - 24t^2 + 3t^5 - t^3 + 12t^2 - 4}{t^3} \div \frac{2}{t}(t^3 - 2)^{\frac{1}{2}}$$

$$= \frac{15t^5 - t^3 - 12t^2 - 4}{2t^2(t^3 - 2)^{\frac{1}{2}}}.$$

EXAMPLE 11. Find $(d/dr)[(3r^2 - 2r)/\sqrt{4r + 1}]$ at the point where $r = 2$.

Solution: Here the main structure is the quotient formula. The derivative is

$$\frac{\sqrt{4r + 1}\,(6r - 2) - (3r^2 - 2r)\frac{1}{2}(4r + 1)^{-\frac{1}{2}}\,4}{4r + 1}.$$

When $r = 2$, the derivative is

$$\left[\sqrt{9}\,(10) - \frac{8(\frac{1}{2})(4)}{\sqrt{9}}\right] \div 9 = \left(30 - \frac{16}{3}\right)\frac{1}{9} = \frac{74}{27}.$$

When the derivative is to be evaluated at a specific point, there is no reason to simplify the original form of the derivative, since the simplification is more easily handled arithmetically, with numbers substituted.

EXERCISES

1. Find $(d/dx)x^4$ by expressing it as $(d/dx)(x^2)^2$ and using the method of a composite function.

2. Use three distinct methods to find $(d/dx)(2x - 1)^2$.

Find the following derivatives. (*Note:* In most of these exercises, there are at least two possible solutions, using techniques of implicit differentiation, product formula, quotient formula, composite function, etc. The student will acquire facility in calculations if Exercises 3–12 are calculated by at least two methods.)

3. $\dfrac{d}{dx}(2x - 1)^3$.

4. $\dfrac{d}{dx}\dfrac{3}{2x - 1}$.

5. $\dfrac{d}{dx}\sqrt{2x - x^2}$.

6. $\dfrac{d}{dt}\dfrac{1}{\sqrt{t + 2t^{\frac{1}{4}}}}$.

7. $\dfrac{d}{dv}\sqrt{(2v^{\frac{3}{4}} - v^{-\frac{1}{4}})^3}$.

8. $\dfrac{d}{dy}\dfrac{1}{(y^2 - 3y)\sqrt{y^2 - 3y}}$.

9. $\dfrac{d}{d\theta}\dfrac{4\pi}{3(2\theta + \theta^{-1})^2}$.

10. $\dfrac{d}{dx}[(5 - x)^2(x^2 - 5)]$.

11. $\dfrac{d}{dt}\dfrac{3t}{\sqrt{t + 5}}$.

12. $\dfrac{d}{df}(2f^2 - 2)^{\frac{1}{2}}$ at $f = 3$.

13. $\dfrac{d}{dx}\dfrac{x^2 - 4}{x + 2}$ at $x = -52$.

14. $\dfrac{d}{di}\sqrt{5i - 1}\sqrt{i + 3}$ at $i = 2$.

15. $\dfrac{d}{dt}[(8 - 4t)\sqrt{t^2 - 9}]$ at the moment when $t = 5$.

16. $\dfrac{d}{dx}\sqrt{\dfrac{x^2 - 2x}{3x^2 + x}}$.

17. $\dfrac{d}{dR}\dfrac{R}{\sqrt{R^2 + X^2}}$.

18. $\dfrac{d}{dX}\dfrac{R}{\sqrt{R^2 + X^2}}$.

19. $\dfrac{d}{dt}i^2R$, where $i = 4t^{\frac{3}{4}} - 8t^{\frac{1}{4}}$ and R is constant.

20. $\dfrac{d}{dx}[(3x + 1)(x^2 - 3)(2 - 5x)^{\frac{1}{2}}]$.

21. $\dfrac{d}{dt}\dfrac{(3t - 1)(4 + t^2)}{\sqrt{1 - 2t}}$.

22. $\dfrac{d}{dt}\sqrt{\dfrac{(t^2 - 7t)(2 - 3t)}{5 + 6t}}$.

23. $\dfrac{d}{dt}\sqrt{\dfrac{(t^2 - 7t)(2t^2 - 14t)}{t^2}}$ at $t = -\sqrt{2}$.

24. $\dfrac{d^2}{dx^2}\sqrt{x^2 - 2}$.

25. The derivative of $i = \dfrac{E}{\sqrt{R^2 + 4\pi^2f^2L^2}}$ (a) with respect to E, (b) with respect to R, and (c) with respect to f.

26. $\dfrac{d}{dC}\dfrac{E}{\sqrt{R^2 + (2\pi fL - 1/2\pi fC)^2}}$.

27. $\dfrac{d}{d(t^2 - 4t)}\sqrt{t^2 - 4t}$.

6.4 PROBLEMS INVOLVING APPLICATIONS OF THE DERIVATIVE

In this chapter, techniques of differentiation have been broadened to include much more complicated functions than were previously considered. This provides the background to deal with more advanced types of problem.

Types of problems involving differentiation in their solution fall roughly into the following categories: analytic geometry, maximum and minimum, rates and related rates, and effects of small changes. The principles involved and the use of differential calculus in solutions have already been analyzed.

In a worded problem, a clear picture of the quantities that vary and their relation to one another must be formed before any details are carried out. The variable quantity whose value determines the values of all other variables in the problem is then chosen as the independent variable and represented by a letter. All other variables should then be expressed as functions of this letter, and in this way equations can be set up and the principles involved can be applied to these equations.

Another point that may arise is that the independent variable may not be completely independent, that is, its value may be restricted within certain limits. An illustration of this point is given in Fig. 6-2. If the distance from A to C in Fig. 6-2*a* is chosen as the independent variable and called x, then $CB = 100 - x$ only if x has a value between 0 and 100. If x_1 is greater than 100 and equal to AC_1, then the physical distance $C_1B = x_1 - 100$. This restriction must be kept in mind if, say, a maximum y is to be found which depends on the physical distance CB, represented by $100 - x$. Suppose Fig. 6-2*b* represents the variation of this y as a function of x. Mathematically, y reaches a maximum at P; but the value of x at this point lies outside the

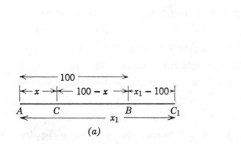

(a)

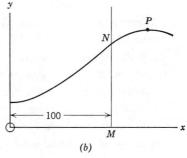

(b)

Fig. 6-2. Illustration of a restriction on the value of the independent variable.

domain 0 to 100 for which the equations are true. Thus, physically, y has a greatest possible value of MN, which is not a mathematical maximum.

Electrical laws frequently used in the solution of problems are listed in Sections 3.4 and 4.3.

EXAMPLE 12.   The effective current in a resistor of R ohms in an a-c circuit is governed by the equation

$$I = \frac{E}{\sqrt{R^2 + (\omega L - 1/\omega C)^2}}.$$

Considering R as the only variable on the right-hand side, find the value of R for maximum power in the resistor, assuming that R is positive.

Solution: Power P is

$$I^2 R = \frac{E^2 R}{R^2 + (\omega L - 1/\omega C)^2}.$$

Using an explicit method and representing the constant quantity $(\omega L - 1/\omega C)$ as X,

$$\frac{dP}{dR} = E^2 \left[\frac{(R^2 + X^2) - 2R^2}{(R^2 + X^2)^2} \right] = 0 \qquad \text{for maximum } P.$$

$X^2 - R^2 = 0$, and $R = \pm X$, the choice of sign being determined by the fact that R must be positive.

The sign of d^2P/dR^2 is the same as the sign of $(d/dR)[E^2(X^2 - R^2)]$, considering only the numerator of dP/dR, as in the analysis of Example 4. Therefore the sign of d^2P/dR^2 is the same as the sign of $-2E^2R$, which is negative for any positive R. Therefore a maximum power occurs when R has a value equal to the absolute value of $\omega L - 1/\omega C$.

EXAMPLE 13.   The horizontal displacement x ft and the vertical displacement y ft of a rocket from its firing point t sec after firing are given by $x = v_0 t \cos \theta$ and $y = v_0 t \sin \theta - 16t^2$, neglecting friction. v_0 is the initial velocity and θ is the angle of projection. Using these equations and using $\cos \theta$ as an independent variable, find the angle of projection to use so that the rocket will land at the farthest possible distance from its starting point, assuming level ground.

Solution: θ is the logical independent variable here. However, the differentiation of $\cos \theta$ and $\sin \theta$ have not yet been covered. This difficulty may be overcome by letting $\cos \theta = u$, say, and treating u as an independent variable. Then $\sin \theta = \sqrt{1 - \cos^2 \theta} = \sqrt{1 - u^2}$. (Note that the absolute value of u here cannot be greater than 1.) t is also variable, but the particular t at which the rocket strikes the ground may be found as a function of u. Thus, when the rocket strikes the ground,

$$y = 0; \qquad \therefore \; v_0 t \sqrt{1 - u^2} - 16t^2 = 0.$$

$$t = 0 \qquad \text{or} \quad t = \frac{v_0 \sqrt{1 - u^2}}{16},$$

the latter value being the one at which the rocket strikes the ground; $t = 0$ is the time of firing, at which, of course, y also $= 0$.

$$\therefore \; x = v_0 \frac{v_0 \sqrt{1 - u^2}}{16} u$$

at the time the rocket strikes the ground. In this equation, u is the only variable on the right-hand side. For maximum horizontal displacement at landing, $dx/du = 0$.

$$\therefore \; \frac{v_0{}^2}{16} [\sqrt{1 - u^2} + u\tfrac{1}{2}(1 - u^2)^{-\frac{1}{2}}(-2u)] = 0.$$

$$\frac{v_0{}^2}{16} \frac{1 - 2u^2}{\sqrt{1 - u^2}} = 0.$$

$$u^2 = \tfrac{1}{2}, \qquad \text{and} \quad u = \frac{1}{\sqrt{2}}.$$

$$\frac{d^2x}{du^2} = \frac{v_0{}^2}{16} \frac{-4u\sqrt{1 - u^2} + u(1 - u^2)^{-\frac{1}{2}}(1 - 2u^2)}{1 - u^2},$$

at $u = +\dfrac{1}{\sqrt{2}}, \dfrac{d^2x}{du^2} = \dfrac{v_0{}^2}{16}\left(-\dfrac{4}{\sqrt{2}}\sqrt{\dfrac{1}{2}} + 0\right)2$, which is negative.

$$\cos \theta = +\frac{1}{\sqrt{2}} \qquad \text{and} \quad \sin \theta = +\sqrt{1 - \tfrac{1}{2}} = +\frac{1}{\sqrt{2}}. \qquad \therefore \; \theta = 45°.$$

Therefore angle of projection for maximum landing distance $= 45°$.

EXERCISES

1. The current through a constant resistance R in an a-c circuit with constant voltage E is given by $I = E/\sqrt{R^2 + X^2}$. Find the value of X for maximum current in the circuit, and find this maximum current.

2. Over a short interval of time t seconds, the charge on a capacitor is given by $q = 4t^2 + \sqrt{1 - 2t}$ coulombs. Find the time for maximum current.

3. Over a short interval of time t seconds, the current in an inductance L of 4.0×10^{-3} henry is given by $i = 1/(3t - 2) + 9t^2$ amperes. Find the time for maximum voltage across the inductance.

4. In an RL series circuit, the current $I = E/\sqrt{R^2 + \omega^2 L^2}$. If E, R, and ω are constant positive values, find the value of L for maximum power in the circuit and find this maximum power.

5. In the circuit of Exercise 4, if E, ω, and L are constant positive values, and R may be varied, find the value of R for maximum power and find this power.

6. The plate current i_p of a triode is given by $i_p = K(v_g + v_p/c)^{3/2}$, where v_g and v_p are the grid voltage and plate voltage, respectively. If K and c are constant, find:

    (a) The mutual conductance g_m, which equals di_p/dv_g with constant v_p.

    (b) The plate resistance r_p, which equals dv_p/di_p with constant v_g.

    (c) The amplification factor, which equals $-dv_p/dv_g$ with constant i_p.

    (d) The value of $r_p g_m$.

7. A low-pass filter has the following characteristics for a frequency f in the interval $f = f_0 - f_1$ to $f = f_0 + f_1$: $E = E_0[1 - (f - f_0 + f_1)^{3/2}]$, E_0 being constant and f, f_0, and f_1 being in kilocycles per second. If $f_0 = 20$ and $f_1 = f_0 \times 10^{-2}$, find the rate of change of E with respect to f when (a) $f = f_0 - f_1$, and (b) $f = f_0$.

8. A circuit consists of a resistance of R ohms, an inductance of L henries and a capacitance of C farads in series with an applied voltage. The voltage v across the capacitance is related to the time t seconds by the equation $v = 2/\sqrt{4 + t}$ volts. Find expressions for (a) charge on the capacitance, (b) current in the circuit, (c) voltage across the inductance, (d) voltage across the resistance, and (e) total applied voltage.

9. A circuit consists of a resistance of 2 ohms in parallel with a capacitance of $\frac{1}{3}$ farad, and, over a short interval of time t seconds beginning at $t = 1$, the voltage across the parallel circuit is given by $v = \sqrt{2t^3 - 1}$ volts. Find as functions of t (a) the charge on the capacitance, (b) the current in the resistance, (c) the current in the capacitance, and (d) the total current flowing in this circuit.

10. For the circuit of Exercise 9, find the minimum current in the capacitive branch.

11. An electron tube has a constant plate resistance of R_p ohms and a variable load resistance r ohms in series with the plate resistance. The voltage applied to this series circuit is constant. For what value of r is the power dissipated in the load resistance a maximum?

12. The output of a transformer is $Vi \cos \phi$, where V and $\cos \phi$ are constant and i is the variable output current. The core loss in the transformer is constant at C watts and the copper loss is i^2R watts, where R is the constant equivalent secondary resistance. If the efficiency of the transformer is expressed as

$$\frac{\text{output}}{\text{output plus core and copper losses}},$$

find the value of current for which the efficiency is greatest.

13. The current in an a-c circuit consisting of a resistance R ohms in series with a capacitance C farads is given by

$$I = \frac{2\pi fCE}{\sqrt{1 + 4\pi^2 f^2 C^2 R^2}} \text{ amperes.}$$

Find the value of R for maximum power in this circuit, and find this maximum power. Assume R to be the only variable on the right-hand side.

14. The current in a circuit is given by $i = 4\sin\theta + \sin 2\theta$. Find the maximum current. (*Hint:* Let $\sin\theta = x$, and use the formula $\sin 2\theta = 2\sin\theta\cos\theta$. Express $\cos\theta$ as a function of x as in Example 13.)

15. A periodic current consists of a fundamental and a third harmonic such that current $i = 4\sin\theta + \sin 3\theta$. Find the maximum positive current and the maximum negative current. ($\sin 3\theta = 3\sin\theta - 4\sin^3\theta$.)

16. Find the maximum value of the function $5\cos\theta + 12\sin\theta$.

17. Find maximum and minimum i, given $i = 6\cos\theta + \sin 2\theta$.

18. The foot of a 13-ft ladder is pulled horizontally outwards along the floor at a rate of 1.5 ft per second while the top slides down a vertical wall. (a) Find the rate at which its top moves down the wall when the foot of the ladder is 5.0 ft from the wall.

(b) Find the theoretical velocity at which the top of the ladder strikes the floor.

19. A boat is being pulled toward a dock by a rope attached to the prow of the boat and running over a pulley on the dock. If the pulley is at a level 7.0 ft higher than the prow of the boat, and the rope is pulled in at a rate of 2.0 ft per second, find the rate at which the boat is approaching the dock when its prow is 24 ft out.

20. A motorboat is 10 miles out from the nearest point on a straight shore, and a passenger wishes to reach an airport 5 miles along the shore in the quickest possible way. If the boat can travel 15 miles per hour and a taxi taken at any point on the shoreline can travel at 60 miles per hour on a straight road to the airport, toward what point on the shore should the boat head for the quickest trip to the airport?

21. If the taxi in Exercise 20 can only go 30 miles per hour, for what point on shore should the boat head?

22. A constant 6-volt battery supplies voltage to a circuit consisting of a variable resistor in series with a parallel arrangement of two resistances of 2 ohms and 4 ohms. Find the setting of the variable resistor for maximum power consumption in this variable resistor.

23. The power in an a-c circuit may be expressed as $P = EI \cos \phi$, where $\tan \phi = X/R$ and $I = E \sin \phi / X$. If E and X are constants and R varies, find the setting of R for maximum power.

24. The amount of illumination from a light source on a level surface varies inversely with the square of the distance from the source and directly with the sine of the angle at which the light ray strikes the surface. Find how far above the edge of a road of width w a light should be placed to supply greatest illumination at the center of the road.

25. The attentuation constant α of a transmission line is expressed by the equation $2\alpha^2 = \sqrt{(R^2 + \omega^2 L^2)(G^2 + \omega^2 C^2)} + RG - \omega^2 LC$. If L is the only variable on the right-hand side of this equation, determine the value of L when the attenuation constant is least.

26. Two resistances of $\frac{1}{2}$ ohm and 2 ohms are connected in series. A uniform 8-in. long wire with a resistance of 1 ohm per inch is tapped so that x in. are connected in parallel with the $\frac{1}{2}$-ohm resistance and the remainder in parallel with the 2-ohm resistance. If each of the two extreme ends of the circuits are connected by resistanceless connections, find the value of x that will give a maximum resistance to the complete circuit. (The equivalent resistance of two resistances in parallel is equal to their product divided by their sum.)

CHAPTER 7

Integration as Inverse of Differentiation

7.1 ANTI-DIFFERENTIATION

Integration and *integral calculus* involve finding what function has a given derivative. For an integration, differentiation is reversed, and an anti-derivative is found. This process, being essentially a backwards reasoning, often becomes quite difficult. However, there are many examples in which it is very simply done.

> EXAMPLE 1.   Find the function whose derivative with respect to x is $6x^2 - 2$.
> *Solution:* The derivative of a power of x has a power of x one lower than the original power. Therefore the original powers of the two terms here must be the cube and the first power. By a process of trial and error, the coefficients of these two terms may be determined; for example, the coefficient of x^3 must be 2 in order that its derivative will be $6x^2$. An antiderivative of $6x^2 - 2$ is therefore $2x^3 - 2x$.

If Example 1 is analyzed a little more thoroughly, we may see that the answer given is only one of many answers. Essentially, we are required to find the equation of a curve whose slope at any point (x,y) is $6x^2 - 2$. This information does not determine the curve exactly, since any curve can be transposed vertically up or down without affecting the slope at any point on the curve. Figure 7-1 shows four curves with different equations but with identical slopes at any value of x. There are actually an infinite number of curves with a given slope at the point (x,y). The equations of these curves have identical variable terms, but the value of y corresponding to any value of x will differ by a constant amount for any two curves.

111

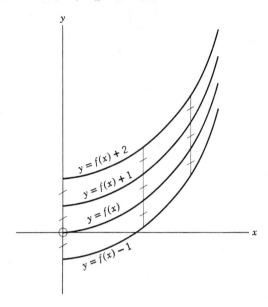

Fig. 7-1. Graphs with identical slopes at any value of x.

In Example 1, the antiderivative of $6x^2 - 2$ could be $2x^3 - 2x + 3$, $2x^3 - 2x - 7$, etc., and in general the antiderivative is $2x^3 - 2x + C$, where C may have any constant value, positive, negative, or zero. The derivative of the constant term is zero, so that the derivative of $2x^3 - 2x + C$ is $6x^2 - 2$, whatever the value of C may be. This extra constant term is called the *integration constant*, and it will appear in all antiderivatives.

EXAMPLE 2.   Find the equation of a curve whose slope is given by the function $3t^5 - 2/t^2 + 1/\sqrt{t}$.
Solution: The powers of t in the successive terms of the answer will be one higher than those in the given function. Therefore there will be terms in t^6, t^{-1}, $t^{\frac{1}{2}}$, and the integration constant. By a process of trial and error and verification, the required function is found to be

$$y = \tfrac{1}{2}t^6 + 2t^{-1} + 2t^{\frac{1}{2}} + C.$$

EXAMPLE 3.   The current flowing to a capacitor of 0.4 farad is related to time t seconds by the equation $i = 2t^{\frac{3}{2}} - 5t^{\frac{1}{2}} + 3$ amperes. Find the voltage across the capacitor as a function of t.

Solution:

$$i = C\frac{dv}{dt}. \qquad \therefore \; \frac{dv}{dt} = \frac{i}{C} = 5t^{3/2} - 12.5t^{1/2} + 7.5.$$

Therefore v is the antiderivative of this function. By trial and error and verification, $v = 2t^{5/2} - \frac{25}{3}t^{3/2} + \frac{15}{2}t + K$, where K is any constant.

EXERCISES

Find y as a function of the independent variable, given the following:

1. $\dfrac{dy}{dx} = 6x^2 + 2x - 7.$

2. $\dfrac{dy}{dt} = 3t^{-1/2} - 6t^{1/2} + 5t^{3/2}.$

3. $\dfrac{dy}{dv} = 7v - \dfrac{2}{\sqrt{v}} - \dfrac{2}{\sqrt[3]{v}}.$

4. $\dfrac{d^2y}{dx^2} = kx^3 + mx - n.$

Solve the following equations for i.

5. $\dfrac{di}{dt} = 7.$

6. $di = (0.5t - 0.4t^{1/2} + 0.9t^{1/4})\, dt.$

7. $di = (t - 2\sqrt{t})^2\, dt.$

8. If the velocity of a body at t seconds is given by $v = 3t + 8$ ft per second, find, as a function of t (a) the displacement and (b) the acceleration.

9. If the acceleration of a body at t seconds is given by $a = 2\sqrt{t} + 3$ ft per second per second, find, as a function of t (a) the velocity and (b) the displacement.

10. The current flowing to a 2.0×10^{-2}-farad capacitor at t seconds is given by $i = 15\sqrt{t} - 20/\sqrt{t}$ amperes. Find as functions of t (a) the charge on the capacitor, (b) the voltage across the capacitor, and (c) the voltage across a 0.2-henry inductance in series with the capacitor.

11. The slope of a curve is given by the function $3 - 5x^{-1/4}$. Find the exact equation of the curve if it passes through the point $(0, -4)$.

12. A circuit consists of an inductance of 0.10 henry in series with a capacitance of 0.050 farad. The voltage across the inductance at time t sec is $t^3 - 1/10t\sqrt{t}$ volts. Find the total applied voltage as a function of t.

7.2   INTEGRATION OF POWER FUNCTIONS BY RULE

The mathematical operation of finding a function whose differential is given is known as *integration*. Integration is represented by the symbol $\int$, which may be interpreted as *the integral of*. The expression following this symbol is always a differential, not a derivative. Thus $\int 3x^2\, dx$ is interpreted to mean "the function whose differential is $3x^2 \ dx$", the answer being $x^3 + C$.

There are reasons why the differential form, rather than the derivative form, is used after the integration sign; these reasons will become more apparent when the meaning of an integral has been more thoroughly investigated. One reason for the use of the differential form is that it indicates the letter representing the independent variable. The preceding example indicates an integration with respect to x, the factor dx making this clear. An expression such as $\int 3uv$ would be ambiguous, since either u or v may be the independent variable. On the other hand $\int 3uv \ dv$ designates v as the independent variable. The function appearing between the complementary symbols $\int$ and dx is called the *integrand.*

After working the exercises following Section 7.1, the student may have found a quick rule for finding integrals of power functions without resorting to trial and error. The rules for finding a differential must be reversed, and this reversal will also apply to the order in which operations are performed. The rules for differentiating a single-termed power function are (1) multiply by the index and (2) lower the index by 1. Completely reversing these operations, the rules for integrating a single-termed power function are (1) raise the index by 1 and (2) divide by the resulting index.

There is one power of the independent variable for which this rule cannot be used. There is no power function of x for which the derivative is a term in x^{-1}. The derivative of x^0 (that is, 1) is 0. Therefore, any term where the independent variable is raised to the power -1 cannot be integrated by the rule for integrating power functions. The integration of this particular power function will be discussed in Chapter 11.

Thus the rule for integrating a power function may be stated as

$$\int ku^n \ du = \frac{ku^{n+1}}{n+1} + C, \qquad \text{provided } n \text{ is not equal to } -1.$$

The letter x or any other letter representing a variable could have been used instead of u throughout this formula. u is used here rather than x because u may be itself a function of x.

Many problems involving integration will contain enough information to enable us to evaluate the integration constant. An integration involves finding the equation of a curve whose slope at any point is given. This produces a "family" of curves (see Fig. 7-1) spaced out vertically at various positions corresponding to different values of the integration constant. In order to limit the solution to one particular curve in this family, some information must be given to fix the exact vertical position of the curve. This is most often given in the form of a particular point through which the curve passes.

EXAMPLE 4. The slope of a straight line is -3 and the line passes through the point $(-2, -1)$. Find the equation of the line.

Solution: This may be done by methods of analytic geometry, as in Example 5 of Chapter 2. Another method is to use calculus. Given $dy/dx = -3$.

$$\therefore y = \int (-3)\, dx = -3x + C.$$

Since the line passes through $(-2, -1)$, $-1 = -3(-2) + C$. $C = -7$. Therefore the equation of the line is $y = -3x - 7$. This may be verified by methods of analytic geometry.

EXAMPLE 5. Find $\int (3x - 2 - 4/x^2)\, dx$, assuming it has a value of 3 when $x = 2$.

Solution:

$$\int \left(3x - 2 - \frac{4}{x^2}\right) dx = \frac{3x^2}{2} - 2x + \frac{4}{x} + C.$$

The additional information is used to solve for the particular integration constant to fit this case. Substituting $x = 2$ in the function representing the integral,

$$\frac{3 \times 4}{2} - 4 + 2 + C = 3.$$

$$C = 3 - 6 + 4 - 2 = -1.$$

Therefore the required integral is $3x^2/2 - 2x + 4/x - 1$.

Considering the linear motion of a body with displacement s, velocity v, and acceleration a, since $v = ds/dt$, $s = \int v\, dt$. Similarly, since $a = dv/dt$, $v = \int a\, dt$. Rotational motion is analogous to this; if angular displacement is θ, angular velocity ω and angular acceleration α, $\omega = d\theta/dt$ and $\alpha = d\omega/dt$. Therefore $\theta = \int \omega\, dt$ and $\omega = \int \alpha\, dt$. These principles are shown briefly in the following arrangement.

Linear Motion	Rotational Motion
$\int v\, dt = s.$	$\int \omega\, dt = \theta.$
$\int a\, dt = v = \dfrac{ds}{dt}.$	$\int \alpha\, dt = \omega = \dfrac{d\theta}{dt}.$
$a = \dfrac{dv}{dt}.$	$\alpha = \dfrac{d\omega}{dt}.$

The arrows in the above arrangement indicate the order of reasoning from given information to required information.

EXAMPLE 6.   The acceleration of a body moving in a straight line is proportional to the time elapsed since the start of the motion. At time 0.5 sec, the acceleration is -1.5 ft per second per second. If the initial velocity is 20 ft per second, find how far it travels before coming to rest.
Solution: $a = kt$ and at $t = 0.5$, $a = -1.5$.

$$\therefore \ -1.5 = 0.5k; \qquad k = -3. \qquad \therefore \ a = -3t.$$

$$v = \int a\, dt = \int (-3t)\, dt = \frac{-3t^2}{2} + C.$$

At $t = 0$, $v = 20$. Therefore $20 = 0 + C$; $C = 20$.

$$\therefore \ v = \frac{-3t^2}{2} + 20.$$

$$s = \int v\, dt = \int \left(\frac{-3t^2}{2} + 20 \right) dt = \frac{-t^3}{2} + 20t + C_1.$$

Since t is measured from the start of the motion, at $t = 0$, $s = 0$.

$$\therefore \ C_1 = 0 \qquad \text{and} \quad s = \frac{-t^3}{2} + 20t.$$

When the body comes to rest, $v = 0$; $-3t^2/2 + 20 = 0$; $t = \sqrt{40/3}$ $= 3.65$. Displacement at this time $= -(3.65)^3/2 + 20(3.65) = 48.7$ ft.

EXERCISES

1. ·Calculate the following:

(a) $\int du.$ (b) $\int d(x^2 - 3x).$ (c) $d \int du.$ (d) $\dfrac{d}{dt} \int d\left(t^2 - \dfrac{5}{t} \right).$

Find the following integrals.

2. $\int (3x^2 - 2)\, dx.$

3. $\int \left(2x + \dfrac{3}{x^2} \right) dx.$

4. $\int (k - 3\sqrt{t})\, dt.$

5. $\int \dfrac{3\sqrt{t} - 2t\sqrt{t}}{t}\, dt.$

6. $\int \left(\dfrac{3}{\sqrt{t}} - \theta^2 \right) dt.$

7. $\int \left(uv + u^2v^2 - \dfrac{u^2}{v^2} \right) du.$

8. $\int \left(uv + u^2v^2 - \dfrac{u^2}{v^2} \right) dv.$

Find y as a function of x for the following:

9. $\dfrac{dy}{dx} = 7x^{3/4} - 5x^{1/4}.$

10. $y = \int m\, dx$, and $y = 3$ when $x = 0$.

11. $y'' = x^{1/2} + x^{-1/2}.$

Find y as an exact function of the independent variable for the following:

12. $y = \int (x^2 - 3)\, dx$, if, when $x = 15$, $y = 1000$.

13. $y = \int \left(2x - \dfrac{1}{x^2} \right) dx$, if, when $x = 1$, $y = 1$.

14. $y = \int \left(5 - \dfrac{3}{\sqrt{t}} \right) dt$, if, when $y = 0$, $t = 1$.

15. $y = \int (\sqrt{p} + 1)(\sqrt{p} - 3)\, dp$, if, when $p = 4$, $y = 10$.

Find the equations of the curves with the following characteristics:

16. Slope is -2, passing through the origin.

17. Slope is $3x - 5$, passing through $(-1,0)$.

18. Slope is twice the horizontal displacement from the origin, with a y-intercept of -3.

19. Slope is $2x^2 + 1/\sqrt{x}$, passing through $(9, -1)$.

20. The acceleration of a body is constant at 2.0 ft per second per second. If displacement is measured from time $t = 0$, and the velocity at this time is 12 ft per second, find by calculus an equation to represent displacement.

21. The acceleration of a body is constant, represented by a. The initial velocity is v_0 and displacement is measured from its position at time $t = 0$. Find, as functions of time (a) its velocity and (b) its displacement.

22. The acceleration of a body depends on time t such that $a = 30 - 4\sqrt{t}$, the units being feet and seconds. If it starts from rest at $t = 0$, find:

(a) How far it is from the starting point after 4 sec.

(b) Its velocity when it returns to the starting point.

23. For a body moving from rest in a straight line, the acceleration is directly proportional to the time of motion. If it moves 12 ft in the first 2 sec, find its displacement as a function of time.

24. Calculate the missing entries in the following tabulation of motion of a body, assuming v is proportional to t:

Time t sec	0	1	2	3	4
Distance s ft	2				
Velocity v ft/sec	0	4			

25. The velocity v of a body depends on its displacement s such that $v = \sqrt{400 - 16s}$, units being feet and seconds. Assuming that, at $t = 0$, $s = 0$, and v is positive, find (a) the acceleration, (b) the initial velocity, (c) the velocity as a function of time, using the acceleration of (a), and (d) the displacement as a function of time.

26. A plane starts from rest along a runway with a constant acceleration of 15 ft per second per second and is airborne at a velocity of 180 ft per second. Find the distance it travels along the runway.

27. A plane lands at 150 ft per second 900 ft from the far end of the runway. Find the minimum constant deceleration that must be maintained so that the plane will stop while still on the runway.

28. The angular acceleration of a rotating machine is given by $\alpha = 5 - t$ rad per second per second. If, at $t = 0$, angular displacement $= 0$ and angular velocity $= 12$ rad per second, find:

   (a) The angular displacement as a function of t.

   (b) The angular displacement when the machine reverses direction.

   (c) The time at which the velocity is -7.5 rad per second.

29. Express y as a function of x, if $d^2y/dx^2 = 3x - 2$, (a) giving a general solution with two unknown constants, and (b) giving an exact solution, if, at $x = 1$, $y = 0.5$, and $dy/dx = 1.5$.

30. A stone is thrown horizontally outwards from the top of a building 225 ft high at a velocity of 20 ft per second. If x and y are its horizontal and vertical displacements in ft from the top of the building at any time t seconds, the equations of motion of the stone are $d^2x/dt^2 = 0$ and $d^2y/dt^2 = -32$. Using calculus, find the following: (a) x as a function of t, (b) y as a function of t, (c) y as a function of x, and (d) how far from the foot of the building the stone strikes the ground.

31. For t seconds after the power is shut off, the angular velocity of a motor armature is $\omega = 22.5 - 3t + 0.1t^2$ rad per second.

   (a) How many revolutions does the armature make in the first 4 sec?

   (b) After how many revolutions is the rate of slowdown 2 rad per second per second?

7.3 ELECTRICAL PROBLEMS

In Section 3.4, several electrical quantities which are determined by taking derivatives of other quantities were listed. It follows that each of these other quantities is determined by taking an integral of the first quantity. For reference, these laws are restated here in both derivative and integral form. The standard unit of measurement for each quantity is assumed.

Current and charge:	$i = \dfrac{dq}{dt}.$	$\therefore \; q = \int i \, dt.$
Current in and voltage across an inductance L:	$v = L\dfrac{di}{dt}.$	$\therefore \; i = \dfrac{1}{L}\int v \, dt.$
Current to and voltage across a capacitance C:	$i = C\dfrac{dv}{dt}.$	$\therefore \; v = \dfrac{1}{C}\int i \, dt.$
Induced voltage v in a coil of N turns linked by flux ϕ:	$v = N\dfrac{d\phi}{dt}.$	$\therefore \; \phi = \dfrac{1}{N}\int v \, dt.$
Voltage induced in one coil due to current in another:	$v_2 = M\dfrac{di_1}{dt}.$	$\therefore \; i_1 = \dfrac{1}{M}\int v_2 \, dt.$

EXAMPLE 7. The current in the capacitive branch of a circuit consisting of a resistance of 0.5 ohm in parallel with a 0.25-farad capacitor varies with time t seconds such that $i = t^3 - 2t\sqrt{t}$ amperes, and the voltage across the circuit is 1.0 volt at $t = 0$. Find the total current flowing to the circuit at $t = 2$.

Solution: The current in the resistor must be found. This current is v/R, where $v = 1/C \int i \, dt$

$$= \frac{1}{0.25}\int (t^3 - 2t^{3/2})\, dt = 4.0\left(\frac{t^4}{4} - 2\times\frac{2}{5}t^{5/2} + k\right).$$

At $t = 0$, this voltage $= 1.0$.

$$\therefore \; v = t^4 - 3.2t^{5/2} + 1.0.$$

Thus current in the resistor is

$$\frac{v}{R} = \frac{t^4 - 3.2t^{5/2} + 1.0}{0.5} = 2t^4 - 6.4t^{5/2} + 2.0.$$

Total current in the two branches is

$$2t^4 + t^3 - 6.4t^{5/2} - 2t^{3/2} + 2.0.$$

At $t = 2$, total current is

$$32 + 8 - 6.4(4\sqrt{2}) - 2(2\sqrt{2}) + 2.0 = 0.1 \text{ ampere}.$$

EXERCISES

1. A capacitor, after discharge, is fed by a current of 0.002 ampere for 5 sec. Determine the final charge on the capacitor.

2. If the current in Exercise 1 is $0.002t$ ampere at t seconds, find the final charge on the capacitor.

3. A current $i = 0.002\sqrt{t}$ amperes flows to an uncharged 50-microfarad condenser for an interval $t = 3$ sec. Find the final voltage across the condenser.

4. Find the flux linking a 200-turn coil at $t = 4.0$ sec if the coil voltage is given by $v = t - t^{1/4}$ volts and the flux is 0.01 weber at $t = 0$.

5. Two coils have a mutual inductance of 5.0×10^{-2} henry. Find an expression for the current in one coil inducing a voltage of $0.4 + 2t\sqrt{t}$ volts at t seconds in the other, if, at $t = 0$, this current is -20 ampere.

6. The current flowing to a capacitor at t seconds is $2\sqrt{t}$ amperes. Find the charge on the capacitor at $t = 0.16$ sec, if it is charged to 0.136 coulomb at $t = 0.09$.

7. The current flowing to a capacitor at t seconds is $0.02t - 0.01t^2$ ampere. Find the time of maximum charge.

8. The voltage across a 0.40-henry inductor is given by $v = 1 + t^{3/2}$ volts, t being in seconds. Find the current in the inductor at 1.21 sec, if the current at 1 sec is 3.5 amperes.

9. For a short interval of t seconds, the voltage across a circuit consisting of a 0.5-ohm resistance and a 0.4-henry inductance in parallel is given by $v = 1/\sqrt{t} - 1/t\sqrt{t}$ volts. At $t = 1$, the current in the inductance is 6 amperes. Find (a) the current in the inductance as a function of time and (b) the total current flowing to this circuit at $t = 0.64$ sec.

10. A voltage is applied to a circuit consisting of a 5.0-ohm resistance in series with a 0.10-farad capacitance. After t seconds, a current of $3 - t^{1/4}$ amperes flows. If the voltage across the capacitor is 2 volts at $t = 0$, find the applied voltage at $t = 1$.

11. The work done in separating two electrically charged spheres with centers r cm apart is expressed as $\int (30/r^2)\,dr$ dyne-cm. If their centers are 2.5 cm apart, calculate the work done in separating them (a) an additional 3.5 cm and (b) an infinite distance.

12. The current through a long cable initially uncharged with a capacitance of C farads per mile and a resistance of R ohms per mile t seconds after the circuit is closed is given by $i = E\sqrt{C/\pi Rt}$ amperes. Find the charge on the cable as a function of t.

13. A voltage is applied to a series circuit consisting of a 2.0-ohm resistor, a 0.50-henry inductor and a 0.40-farad capacitor. After t seconds, a current of $3t - 2$ amperes flows. If, at $t = 0$, the capacitor is charged to 0.60 coulomb, find the applied voltage at $t = 0.80$ sec.

14. A three-branch parallel circuit consists of a 0.5-ohm resistor, a 0.4-farad capacitor, and a 0.2-henry inductor. After t seconds, the voltage across this circuit is $3t^2 + 4t$ volts. Find the total current flow to this circuit at $t = 0.5$ sec if the total current is 3 amperes at $t = 0$.

CHAPTER 8

Integration as Summation

8.1 AREAS

Methods of calculating areas of figures bounded entirely by straight lines have been known since ancient times; they involve applications of basic geometry. The Greek mathematicians also developed approximate methods of calculating areas bounded in part or entirely by curves. Archimedes in particular devised a means of finding areas by summing up narrow rectangles within the area which could be made to fit the area more and more closely by making the width of each rectangle smaller and smaller; that is, letting the width approach zero.

With the discovery of differential calculus in the seventeenth century, it was found that this summing-up process involved antidifferentiation or integration, and by using calculus methods, such areas could be calculated exactly. Formulas for the more familiar areas bounded by curves were then developed. It is this use of integration as a summation that we will investigate in this chapter.

Consider the area under the curve of Fig. 8-1. If we think of a vertical line OB along the y-axis moving toward the right and always extending from the x-axis to the curve, it sweeps over an area which increases as x itself increases; in this way the area swept out may be treated as variable, a function of x. When this vertical line has reached the position PM, its length y and the area it has moved across ($BOMP$) are both dependent on x. Now let x take on an increment Δx, y and A taking on corresponding increments Δy and ΔA. ΔA is the area of $QPMN$, QP being the curved path, and ΔA lies between area of rectangle $RPMN$ and area of rectangle $QSMN$. That is, ΔA lies between

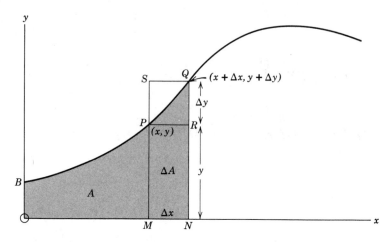

Fig. 8-1. Calculus applied to the area under a curve.

$y \, \Delta x$ and $(y + \Delta y) \, \Delta x$; or $\Delta A/\Delta x$ lies between y and $y + \Delta y$. Now let $\Delta x \to 0$. $\Delta y \to 0$, and $\Delta A/\Delta x \to dA/dx$. Therefore dA/dx lies between y and y.

$$\therefore \ \frac{dA}{dx} = y. \qquad \therefore \ A = \int y \, dx.$$

In order to see exactly how integration is applied to the calculation of an area, a particular example will be studied.

EXAMPLE 1.   Find the area between the curve $y = x^2 + 2$ and the x-axis bounded by the lines $x = 2$ and $x = 4$.
Solution: (See Fig. 8.2.)

$$A = \int y \, dx = \int (x^2 + 2) \, dx = \frac{x^3}{3} + 2x + C.$$

Figure 8-2*b* shows a graph of area A plotted vertically against x, and for every value of the integration constant C there is a corresponding curve of A against x, so that we get a family of curves at different levels, of which only a few are depicted in this diagram. From any one of these curves we can read directly the area swept out in Fig. 8-2*a* by a vertical line from the x-axis to the curve as it moves from left to right; the area swept out to any value of x in (*a*) is the ordinate of the curve of (*b*) at that value of x. Which of the curves of (*b*) do we use? For the problem stated, the sweep of area starts at $x = 2$. Therefore, at $x = 2$, $A = 0$,

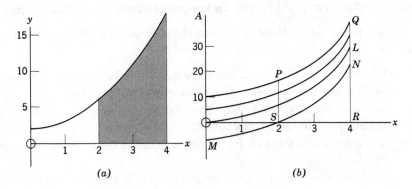

Fig. 8-2. Area under the curve $y = x^2 + 2$.

$$A = \frac{x^3}{3} + 2x + C.$$

which indicates that the curve MN fits our case. If the area were to start at $x = 0$, the curve OL would have to be used. Thus the left boundary of the area provides the information necessary to solve for the integration constant. The total area swept out as far as $x = 4$ is thus the length RN as measured on the A scale. The completion of the calculation is as follows:

$$A = \frac{x^3}{3} + 2x + C. \qquad \text{At } x = 2,\ A = 0.$$

$$\therefore \frac{2^3}{3} + 2(2) + C = 0. \qquad C = -\left(\frac{2^3}{3} + 2 \times 2\right) = -\frac{20}{3}.$$

$$A = \frac{x^3}{3} + 2x - \frac{20}{3} \text{ at } x = 4.$$

$$A = \frac{4^3}{3} + 2 \times 4 - \frac{20}{3} = \frac{68}{3} \qquad \text{or 22.7 square units.}$$

The concept of the integral as a summation now is beginning to emerge. If the area under the curve up to a certain point in Fig. 8-1 were split up into a series of rectangles such as $PMNR$ each with width Δx and height y matching the height of the curve at each left edge, the area could be approximated by summing the areas of the individual rectangles. Using the symbol Σ to represent a sum of several terms of the form indicated, the area equals approximately $\Sigma y \,\Delta x$. But the exact area is $\int y \, dx$, which then can be thought of as $\lim_{\Delta x \to 0} \sum y \, \Delta x$. Thus the integral performs a continuous summation of graphical areas whose heights match the height of a curve continuously and whose width has become infinitely small.

EXAMPLE 2.   Find the area of a trapezoid with sides 3 in. and 5 in., and width 6 in., using calculus. Verify the answer by formula.

Solution: The trapezoid may be fitted into a graphical coordinate system most simply by choosing its left lower vertex as the origin and fitting the sides extending from this vertex along the axes, as shown in Fig. 8-3.

The method of finding areas by integration is perfectly general, and may be applied to simple instances where the boundaries of the area are all straight lines, as in this example. To use calculus, we must know the equation of the upper boundary of the area. Analytic geometry or calculus may be used to obtain this equation. The upper boundary is a straight line of slope $\frac{2}{6}$;

$$\therefore \frac{dy}{dx} = \frac{1}{3} \qquad y = \int \frac{1}{3} \, dx \qquad y = \frac{x}{3} + C.$$

At $x = 0, y = 3,$ $\therefore 3 = C.$ $\therefore y = \frac{x}{3} + 3.$

$$\text{Area} = \int \left(\frac{x}{3} + 3 \right) dx = \frac{x^2}{6} + 3x + k.$$

At $x = 0, A = 0.$

$$\therefore 0 = 0 + 0 + k. \qquad k = 0 \quad \text{and} \quad A = \frac{x^2}{6} + 3x \text{ at } x = 6.$$

$$A = \frac{6^2}{6} + 3 \times 6 = 6 + 18 = 24 \text{ in.}^2$$

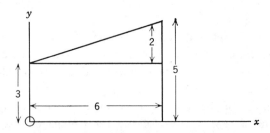

Fig. 8-3. Area of a trapezoid by integration.

The area of a trapezoid, by formula, is width times average height.

$$A = 6 \times \frac{3 + 5}{2} = 6 \times 4 = 24 \text{ in.}^2$$

The answer is verified.

EXAMPLE 3. Find the area bounded by the curve $y = 1 + 5x - x^2$ and the line $y = 5$.

Solution: A diagram is necessary here, as in many cases, to interpret the question. Figure 8-4 shows graphs of the two given equations that completely enclose the required area. It is necessary to determine the values of x at the left and right extremes of the area. We see that these are the points of intersection of the graphs, and the values of x may be found by algebraic methods. Substituting $y = 5$ in the other equation,

$$5 = 1 + 5x - x^2; \qquad x^2 - 5x + 4 = 0.$$
$$(x - 1)(x - 4) = 0; \qquad x = 1 \quad \text{or} \quad 4.$$

Thus the area extends from $x = 1$ to $x = 4$. Rectangle *KLMN* represents approximately an element of the total area. The height of this rectangle is not y in this case but $y_1 - y_2$, where y_1 is the y of the first equation

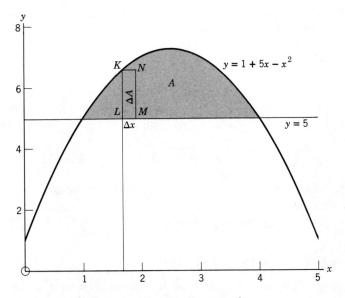

Fig. 8-4. Area between two graphs.

and y_2 is the y of the second. The rectangular element of area is therefore $(1 + 5x - x^2 - 5)\,\Delta x = (5x - 4 - x^2)\,\Delta x$. By using integration as exact summation, the area is

$$\int (5x - 4 - x^2)\, dx = \frac{5x^2}{2} - 4x - \frac{x^3}{3} + C.$$

At $x = 1$, $A = 0$. Therefore $0 = \frac{5}{2} - 4 - \frac{1}{3} + C$. $C = \frac{11}{6}$. Substituting $x = 4$, the area is

$$\frac{5(4)^2}{2} - 16 - \frac{64}{3} + \frac{11}{6} = \frac{9}{2} \text{ square units.}$$

EXERCISES

1. Find the area bounded by the graph of $y = x^2$, the x-axis, and the line $x = 3$.

2. Find the area bounded by the graph of $y^2 = x$, the x-axis, and the line $x = 9$.

3. Any point (x,y) is taken on the graph of $y = x^2$, and from this point a horizontal and a vertical line are drawn to form a rectangle with the two axes. Find the area bounded by the graph, the x-axis, and the vertical line (a) as a function of x and (b) as a fraction of the area of the rectangle.

4. Repeat Exercise 3 for the graph of $y = x^n$ for any constant value of $n \geq 0$.

5. Find the area bounded by the given curve, the x-axis, and the given lines for the following functions.
   (a) $y = 0.5x^2 - 1$, $x = 2$, $x = 4$.
   (b) $y = (x - 3)^2$, $x = 3$, $x = 6$.
   (c) $y = 2/x^2$, $x = 1$, $x = 4$.
   (d) $y = 2/x^3 + 2x$, $x = 0.01$, $x = 1$ (illustrate).

6. Find the area bounded by the graph of $y = (x - 5)(2 - x)$ and the x-axis.

7. Find the area bounded by the graph of $y = -x^2 + 6x - 3$ and the line $y = 2$.

8. Find the area bounded by the graph of $y = x^2 - 4x + 6$ and the line $y = 3$.

9. Find the area in the first quadrant bounded by the graph of $y = 4x - x^3$ and the x-axis.

10. Find the area bounded by the graphs of $y = -x^2 + 8x - 8$ and $y = 2x$.

8.2 THE DEFINITE INTEGRAL

In the calculation of areas it will be seen that the value of the integration constant required to fit the problem is always the negative of the value of the variable part of the expression for the integral at the left-hand boundary of the area. In Example 1, for instance, the expression for the integral is $x^3/3 + 2x + C$, and the desired value of C is $-(2^3/3 + 2 \times 2)$, 2 being the value of x at the left-hand boundary of the area. Essentially the calculation does not involve an integration constant, since we use the value of the variable part of the integral at the right-hand boundary and subtract its value at the left-hand boundary.

Figure 8-2*b* will further clarify this part of the calculation. Each of the graphs has a different integration constant C. The actual area is represented by the length RN as measured on the A scale. However, there is no reason why we could not use the curve PQ from which we can obtain the correct area by taking the length RQ and subtracting the length NQ. But every point on the curve PQ is an equal distance above the corresponding point on the curve SN, and therefore $NQ = SP$. Thus, using the curve PQ, the correct area is $RQ - SP = 4^3/3 + 2 \times 4 + C - (2^3/3 + 2 \times 2 + C)$, where both C's are identical, the value of C for the curve PQ. In the calculation, therefore, C vanishes and, irrespective of its particular value, the area is $4^3/3 + 2 \times 4 - (2^3/3 + 2 \times 2)$. In the same way any of the other curves could be used, the particular C for each curve disappearing in the calculation.

When an integral is used to find a graphical area, it is called a *definite integral*, since no undetermined integration constant appears in the numerical answer to the integral. The left-hand and right-hand boundaries of the area are called the *lower* and *upper limits*, respectively, and they are usually indicated in the integration symbol. For example, the integral used to calculate the area in Example 1 is indicated briefly as $\int_2^4 (x^2 + 2) \, dx$, and is calculated as

$$\left[\frac{x^3}{3} + 2x\right]_2^4 = \frac{4^3}{3} + 2 \times 4 - \left(\frac{2^3}{3} + 2 \times 2\right) = \frac{64}{3} + 8 - \frac{8}{3} - 4 = \frac{68}{3}.$$

In determining areas, it is advisable to draw a graphical diagram; otherwise the interpretation of the areas involved is open to error. The following examples will illustrate further instances where diagrams are necessary.

EXAMPLE 4. Find the total area enclosed by the x-axis and the graph of $y = (x^2 - 1)(4 - x)$.
Solution: Figure 8-5 shows that one portion of the area is above the x-axis and another portion below the x-axis. If we calculate areas below the x-axis using $\int y \, dx$, with lower limit on the left and upper limit on the right, we are summing up infinitely thin rectangles whose heights y ·

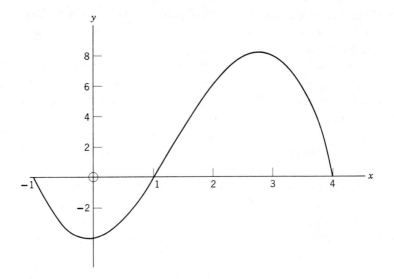

Fig. 8-5. Area between the x-axis and the curve $y = (x^2 - 1)(4 - x)$.

are all negative and whose widths dx are positive; thus the area of each rectangle will be a negative value, and the area calculated by an integral will be negative. However, the area is a positive physical quantity. Here, if we use one integral with limits -1 and 4, we will get a value equal to the physical area above the x-axis minus the physical area below the x-axis.

To find the total physical area, we must determine the physical area of the two portions separately. Area of the portion between -1 and 1 is

$$\int_{-1}^{+1} (x^2 - 1)(4 - x) \, dx = \int_{-1}^{+1} (-x^3 + 4x^2 + x - 4) \, dx$$

$$= \left[\frac{-x^4}{4} + \frac{4x^3}{3} + \frac{x^2}{2} - 4x\right]_{-1}^{1}$$

$$= -\frac{1}{4} + \frac{4}{3} + \frac{1}{2} - 4 - \left(-\frac{1}{4} - \frac{4}{3} + \frac{1}{2} + 4\right)$$

$$= \frac{-16}{3}.$$

This is a physical area of $16/3$ below the x-axis.

Area of the portion between 1 and 4 is

$$\left[\frac{-x^4}{4} + \frac{4x^3}{3} + \frac{x^2}{2} - 4x\right]_1^4$$

$$= -64 + \frac{256}{3} + 8 - 16 - \left(-\frac{1}{4} + \frac{4}{3} + \frac{1}{2} - 4\right)$$

$$= \frac{40}{3} + \frac{29}{12} = \frac{189}{12} = 15.75.$$

Therefore total physical area in the two portions = 5.33 + 15.75 = 21.08 square units.

(Notice that in Fig. 8-5 the y-axis is contracted and the actual area is contracted as a result.)

The single integral $\int_{-1}^{4} (-x^3 + 4x^2 + x - 4)\, dx$ would give $-5.33 + 15.75 = 10.42$.

EXAMPLE 5. Find the area enclosed by the graph of $y = x^2$ and the lines $y = 2$ and $y = 4$.

Solution: (See Fig. 8-6.) It is inconvenient in this case to consider an element of area as a rectangle of height y and width Δx approaching zero. Here it is easier to take the element of area as $KLMN$, with length x and width Δy approaching zero; then, using integration as continuous summation, the area to the right of the y-axis may be represented as

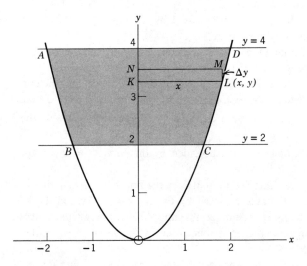

Fig. 8-6. Area between $y = x^2$, $y = 2$, and $y = 4$.

$\int x \, dy$. In this form, y is treated as the independent variable which implies that x must be expressed as a function of y and that the lower and upper limits must be the values of y at the lower and upper boundaries of the area. Since the curve has even symmetry, the total area is double the portion to the right of the y-axis. Thus the required area is

$$2 \int_0^2 x \, dy = 2 \int_2^4 y^{1/2} \, dy = \tfrac{4}{3}\Big[y^{3/2} \Big]_2^4 = \tfrac{4}{3}[4^{3/2} - 2^{3/2}] = 6.9 \text{ square units.}$$

EXERCISES

Calculate the values of the following definite integrals:

1. $\int_0^2 x^2 \, dx.$ 2. $\int_0^3 (x^3 - 1) \, dx.$

3. $\int_0^4 \sqrt{x} \, dx.$ 4. $\int_0^4 \dfrac{dx}{\sqrt{x}}.$

5. $\int_1^3 \dfrac{dx}{x^2}.$ 6. $\int_1^\infty \dfrac{dx}{x^2}.$

7. $\int_{-1}^1 x \, dx.$ 8. $\int_{-2}^2 (1 - t^2) \, dt.$

9. $\int_{-3}^{-1} \dfrac{3}{t^2} \, dt.$ 10. $\int_0^2 \left(\dfrac{4}{\sqrt{y}} - 2\sqrt{y} \right) dy.$

11. Calculate the area enclosed by the curve $y = 0.8x^3 + 1$, the y-axis, the x-axis, and the line $x = 2$.

12. Calculate the area completely enclosed by the curve $y = 0.8x^2 - 3.2$, the x-axis, and the line $x = 3$.

13. Calculate the area enclosed by the line $y = 3$ and the curve $y = 4 - x^2$.

14. Calculate the area enclosed by the curve $y = x^2 + 3x$ and the line $y = 4$.

15. Calculate the area enclosed by the curve $y = 1/\sqrt{x}$, the x-axis, and the lines $x = 1$ and $x = 4$.

16. Calculate the area bounded by the curve $y = 2/\sqrt{x}$, the y-axis, and the lines $y = 1$ and $y = 4$.

17. Repeat Exercise 16, replacing the line $y = 1$ by the line $y = 0$.

18. Find the area enclosed by the curve $y^2 = 2x$ and the line $x = 2y$, using (a) x as the independent variable and (b) y as the independent variable.

19. Find the smaller area enclosed by the curve $y^2 = 2x$, the line $x = 8$, and the line $y = 2$.

20. Find the area enclosed by the curve $y = 2x^2$ and the lines $y = 2$, $y = 8$, and $x = 8$.

21. For the curve $y = x(2 - x)$, find (a) the area above the x-axis, (b) the area below the x-axis between $x = 2$ and $x = 4$, and (c) the vertical line to the right of $x = 2$ which must be used as a boundary so that the area below the x-axis equals the area above the x-axis.

22. A tangent is drawn from the maximum point of the graph of $y = (2 + x)^2(1 - x)$ and extended to cut the curve again. Find:

    (a) The coordinates of the point of contact.

    (b) The coordinates of the point where the tangent crosses the curve.

    (c) The area enclosed by the curve and the x-axis.

    (d) The area enclosed by the curve and the tangent.

23. What physical quantity does the area between the curve and the horizontal axis represent for a graph of (a) velocity against time, (b) acceleration against time, and (c) current against time?

8.3 PHYSICAL APPLICATIONS

In engineering and science, a graph seldom represents one numerical value varying with another; each of the variables almost always is a specified physical quantity. The areas we have been analyzing will not be measured simply in square units but in units that are the product of two physical quantities.

If a current i is plotted against time t, for example, and the current remains constant, the product of the constant current and the time will give the total charge q that has passed during the time interval. If the current is not constant, we can consider the charge passing during an increment of time Δt as being approximately $i\,\Delta t$, which is the graphical area of a rectangle of height i and width Δt. The total charge passing during any interval of time will then be approximated as the sum of all these rectangles of height i to match the graph for each rectangle. From our analysis of areas, we see that the exact charge passing during any interval of time will be the exact area under the graph, which can be calculated by the definite integral $\int_{t_1}^{t_2} i\,dt$, where t_1 and t_2 are the lower and upper limits of the time interval.

In general, any quantity that is calculated as the product of a dependent variable and an independent variable (a graphical rectangular area) when the dependent variable remains constant, is calculated as the integral of the dependent variable with respect to the independent variable (a graphical area under the curve) when the dependent variable does not remain constant.

A partial list of quantities to which this analysis applies is given in the following tabulation.

Independent Variable	Dependent Variable	If Dependent Variable Remains Constant	If Dependent Variable Varies
Time t	Velocity v	Displacement $s = vt$	$s = \int v \, dt$
Time t	Acceleration a	Velocity $v = at$	$v = \int a \, dt$
Time t sec	Current i amp	Charge $q = it$ coulombs	$q = \int_{t_1}^{t_2} i \, dt$
Displacement s	Force F	Work done $W = Fs$ (or energy used)	$W = \int_{s_1}^{s_2} F \, ds$
Time t	Power P	Work done $W = Pt$ (or energy used)	$W = \int_{t_1}^{t_2} P \, dt$

The right-hand sides of the equations in the last two columns are all graphical areas, those in the third column being rectangular and those in the fourth column having a variable top boundary. Each equation in the fourth column is general enough to cover the simpler case of the third column. For instance, $\int_{t_1}^{t_2} i \, dt$, where i is constant equals $i[t]_{t_1}^{t_2} = i(t_2 - t_1)$. This amounts to a multiplication of the constant i and the time interval represented by t in third column. Thus the simpler cases of the third column are particular cases of those in the fourth column.

The use of the definite integral as the summation of elements, even in simple cases, can be shown for the simplest instance of all, the length of a straight line. Suppose the length of a straight line is k. If this length is divided up into several elements of length Δx, approaching zero, the total length is $\int_0^k dx = [x]_0^k = k - 0 = k$.

EXAMPLE 6.   Two equally charged particles of opposite polarity attract one another with a force inversely proportional to the square of their distance apart. When they are 2.0 cm apart, the force of attraction is 2.5 dynes.

Find the work done in moving one particle from a position 2.0 cm from the other particle to an infinite separation.

Solution: $F = k/s^2$, where s is the distance between the particles. From information in the problem, $2.5 = k/4$; therefore $k = 10$. Thus the variation of force with distance of separation is given by $F = 10/s^2$. This function is represented graphically in Fig. 8-7.

$$\text{Total work done} = \int_2^\infty F \, ds = \int_2^\infty 10s^{-2} \, ds = -10\left[\frac{1}{s}\right]_2^\infty$$

$$= -10[0 - \tfrac{1}{2}]$$

$$= 5 \text{ dyne-cm.}$$

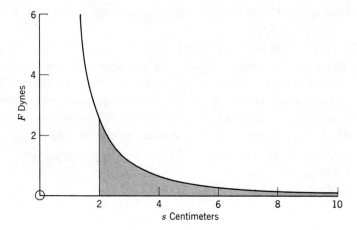

Fig. 8-7. Variation of force between charged particles with separation.

The total work done is represented graphically by the shaded area in Fig. 8-7. An interesting feature of this problem is that the graphical area has an infinite width and yet its area is not infinite. This may often result from an asymptotic condition where the graph approaches the axis of the independent variable as this variable approaches infinity.

EXAMPLE 7. A current of $0.2t^2(3 - t)$ milliamperes flows to a capacitor over a short interval of time t seconds. The capacitor is uncharged at $t = 0$. Find (a) the charge at $t = 1$, (b) the time of maximum positive charge and evaluate it, and (c) when, after $t = 0$, the capacitor is again uncharged.

Solution: (a) Charge q is

$$\int_0^1 i \, dt = \int_0^1 (0.6t^2 - 0.2t^3) \, dt = \left[0.2t^3 - 0.05t^4\right]_0^1$$

$$= 0.15 \text{ millicoloumb.}$$

(b) For maximum positive q, $dq/dt = 0$. In other words, $i = 0$.

$$0.2t^2(3 - t) = 0. \qquad t = 0 \text{ or } 3.$$

$$\frac{d^2q}{dt^2} = \frac{di}{dt} = \frac{d}{dt}(0.6t^2 - 0.2t^3) = 1.2t - 0.6t^2.$$

$$\frac{d^2q}{dt^2} = 0 \text{ at } t = 0 \qquad \text{and} \qquad \frac{d^2q}{dt^2} = -1.8 \text{ at } t = 3.$$

Thus at $t = 3$ there is a maximum positive charge of $0.2(27) - 0.05(81)$ $= 1.35$ millicoulombs.

(c) The area between the graph of current and the t-axis represents the charge on the capacitor. In Fig. 8-8, this area builds up positively until $t = 3$ and negatively after $t = 3$. Positive area therefore represents charging and negative area discharging. The capacitor will be uncharged again when the amount of discharge equals the maximum charge at $t = 3$. Graphically, the negative area equals the positive area at this time; that is, the net area is 0. For an uncharged capacitor, $\int_0^t i\, dt = 0$. Therefore

$$\left[0.2t^3 - 0.05t^4\right]_0^t = 0.$$

$$t^3(0.2 - 0.05t) = 0.$$

Thus at $t = 0$, the capacitor is uncharged, and at $t = 4$, the capacitor is again uncharged. If the current continues beyond 4 sec, the capacitor rapidly acquires a negative charge.

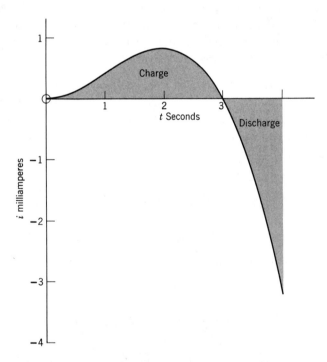

Fig. 8-8. Charge on a capacitor where $i = 0.2t^2(3 - t)$.

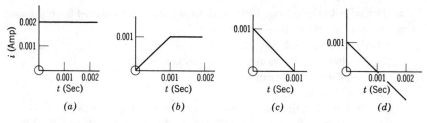

Fig. 8-9.

It should be noted that a definite integral of current with respect to time represents the accumulation of charge on a capacitor for the interval of time. Thus, if a capacitor has a charge Q at $t = 0$, its charge at any time $t = Q + \int_0^t i \, dt$.

EXERCISES

1. The following currents flow to a capacitor, initially uncharged, for t sec. For each value of i, draw a graph of the current and the charge on the capacitor from $t = 0$ to $t = 2$. (a) $i = \underline{0.2, \text{ constant}}$, (b) $i = 0.2t$, and (c) $i = -0.2t$.

2. For each of the graphs of current flowing to an uncharged capacitor in Fig. 8-9, draw the graph of charge against time.

3. In the circuit of Fig. 8-10, a method of generating a low-output voltage e_0 with a wave form different from that of a higher input voltage e_i is illustrated. $e_0 = 1/C \int_0^t i \, dt$, and, if e_0 is small compared to e_i, i is approximately equal to e_i/R. Thus e_0 is approximately equal to $1/RC \int_0^t e_i \, dt$. If $R = 5$ and $C = 0.02$, plot e_0 against time from $t = 0$ to $t = 0.002$ for each of the following input voltages: (a) $e_i = $ a constant 0.5 volt and (b) $e_i = 500t$.

4. In a solution of silver nitrate, a charge of one coulomb will deposit 1.118 mg of silver. Find the amount of silver deposited in 10 sec by a current of $0.15t^2$ ampere, t being in seconds.

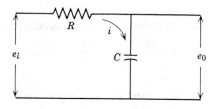

Fig. 8-10.

5. The velocity v of a body varies with time t seconds measured from the start of motion such that $v = 3t - t^2$ ft per second.
   (a) Sketch the graph of v against t.
   (b) At what time is the displacement a maximum?
   (c) Find the displacement at $t = 3$.
   (d) Find the acceleration at $t = 3$.
   (e) Find at what time it returns to its starting point.
   (f) Interpret (e) from a standpoint of graphical areas in the graph of (a).
   (g) Find at what time its displacement is $\frac{7}{6}$.

6. A spring hangs vertically and is fixed at its highest point. When its free end is pulled downwards from its point of equilibrium, the force required is directly proportional to the distance of the free end from its position of equilibrium. When this distance is 3.0 in., the force is 4.5 lb. Find the potential energy of the spring when its free end has been extended 5.0 in. (The potential energy is the work done in moving any object from its position of equilibrium.)

7. Two oppositely charged particles attract one another with a force that is inversely proportional to the square of their distance of separation. Find the work done in separating them from 0.5 cm to 5.0 cm apart if the force of attraction at 0.5-cm separation is 24 dynes.

8. Find the energy E expressed by $E = \int_m^\infty \frac{e^2}{4x^2}\,dx$.

9. Two similarly charged particles repel one another with a force that is inversely proportional to the square of their distance of separation. If the force of repulsion at 1.0-cm separation is 4.0 dynes, find the energy required to move them (a) from a separation of 2.0 cm to a separation of 1.0 cm, and (b) from infinite separation to a separation of 2.0 cm. (*Note:* $\int F\,ds$ represents energy expended when F is in the direction of increasing s. If s represents separation, in this case F is in the direction of decreasing s, energy expended equals $-\int F\,ds$, and the lower limit for s is greater than the upper limit. This is equivalent to $+\int F\,ds$ with the limits reversed.)

10. A rocket weighing 2.0 tons at the earth's surface is fired vertically. The force of gravity (that is, its weight) varies inversely as the square of its distance from the earth's center. Neglecting air friction, find the energy required to fire it to a height equal to the earth's radius, 3960 mi.

11. Force on a unit charge outside a spherical conductor is given by

$$F = \frac{\pi a \sigma}{c^2} \int_{c-a}^{c+a} \frac{r^2 + c^2 - a^2}{r^2}\,dr.$$

Find F in terms of the constants.

12. The current i in an RLC series circuit varies with time t seconds such that $i = 3t^{3/2} - 2t^{5/2}$ milliamperes, and at $t = 0$, the capacitor is uncharged.
   (a) Sketch i against t.

(b) Find the charge on the capacitor at $t = 1$.

(c) Find the charge on the capacitor at $t = 4$.

(d) Find the first time after $t = 0$ when the capacitor is uncharged.

(e) Find the time of maximum charge.

13. A capacitor initially charged to 0.02 coulomb is fed by a current of $0.16t - 0.1$ ampere for t seconds.

(a) Find when the capacitor is completely discharged for the first time.

(b) In what interval of time is the charge negative?

14. The voltage across an inductor in a circuit is $0.3t^2 - t^3$ volts at t sec. Find the time at which the current is a maximum.

15. The power in a circuit varies with time t sec such that $P = 3.5t^{1/2} - 1.5t^{5/2}$ watts. Find (a) the energy dissipated from $t = 0$ to $t = 1$, (b) the energy dissipated from $t = 1$ to $t = 4$, and (c) the time at which net energy dissipated equals zero.

16. If $P = 96/V^{1.2}$ lb/in.^{2}, V being in cubic feet, find the area enclosed between this graph, the line $V = 1$ and the line $P = 1.5$, expressing the answer in foot-pounds.

17. The current through a resistance of 2.0 ohms at t seconds is $3.0t^{1/2} - 1.5t$ amperes. Find the energy dissipated from $t = 0$ to $t = 0.5$.

18. After the free end of the spring in Exercise 6 is extended 5.0 in., it is released. Neglecting friction, the velocity of the free end, when plotted against time, is a sine wave, as shown in the solid curve of Fig. 8-11, positive velocity representing upward motion and negative velocity downward motion. At $t = 0$, the free end is at its lowest point 5.0 in. below its equilibrium position. From the graph:

(a) Find when the free end first reaches its highest point.

(b) Find when the free end first returns to its lowest point.

(c) If its highest point is 5.0 in. above its equilibrium point, find when the free end is first at its point of equilibrium.

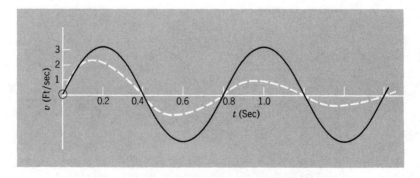

Fig. 8-11. Velocity of the free end of an oscillating spring.

(d) What is the acceleration of the free end at its position of equilibrium?

(e) Is the acceleration positive or negative at the highest point?

19. In actual fact, because of spring and air resistance, the spring of Exercise 18 has the velocity shown in the dotted curve of Fig. 8-11, in which the amplitude of the wave decreases continuously with time. Using relative areas to determine answers:

(a) Will the free end ever reach a height of 5 in. above its equilibrium point? Explain graphically.

(b) What physical quantity is represented by the net area from $t = 0$ to $t = 0.8$.

(c) When it reaches its lowest point again for the first time, where is the free end in relation to its starting point at $t = 0$? Explain graphically.

20. The graph of an alternating current against time is a sine wave as shown in Fig. 8-12. If this current flows through a capacitor and an inductor in series:

(a) What are the first two times at which the charge on the capacitor reaches a maximum?

(b) If the capacitor is uncharged at $t = 0$, when is its charge next equal to zero?

(c) When is the first time the voltage across the inductor is zero?

(d) At the time of (c), to what portion of its maximum charge is the capacitor charged?

(e) When the voltage across the capacitor is at its first maximum, is the voltage across the inductor positive, negative, or zero?

21. The current in an *RLC* circuit varies with time as illustrated in Fig. 8-13. In its later stages the current gets closer and closer to zero and the graph gradually flattens out. Theoretically the current approaches zero as t approaches ∞.

(a) At what time does the capacitor have its maximum charge?

(b) What is the voltage across the inductance at this time?

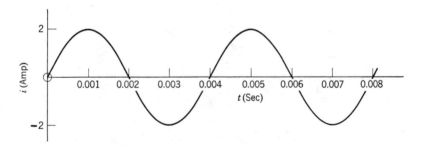

Fig. 8-12. Graph of an alternating current with an amplitude of 2.

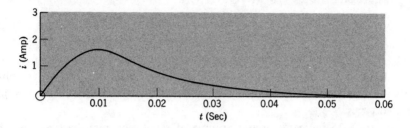

Fig. 8-13. Transient current surge in an *RLC* circuit.

(c) Is there any other time when the voltage across the inductance is the same as in (b)?

(d) At what state is the voltage across the resistance at the time of (c)?

22. The current in an *RLC* series circuit is a damped oscillation, as graphed in Fig. 8-14.

(a) At what time does the capacitor have its maximum charge?

(b) Assuming the capacitor uncharged at $t = 0$, is the charge on the capacitor zero, positive, or negative at $t = 0.02$?

(c) Is this charge increasing or decreasing at the beginning of the interval starting at $t = 0.02$?

(d) If the amplitude of the current continuously diminishes to zero, will the capacitor finally have a positive, negative, or zero charge?

(e) At point *A*, what is the state of the voltage across the inductor?

(f) At point *A*, what is the state of the voltage across the resistor?

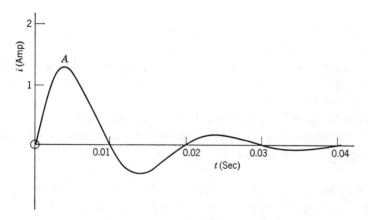

Fig. 8-14. A damped oscillation.

8.4   VOLUMES OF REVOLUTION

A volume of revolution is a volume for which all cross sections at right angles to a central axis are circles with their centers on the axis. The cylinder, the cone, and the sphere are common examples, and well-known formulas exist for calculating their volumes. Definite integrals are used to establish these formulas and also to calculate volumes of revolution of more complicated types.

Figure 8-15 indicates how a volume of revolution may be swept out by revolving the area under any curve $y = f(x)$ about the x-axis through 360°.

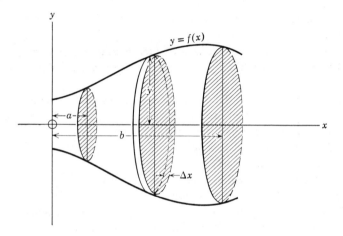

Fig. 8-15. Volume of revolution generated by the area under $y = f(x)$.

The x-axis will be an axis of symmetry for this volume, and all cross sections at right angles to the x-axis will be circular.

As for previous summations, to find a total volume we consider an element of the volume, a thin disk whose circular cross-sectional area can be determined from the formula for the area of a circle as πy^2 and whose element of thickness is Δx, approaching zero. The y varies to fit the curve as x increases. Using an integral as a continuous summation, we see that the total volume between $x = a$ and $x = b$ is $\int_a^b \pi y^2 \, dx$, or $\pi \int_a^b [f(x)]^2 \, dx$.

EXAMPLE 8.   Liquid is poured into a hemispherical container of radius 17 in. until the surface of the liquid has a diameter of 30 in. Find the volume of liquid in the container.

Solution: From Fig. 8-16, it should be apparent that the volume we wish to calculate may be generated by rotating the area QLN through $360°$ about the y-axis. A general cross section of this volume at right angles to the y-axis will be a circle of radius x in this case and its area will be πx^2. An element of the total volume is then formed by giving this circle an element of thickness Δy, the element of volume being the volume of a thin disk, $\pi x^2 \, \Delta y$, and the total volume will be $\pi \int x^2 \, dy$. In this treatment, y is taken as the independent variable, and x must be expressed as a function of y using the equation of the circle with center at the origin and radius 17; in addition, lower and upper limits of y must be established. For any point on the circle with coordinates (x,y),

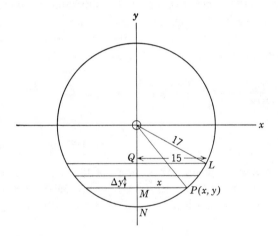

Fig. 8-16. Volume of liquid in a hemispherical container.

$x^2 + y^2 = 17^2$, and therefore $x^2 = 289 - y^2$. Also, in the triangle OQL, $OQ = \sqrt{17^2 - 15^2} = \sqrt{289 - 225} = 8$. Therefore the upper limit of y is the ordinate of $Q = -8$, and the lower limit is the ordinate of $N = -17$. Thus

$$\text{volume} = \pi \int_{-17}^{-8} (289 - y^2) \, dy = \pi \left[289y - \frac{y^3}{3} \right]_{-17}^{-8}$$

$$= \pi \left[289(-8 + 17) - \frac{-512 + 17^3}{3} \right]$$

$$= \pi \left[289 \left(-8 + 17 - \frac{17}{3} \right) + \frac{512}{3} \right] = \frac{\pi(2890 + 512)}{3}$$

$$= 1134\pi = 3560 \text{ in.}^3.$$

8.5 MEAN AND ROOT MEAN SQUARE VALUES

The method of finding a mean or average of several individual numerical values is already well known; the sum of all the numerical values is divided by the number of values occurring. We are concerned now with finding the mean or average value of a variable that is continuously changing over an interval. It will be realized by this time that some form of calculus must be used in finding this kind of mean since calculus deals with continuous change.

The problem is to find the mean value of a continuously changing y which is equal to any function of x, as in Fig. 8-17, in the interval from $x = a$ to $x = b$. An approximate answer may be found if we divide the interval of x into a number of smaller intervals Δx, find the value of y at the midpoint of each of these intervals, and average these individual numerical values. In Fig. 8-17, the whole interval has been divided into ten equal parts of width Δx, and the values of y, $(y_1, y_2, y_3 \cdots)$ at the midpoints of each small interval are determined. Then the mean y will approximately be

$$\frac{y_1 + y_2 + y_3 \cdots + y_{10}}{10} = \frac{(y_1 + y_2 + y_3 \cdots + y_{10})\, \Delta x}{10\, \Delta x}$$

$$= \frac{y_1\, \Delta x + y_2\, \Delta x + y_3\, \Delta x \cdots + y_{10}\, \Delta x}{MN}.$$

The numerator here is the sum of the areas of the ten rectangles of variable height y and width Δx. If the interval MN is divided into more and more intervals of smaller width, we get closer and closer to the exact mean y, and

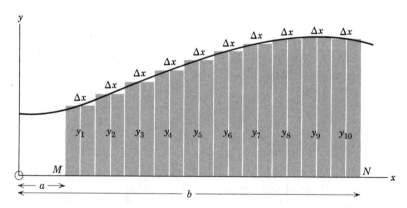

Fig. 8-17. The mean value of $y = f(x)$ over an interval of x.

the numerator of the resulting expression approaches closer to the area under the curve. Thus

$$\text{mean } y = \lim_{\Delta x \to 0} \frac{y_1 \Delta x + y_2 \Delta x + y_3 \Delta x \cdots + y_n \Delta x}{n \Delta x}$$

$$= \frac{\int_a^b y \, dx}{MN} = \frac{\int_a^b y \, dx}{b - a}.$$

Therefore the mean value of a dependent variable over an interval of the independent variable is the area under the graph for this interval divided by the interval.

EXAMPLE 9.   Find the mean current from $t = 0$ to $t = 0.25$ for the current $i = 4t^{1/2} - t^{3/2}$ amperes.

Solution:

$$\text{Mean } i = \frac{\int_0^{0.25} (4t^{1/2} - t^{3/2}) \, dt}{0.25}$$

$$= \frac{1}{0.25} \left[\frac{8}{3} t^{3/2} - \frac{2}{5} t^{5/2} \right]_0^{0.25} = 4.0 \left[\frac{8}{3}(0.125) - \frac{2}{5}(0.03125) \right]$$

$$= 1.3333 - 0.0125 = 1.32 \text{ amperes.}$$

It is seen that, in the case of a current, the mean value over an interval of time is simply the charge that has passed over that interval divided by the time interval, since $\int i \, dt$ is charge q. If, for instance, a variable current flows to a capacitor, the mean current over a time interval starting when the capacitor is uncharged will be the charge on the capacitor at the end of the interval divided by the time.

For an alternating current which changes its direction so that charge flows as much in one direction as in the other over one cycle, the mean current over a cycle is zero, since current in one direction would deduct from current in the opposite direction when the over-all effect is considered. Yet this type of current may produce considerable power.

To determine the power produced by a current, the mean current is of no significance. The power of a current i flowing through a resistance R is $i^2 R$ at any instant; and if i varies and R is constant, the power varies with the square of the current, which is positive for either positive or negative current. Thus to find mean or average power, we determine the mean of the square of the variable current, called the *mean square current*, and multiply by the constant resistance. The power could also be determined by using e^2/R, where e is the variable voltage across the resistance; in this case the average power could be calculated by determining the *mean square voltage* and dividing by the constant resistance. In either case, a *mean square* value, that is, the mean value of the square of a variable, must be determined. It is desirable still to use formulas I^2R and E^2/R to calculate average power, where I and E are fixed values in amperes and volts, respectively. The correct fixed value of each must be the one which, when squared, gives the mean square value. This is known as the *root mean square* value.

To find the root mean square, often abbreviated to rms, of a dependent variable involves finding the mean of the square of the variable and taking the square root of this quantity.

EXAMPLE 10.   (a) Find the rms current from $t = 0$ to $t = 0.25$ for the current $i = 4t^{1/2} - t^{3/2}$ (the same current as in Example 9).

(b) Find the average power when the current of (a) flows through a 3.0-ohm resistor.

Solution: (a) First we must square the function, then find the mean of this squared function, and finally determine the square root of this mean.

$$i^2 = (4t^{1/2} - t^{3/2})^2 = 16t - 8t^2 + t^3.$$

$$\text{mean square } i = \int_0^{0.25} \frac{(16t - 8t^2 + t^3)\, dt}{0.25} = 4.0\left[8t^2 - \frac{8}{3}t^3 + \frac{t^4}{4}\right]_0^{0.25}$$

$$= 4.0\left[8(0.0625) - \frac{8}{3}(0.0156) + \frac{0.0039}{4}\right]$$

$$= 1.84 \text{ ampere}^2.$$

$$\text{rms } i = \sqrt{1.84} = 1.36 \text{ ampere}.$$

(The rms value is never less than the mean value.)

(b) Average power = mean square $i \times R = 1.84 \times 3.0 = 5.52$ watts.

EXERCISES

1. Find the mean and rms values of the following individual numbers:
   (a) 3, 5, 7, 9.   (b) 6, 4, 0, -4, -6.   (c) 3, 3, -4, -4.   (d) 5, 5, 5, 5.
2. For what type of function is the mean value equal to the rms value?
3. Sketch graphs for which the following conditions are true:
   (a) The mean value of the dependent variable is zero, although this variable is only equal to zero once during the interval.
   (b) The mean value of the dependent variable is negative for any interval.
   (c) The mean value of the dependent variable is positive for short intervals, zero for one particular interval, and negative for intervals greater than this.

4. Establish by calculus the volume of a sphere of radius R.

5. (a) Sketch the complete graph of $4x^2 + 9y^2 = 36$.

(b) Find the volume of revolution resulting from a rotation of the area of this complete graph about the x-axis.

(c) Find the volume of revolution resulting from a rotation of this complete area about the y-axis.

6. The equation of an ellipse centered at the origin is $x^2/a^2 + y^2/b^2 = 1$. Assuming a and b as constants and $a > b$:

(a) Sketch the complete graph, indicating the lengths of a and b.

(b) Find the volume of revolution of this complete area about the x-axis.

(c) Find the volume of revolution of this complete area about the y-axis.

7. Find the volume of revolution for the graph of $xy^{3/2} = 1$ with (a) a rotation of the area in the first quadrant bounded by the graph, the y-axis, and the lines $y = 2$ and $y = 8$ about the y-axis and (b) a rotation of the area in the first quadrant bounded by the graph, the x-axis, and the lines $x = 2$ and $x = 8$ about the x-axis.

8. The vertical cross section through the axis of symmetry of a solid volume has an upper boundary given by the equation $y = 0.5 + 0.2x^2$, where x and y are in feet and the axis of symmetry is the x-axis. All vertical cross sections at right angles to the axis of symmetry are squares with their centers on the x-axis. Find the volume of the solid if its axis extends from $x = 0$ to $x = 2$.

9. Find the volume remaining when a hole of 1.0-in. diameter is drilled through the center of a solid sphere of 1.3-in. radius from one surface to the other.

10. A variable current flows to a capacitor. Find the mean current for the following situations:

(a) The capacitor is charged from 0 to 0.4 coulomb in 0.05 sec.

(b) The capacitor is charged from 0.3 coulomb of negative polarity to 0.2 coulomb of positive polarity in 2.0 sec.

(c) The capacitor has 0.6 coulomb of positive charge and is completely discharged in 0.04 sec.

11. Find the mean value of the following over the interval indicated:

(a) $i = 1.5t$ from $t = 0$ to $t = 0.6$.

(b) $i = 0.8 - 4t$ from $t = 0$ to $t = 0.4$.

(c) $v = 2\sqrt{t}$ from $t = 1$ to $t = 3$.

(d) $i = t^2 - t$ from $t = 0.3$ to $t = 0.5$.

(e) $i = 5t^{1/4} - 2t^{3/4}$ from $t = 0$ to $t = 0.0016$.

12. Find the rms values for all parts of Exercise 11.

13. A current $i = 2t - 4t^2$ amperes flows through a resistance of 3.0 ohms from $t = 0$ to $t = 0.5$ sec. Find the average power developed.

14. A voltage $v = 6t^{1/2} - 24t^{3/2}$ is applied across a resistance of 0.5 ohm. Find the average power developed during the time that v is positive.

CHAPTER 9

Transcendental Functions

The discussion of functions so far has been limited to the power function. It is now necessary to examine other types of functions, which are classified broadly as *transcendental functions*. These include trigonometric, logarithmic, exponential, hyperbolic, and inverse functions.

9.1 THE TRIGONOMETRIC FUNCTIONS

This heading includes any expressions involving any of the six trigonometric functions of variables. These are extremely important in the electrical theory of the a-c circuit and in mechanical problems dealing with rotational motion. The expression for instantaneous current in an a-c system, $i = A \sin \omega t$, is a trigonometric function of t, A and ω being constants.

The trigonometric functions are all *periodic functions*, which implies that the patterns of their graphs are repetitive in cycles of equal duration.

Unless otherwise specified, the angle following the trigonometric function is assumed to be in radians. This unit of angular measure allows for greater simplicity in calculus—a fact that will be apparent later. If the angle is intended to be in degrees, the symbol for degrees (°) should be used.

EXAMPLE 1. An alternating voltage at time t seconds is given by $v = 3 \cos 200t$ volts. Evaluate (a) v when $t = 0.006$ and (b) t when $v = 2.4$.

Solution: Figure 9-1 shows the graph of this function, $200t$ representing an angle in radians. (a) At $t = 0.006$ sec,

$$v = 3 \cos (200 \times 0.006 \text{ rad})$$
$$= 3 \cos 1.2 \text{ rad}$$
$$= 3 \cos \frac{1.2 \times 180°}{\pi}$$
$$= 3 \cos 68.7°$$
$$= 3 \times 0.363 = 1.09 \text{ volts.}$$

(b) When $v = 2.4$,

$$3 \cos 200t \text{ rad} = 2.4.$$
$$\cos 200t \text{ rad} = 0.80.$$

From tables, $\cos 36.9° = 0.80$. In each cycle of the cosine function, there are two points where the function has a specified value. Here, $\cos 323.1°$ also equals 0.80, and in successive cycles there are two values of the angle whose cosine is 0.80, so that there are actually an infinite number of solutions to this problem. Limiting the solutions to the first cycle, $200t$ rad $= 36.9°$ or $323.1°$. The units on the two sides of this equation must correspond.

$$200t \text{ rad} = 36.9 \times \frac{\pi}{180} \text{ rad} \quad \text{or} \quad 323.1 \times \frac{\pi}{180} \text{ rad.}$$
$$200t = 0.644 \quad \text{or} \quad 5.639.$$

$t = 3.22 \times 10^{-3}$ sec or 2.82×10^{-2} sec in the first cycle only. A reference to Fig. 9-1 shows that $v = 2.4$ twice in each cycle.

An alternating voltage producing alternating current is generated by the uniform rotation of a generator, and many of the terms used with trigonometric functions are based on the concept of rotational motion. In a perfectly

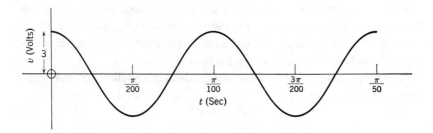

Fig. 9-1. Graph of $v = 3 \cos 200t$.

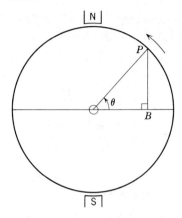

Fig. 9-2. Analysis of the voltage produced by an a-c generator under conditions of uniform rotation.

designed two-pole a-c generator, the voltage produced is proportional to the vertical distance of a point on the circumference of the armature from an axis at right angles to the poles through the center of a circular cross section of the machine. Figure 9-2 represents an armature rotating counterclockwise with a fixed magnetic field. The voltage produced is proportional to distance *PB*. Therefore

$$\text{voltage } e = kPB$$
$$= kOP \sin \theta$$
$$= A \sin \theta, \quad \text{where } A \text{ is constant.}$$

If, for example, the constant rotational velocity of the generator is 50 rad per second, in 2 sec the generator will rotate 100 rad. With simple examples such as this, it is apparent that the angle of rotation θ is ωt, where ω is the constant angular velocity in radians per second and t is time in seconds.

The *frequency* of this rotation and of the voltage and current developed is defined as the number of rotations or cycles per second (cps). Frequency is simply another way of indicating the angular velocity. Since one revolution $= 2\pi$ rad, f revolutions per second are equivalent to $2\pi f$ rad per second. Thus $\omega = 2\pi f$ and $f = \omega/2\pi$, where f is in cps.

The voltage generated by an a-c generator may therefore be expressed in three ways:

$$e = A \sin \theta,$$
$$e = A \sin \omega t,$$

or

$$e = A \sin 2\pi ft.$$

The expressions θ, ωt, and $2\pi f t$ all represent a variable angle in radians. In the first of these equations, θ is the independent variable, even though θ itself depends on t. In the second and third forms, t is the independent variable. The constants are A, ω, π, and f.

The maximum value of the function is A, the amplitude of the voltage, since the sine function has a maximum value of 1.

The *period* T of the sine function is the time in seconds taken for a complete cycle of 2π rad. If the frequency is 60 cps, the period is $\frac{1}{60}$ sec. In general, period $T = 1/f$.

The current flowing from an a-c generator is also a sine function for which A, ω, f, and T are determined by the preceding methods.

EXAMPLE 2.   For the current $i = 4 \sin 1000t$ amperes, t being in seconds, find (a) the amplitude, (b) the angular velocity, (c) the frequency, (d) the period, and (e) the current when $t = 0.0005$.

Solution: (a) Amplitude $= 4$ amperes, the maximum current.

(b) $i = A \sin \omega t$ in general. Therefore, for this example, $\omega = 1000$ rad per second.

(c) $2\pi f t = 1000t$. Therefore $f = 1000/2\pi = 159$ cps.

(d) $T = 2\pi/1000 = 0.0063$ sec.

(e) $i = 4 \sin (1000 \times 0.0005 \text{ rad}) = 4 \sin 0.5 \text{ rad} = 4 \sin 28.6° = 1.9$ amperes.

EXAMPLE 3.   For the voltage $v = 6.0 \sin \theta$ volts, where $\theta = 377t$ at t seconds, find the frequency and sketch a graph of v against θ indicating the times t for significant values of θ.

Solution: $f = 377/2\pi = 60$ cps. The required graph is shown in Fig. 9-3.

In practice, when dealing mathematically with a-c voltages and currents, the rotating machine is forgotten. Actually, in a machine with more than two poles, the angle θ used in expressions for voltage and current is not the angle through which the machine rotates mechanically. The symbol θ represents electrical radians, with one cycle completed by a rotation of the machine from one pole to the next pole of similar polarity.

Figure 9-4 shows that $i = A \cos \theta$ follows the same pattern as $i = A \sin \theta$, but it is displaced 90° or $\pi/2$ rad to the left of the sine function. Both graphs are said to be *sinusoidal*, and the amplitude, frequency, and period of the cosine function are determined in exactly the same way as for the sine function.

Since $\theta = \omega t$, points to the left on either of the graphs of Fig. 9-4 occur earlier in time than points farther to the right. Hence $A \cos \theta$ is said to *lead* $A \sin \theta$ by 90° or $\pi/2$ rad. This introduces the idea of *phase relationship*

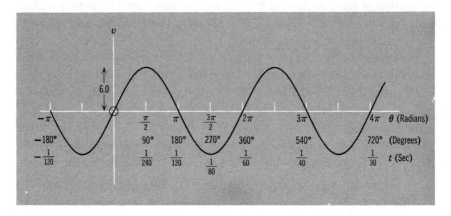

Fig. 9-3. Graph of $v = 6.0 \sin 377t$.

between two sinusoidal waves of the same frequency, which refers to the angle by which one is displaced from the other. This angle is called the *phase angle* of one as referred to the other. It may also be said that $A \sin \theta$ *lags* $A \cos \theta$ by 90°. Of course, a lagging phase angle of 90° is equivalent to a leading phase angle of 270° when referred to the next cycle of the reference function. However, phase relationship is always expressed in the form with the smallest possible phase angle.

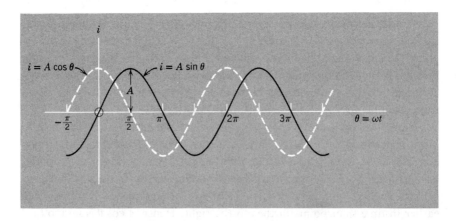

Fig. 9-4. Graphs of $A \sin \theta$ and $A \cos \theta$.

Since the sine and cosine functions are identical except for a 90° phase shift, it must be possible to express $\sin \theta$ as a cosine and $\cos \theta$ as a sine. It is seen from Fig. 9-4 that the sine of any angle is always equal to the cosine of that angle minus 90°. Thus $\sin \theta = \cos (\theta - 90°)$ or $\cos (\theta - \pi/2)$. Also $\sin (\theta + \phi) = \cos (\theta + \phi - \pi/2)$. To take a particular case, $\sin (\theta + \pi/3)$ $= \cos (\theta - \pi/6)$. By the same reasoning, $\cos \theta = \sin (\theta + \pi/2)$ and $\cos(\theta + \phi) = \sin (\theta + \phi + \pi/2)$.

The graph of $\sin (\theta + \pi/3)$ crosses the θ axis when the bracket equals 0. At $\theta = -\pi/3$, $\sin (\theta + \pi/3) = 0$, and to the right of this point on the graph the function is positive. Thus $\sin (\theta + \pi/3)$ leads $\sin \theta$ by $\pi/3$ rad. Also, for example, $\cos (\theta - \pi/4)$ lags $\cos \theta$ by $\pi/4$ rad. Therefore the plus or minus sign in the expression for the angle introduces a leading or lagging phase angle, respectively, when referred to the angle θ. A close study of Fig. 9-4 also shows that $\sin (\theta \pm \pi) = -\sin \theta$ and $\cos (\theta \pm \pi) = -\cos \theta$.

EXAMPLE 3. Find the phase relationship between $v_1 = 3 \sin (\theta + \pi/3)$ and $v_2 = 2 \cos (\theta - \pi/3)$.

Solution: The phase relationship is more apparent if both functions are expressed as sine functions. $v_2 = 2 \cos (\theta - \pi/3) = 2 \sin (\theta + \pi/6)$. A comparison with $v_1 = 3 \sin (\theta + \pi/3)$ now shows that v_1 leads v_2 by $\pi/6$ rad.

When two or more sinusoidal functions of the same frequency are added, the result is a sinusoidal function of the same frequency.

In Fig. 9-5, $i_1 = 3 \sin \theta$ and $i_2 = 4 \cos \theta$ are added graphically by simply superimposing the values of i_1 and i_2 at several values of θ and sketching the resulting graph. The sum of the two sinusoidal functions is a sinusoidal function of amplitude 5 with the same frequency as its two components and with a phase displacement.

The sum of these two functions may be calculated, yielding accurate results, as follows: $A \sin (\theta + \phi) = A \sin \theta \cos \phi + A \cos \theta \sin \phi$, an expression that has the same structure as $3 \sin \theta + 4 \cos \theta$, where $A \cos \phi = 3$ and $A \sin \phi = 4$. The right-angled triangle of Fig. 9-6 incorporates these two equations, and A and ϕ may be determined by a solution of this triangle.

From Fig. 9-6, $A = 5$ and $\tan \phi = \frac{4}{3} = 1.333$. Therefore $\phi = 53.1°$. Therefore $i_1 + i_2 = A \sin (\theta + \phi) = 5 \sin (\theta + 53.1°)$, a sinusoidal function with amplitude 5, leading the standard $\sin \theta$ by 53.1° and with the same frequency as its two components, since the variable angle θ is identical throughout.

The amplitude A and the phase angle ϕ of this function may be calculated

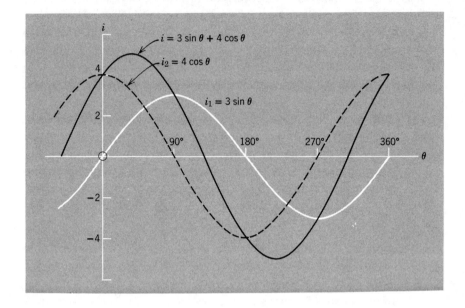

Fig. 9-5. $3 \sin \theta + 4 \cos \theta = 5 \sin (\theta + 53.1°)$.

by solving the two equations $A \cos \phi = 3$ and $A \sin \phi = 4$ without the aid of the right-angled triangle of Fig. 9-6. This is left as an exercise for the student.

In general **$c \sin \theta + d \cos \theta = A \sin (\theta + \phi)$, where** $A = \sqrt{c^2 + d^2}$ **and tan ϕ = d/c.** It is better to work this kind of problem from first principles, as previously shown, than to use this formula.

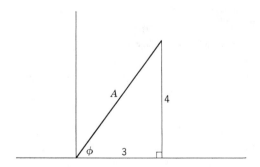

Fig. 9-6. $A \cos \phi = 3$ and $A \sin \phi = 4$.

EXAMPLE 4.   Find the maximum value of the function $v = -3 \cos \theta +$ $4 \sin (\theta - 20°)$ and two values of θ for which this maximum occurs.
Solution: The maximum value of a single sine function is its amplitude, and it occurs when the angle following the sine function is 90° plus or minus any multiple of 360°. If the given function is expressed as a single sine function, the maximum value may be simply determined.

$$v = -3 \cos \theta + 4(\sin \theta \cos 20° - \cos \theta \sin 20°)$$
$$= -3 \cos \theta + 4(0.94 \sin \theta - 0.342 \cos \theta)$$
$$= 3.76 \sin \theta - 4.37 \cos \theta.$$

Thus v may be expressed as $A \sin (\theta + \phi)$, where A and ϕ may be determined from the right-angled triangle of Fig. 9-7.

$$\therefore A = \sqrt{3.76^2 + 4.37^2} = 5.8,$$

and

$$\tan \phi = \frac{-4.37}{3.76} = -1.162. \qquad \phi = -49.3°.$$

Thus $v = 5.8 \sin (\theta - 49.3°)$ and maximum $v = 5.8$, occurring when $\theta - 49.3° = 90°, 450°, -270°$, etc. For maximum v, $\theta = 139.3°$, $499.3°, -220.7°$, etc.

The sum of two or more sinusoidal functions may also be calculated by the method of using a rotating vector representation of sinusoidal functions. As this method is developed in most texts on a-c theory, it is omitted here.

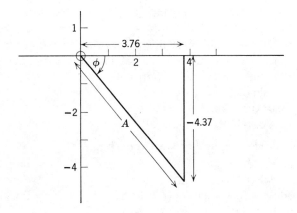

Fig. 9-7.

EXERCISES

1. Find the amplitude, angular velocity, frequency, and period for the following functions:

(a) $v = 7.5 \sin 200\pi t$.

(b) $i = 2.3 \cos 400t$.

(c) $i = 5.2 \sin (377t + \phi)$.

(d) $e = 7.5 \cos (10^4 t - \pi/6) - 2.3 \cos (10^4 t - \pi/6)$.

2. For the function of Exercise 1c, find the value of i when $t = 1/240$, assuming $\phi = -\pi/3$.

3. Change each sine function to an equivalent cosine function and each cosine function to an equivalent sine function in the following:

(a) $3 \sin (\theta + 30°)$.

(b) $5 \cos (\theta - 22°)$.

(c) $4 \sin (300\pi t - \pi/8)$.

(d) $2 \cos (120\pi t + 1.5\pi)$.

4. Find the phase relationship between each of the following, where possible, using a phase angle no greater than 180°:

(a) $i_1 = \sin (\theta - \pi/4)$ and $i_2 = \sin \theta$.

(b) $v_1 = \sin (\theta - \pi/4)$ and $v_2 = 3 \sin (\theta - \pi/3)$.

(c) $e_1 = 0.5 \cos (1000t + \pi/8)$ and $e_2 = 1.2 \cos (1000t - \pi/6)$.

(d) $v_1 = 4 \cos \theta$ and $v_2 = 2 \cos (2\theta - 40°)$.

(e) $v_1 = 3 \cos (\theta + 120°)$ and $v_2 = 7 \cos (\theta - 150°)$.

5. Find the phase relationship between each of the following, using a phase angle no greater than 180°:

(a) $e_1 = \sin (2\pi ft + \pi/4)$ and $e_2 = 3 \cos 2\pi ft$.

(b) $i_1 = 5 \cos (2\theta + \pi/3)$ and $i_2 = 4 \sin (2\theta + \pi/3)$.

(c) $v_1 = 2.5 \cos (500t + \pi/6)$ and $v_2 = 3.8 \sin (500t - \pi/3)$.

(d) $i_1 = 0.5 \sin (120\pi t - \pi/2)$ and $i_2 = 0.5 \cos (120\pi t + \pi/6)$.

6. Verify that $\sin (\theta + 30°) = \cos (\theta - 60°)$ by using a standard expansion for each of the functions and simplifying to the sum of a sine and a cosine.

7. Graph the following functions for at least one cycle, marking clearly significant angles and times along the horizontal axis:

(a) $i = 2.8 \sin (200\pi t + \pi/4)$.

(b) $v = 0.08 \cos (1000\pi t + \pi/6)$.

(c) $v = 1.4 \cos (1000t - \pi/3)$.

Express as a single sine function where possible.

8. $4.0 \sin \theta + 3.0 \cos \theta$.

9. $3.0 \sin \theta + 4.0 \cos (\theta - 30°)$.

10. $3.0 \sin (\theta + 30°) - 4.0 \cos (\theta + 45°)$.

11. $5.0 \sin (\theta + 60°) + 4.0 \cos (\theta - 60°)$.

12. $3.2 \sin (\omega t - 40°) + 2.5 \sin (\omega t + 140°)$.

13. $4.2 \sin 377t + 5.6 \cos (377t - \pi/4)$.

14. $6.0 \sin (\theta - \pi/6) - 4.0 \cos (\theta - 2\pi/3)$.

15. $7.5 \cos 377t - 5.2 \sin (200t - 20°)$.

16. $7.5 \cos 377t + 5.2 \sin (377t - 20°)$.

17. Express $4.0 \cos (500t + 30°) - 3.0 \sin (500t - 150°)$ as a single cosine function.

Find the maximum values of the following functions and the first positive value of the independent variable for which maximum occurs:

18. $5.0 \cos (\theta + 30°) - 12.0 \cos (\theta + 120°)$.

19. $0.8 \sin \theta + 1.5 \sin (\theta + \pi/2)$.

20. $6.0 \cos (\theta + 60°) - 4.5 \sin (\theta - 120°)$.

21. $0.40 \sin (800t - 3\pi/2) - 0.50 \cos (800t - \pi/6)$.

22. $4.0 \cos (\theta - 120°) + 3.0 \sin (\theta + 60°) + 12.0 \cos (\theta + 6.9°)$.

23. $3.2 \sin (2000t + 3\pi/8) - 1.2 \cos (2000t + 7\pi/8) + 2.0 \sin (2000t - 5\pi/8)$.

24. Prove that $\sin^2 \theta - 2\sqrt{3} \sin \theta \cos \theta + 3 \cos^2 \theta = 4 \sin^2 (\theta - \pi/3)$.

9.2   THE LOGARITHMIC FUNCTION

The *logarithmic function*, often abbreviated to log function, includes all expressions involving the log of a variable expression to a constant base. The theory of logs provides a powerful mathematical tool. The most important laws of logs are as follows:

1. If $a^y = x$, then $y = \log_a x$.

2. $\log_a xy = \log_a x + \log_a y$.

3. $\log_a \dfrac{x}{y} = \log_a x - \log_a y$.

4. $\log_a x^y = y \log_a x$.

where base a may be any positive number other than 1.

The principal use of logarithms in mathematics is to simplify operations. Thus, by the second law, a multiplication of x and y is simplified to an addition by using logs. Similarly, a division is simplified to a subtraction, and a power is simplified to a multiplier, which implies that a root is simplified to a divisor. These are the principles of the slide rule, which is marked off on a logarithmic scale.

There are two bases in common usage, 10 and e. Base 10 lends itself best to calculations involving products, quotients, powers, and roots. Base e is the logical base to use in calculus, since it makes calculations much easier.

The meaning of e may be understood from a consideration of compound interest. If one dollar is invested at 5% interest compounded annually, it would amount at the end of 20 yrs, to $(1 + \frac{1}{20})^{20}$ dollars $= 2.65$ dollars. If the compounding is quarterly, it amounts to $(1 + \frac{1}{80})^{80}$ dollars, or if compounded monthly, it becomes $(1 + \frac{1}{240})^{240}$ in 20 years. The shorter the compounding period, of course, the greater the amount will be. e represents the amount resulting if compounding could be continuous; that is if, the instant the deposit is made, it starts increasing, and the instant it increases, the new increase is based on the already increased amount, etc. This type of continuous increase does occur in nature and science. The rate of growth of cell structure, for instance, depends on the number of cells present at any instant.

Mathematically, e may be defined as $\lim_{n \to \infty} (1 + 1/n)^n$. The value of e may be determined to any degree of accuracy by using the binomial theorem as follows:

$$\left(1 + \frac{1}{n}\right)^n = 1 + n\left(\frac{1}{n}\right) + \frac{n(n-1)}{2!}\left(\frac{1}{n}\right)^2 + \frac{n(n-1)(n-2)}{3!}\left(\frac{1}{n}\right)^3 \cdots$$

to an infinite number of terms.

$$= 1 + 1 + \left(1 - \frac{1}{n}\right)\frac{1}{2!} + \left(1 - \frac{3}{n} + \frac{2}{n^2}\right)\frac{1}{3!} \cdots$$

As n becomes infinitely large, any finite number divided by n or any positive power of n approaches zero. Then

$$e = \lim_{n \to \infty} \left(1 + \frac{1}{n}\right)^n = 1 + 1 + \frac{1}{2!} + \frac{1}{3!} + \frac{1}{4!} \cdots \quad \text{to an infinite number of terms.}$$

By taking seven terms of this series and neglecting the remainder, which are relatively small, we find that $e \simeq 2.718$. Note that e can never be calculated exactly, since it is what is known as an irrational number. However, e may be calculated to any necessary degree of accuracy, and any set of tables will give the value of this constant to the degree of accuracy of the tables.

If n were actually replaced by infinity in the expression $(1 + 1/n)^n$, the result is $(1)^\infty$, which is not, as might be supposed, equal to 1; it is an indeterminate expression, and in this case it has been evaluated by putting the expression into a form that is not indeterminate.

Sets of tables usually include logs of numbers to base e. These are called natural, napierian, or hyperbolic logs. $\text{Log}_e x$ is often abbreviated to $\ln x$.

EXAMPLE 5.   Solve for x: $\log (x^4 + 3x^3) - \log (x + 3) + \log 2 - \log 6 = 2 \log x$.

Solution: When all terms involve logs, it is convenient, as in algebra, to arrange all unknown terms on the left-hand side, and known terms on the right. Thus

$$\log (x^4 + 3x^3) - \log (x + 3) - 2 \log x = \log 6 - \log 2.$$

Using laws of logs,

$$\log \frac{x^4 + 3x^3}{x^2(x + 3)} = \log \frac{6}{2} = \log 3.$$

Since all logs are to the same base, when the log of one quantity equals the log of another, we may conclude that the two quantities are equal, as every number has its own distinct log. Therefore

$$\frac{x^4 + 3x^3}{x^2(x + 3)} = 3.$$

By simple algebra, $x = 3$.

EXAMPLE 6.   If $\ln (E - iR) - \ln E = -Rt/L$, find i as a function of t.

Solution: Using log theory,

$$\ln \frac{E - iR}{E} = \frac{-Rt}{L} \quad \text{or} \quad \ln \left(1 - \frac{iR}{E}\right) = \frac{-Rt}{L}.$$

$$\therefore \; 1 - \frac{iR}{E} = e^{-Rt/L}, \qquad \frac{iR}{E} = 1 - e^{-Rt/L}, \qquad i = \frac{E}{R}(1 - e^{-Rt/L}).$$

Since any positive base raised to a positive or negative power will result in a positive number, it follows that we can find the logarithm of positive numbers only. For example,

$$10^{-100,000} = \frac{1}{10^{100,000}}$$

which is a very small positive number. The preceding equation may be written in log form thus:

$$\log_{10} \left(\frac{1}{10^{100,000}}\right) = -100,000.$$

Using mathematical terminology, $\lim_{n \to 0} \log n = -\infty$, or simply $\log 0 = -\infty$, provided the base is greater than one. Graphs of $y = \log_{10} x$ and $\log_e x$ are shown in Fig. 9-8, and we see that x cannot be negative.

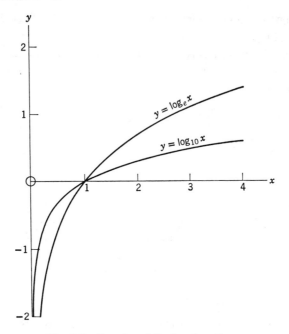

Fig. 9-8. Graphs of the log function.

9.3   THE EXPONENTIAL FUNCTION

The *exponential function* is an expression in which a positive constant base other than 1 is raised to a variable power. It is somewhat similar to the power function; however, it is important to note that in the power function, a variable base is raised to a constant power. Thus $3x^5$ is a power function, whereas 3×5^x is an exponential function.

If $y = e^x$ (an exponential function) then $x = \ln y$, and thus the exponential function is the inverse of the log function. The *inverse* of a function is the function resulting when the independent variable of the original function is expressed in terms of the dependent variable of the original function. More strictly speaking, $y = e^x$ is the inverse of $y = \ln x$, the letters being interchanged in the inverse function to indicate that the single letter on the left is now treated as the dependent variable.

If, as is usual, a base that is greater than 1 is used (10 and e being the most frequent bases), then, for $y = a^x$, we see that y increases as x increases, whereas for $y = a^{-x}$, y decreases as x increases. The exponential function is characteristic of natural growth and natural decay. If the exponent is positive, the function represents natural growth; if the exponent is negative, it represents natural decay.

The growth and decay of cell structures follow exponential laws. In the electrical field, as another example, the growth and decay of current from one fixed value to another, when current cannot suddenly change value due to the presence of an inductor, follow exponential patterns.

It is noted that the value of e^x, when $x = 0$, is 1. Also e^{2x} grows more rapidly with increasing x than e^x. Thus, to represent the most general exponential function, it is necessary to introduce two constants. The function $y = Ae^{mx}$ is general enough to represent any kind of exponential function. The factor A controls the value of y when $x = 0$, and the factor m controls the rapidity of growth (if m is positive) or rapidity of decay (if m is negative).

It is usual to take e as the base. No generality is sacrificed by so doing. If $y = A(10^{kx})$, $y/A = 10^{kx}$.

$$\ln \frac{y}{A} = kx \ln 10 = 2.303 \, kx.$$

$$\frac{y}{A} = e^{2.303kx}.$$

$$y = Ae^{2.303kx}.$$

This may be written $y = Ae^{mx}$. For every value of k there is a fixed value of $m = 2.303k$.

Thus all cases of natural growth or decay may be represented by $y = Ae^{mx}$. Graphs of the growth and decay functions are shown in Fig. 9-9.

For the decay function, as x approaches ∞, $(Ae^{-\infty})$ approaches zero. The graph approaches closer and closer to the x-axis as x increases and gradually flattens out. The x-axis is said to be an asymptote to the curve.

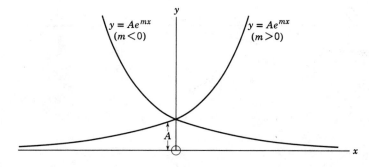

Fig. 9-9. The growth and decay functions (positive and negative exponentials).

EXAMPLE 7.   It is also correct, but not usual practice, to represent a general growth function as $y = e^{mx+k}$, the constant k being used to control the value of y when $x = 0$. Show that this is equivalent to $y = Ae^{mx}$.

Solution: $y = e^{mx+k} = e^k e^{mx} = Ae^{mx}$, where $A = e^k$, and for every value of k there is a distinct positive value of A.

EXAMPLE 8.   Plot the graph of $v = 3(1 - e^{-10t})$ from $t = 0$ to $t = 0.5$.

Solution: The solution will not be given in detail here. Tables of negative exponentials or the slide rule may be used to find individual points on the graph.

The clearest understanding of this graph will be achieved if these three curves are drawn (see Fig. 9-10): (1) $v = 3e^{-10t}$, a decay function; (2) $v = -3e^{-10t}$, the negative of the first, reflecting the first curve through the t-axis; and (3) the second curve adjusted upward three units for all values of t. This gives $v = -3e^{-10t} + 3$, or $v = 3(1 - e^{-10t})$, as required. It will be noted that the curve passes through the origin and rises rapidly at first and more slowly later on, tapering off to a constant value of 3. This contrasts with the form of the natural growth function which rises more and more rapidly as the independent variable increases. The graph of $v = 3(1 - e^{-10t})$ (Fig. 9-10) has a positive first derivative

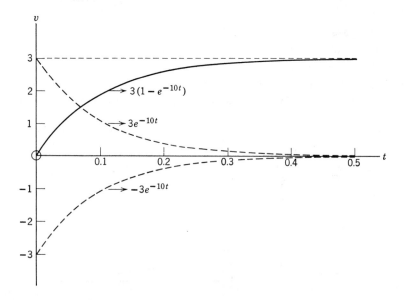

Fig. 9-10. Graph of $v = 3(1 - e^{-10t})$.

and a negative second derivative. For the growth function $v = Ae^{mx}$, both first and second derivatives are positive.

EXAMPLE 9.   Find a function to represent the population of a city as a function of time in years, assuming the present population is 600,000 and that it will grow exponentially, doubling itself in 30 years.
Solution: Let $P = Ae^{mt}$, where A and m must be determined to fit the information in the problem. If the present is taken as $t = 0$, at $t = 0$, $P = 600,000$. By substitution, $600,000 = Ae^0 = A$. At $t = 30$ years, $P = 1,200,000$.

$$\therefore\ 1,200,000 = 600,000e^{30m}.$$
$$e^{30m} = 2.$$
$$30m = \ln 2 = 0.693.$$
$$m = \frac{0.693}{30} = 0.0231.$$

Thus $P = 600,000e^{0.0231t}$, where t is measured in years from the present. The population in 15 years should be $600,000e^{0.3465} = 600,000(1.41) = 846,000$.

EXERCISES

1. If $\log_x 64 = \frac{3}{2}$, find x.
2. If $R = 10^{u/v}$, find an expression for u.
3. Given $\log_a \sqrt{x} = 0.35$, find $\log_a x^2$.
4. Given $\log_a (1/x) = \frac{1}{4}$, find $\log_a x$.
5. Rewrite without exponents or radicals:
   (a) $\log_b 2x^3$.   (b) $\log_b \sqrt[3]{x^2}$.   (c) $\log_b b^2$.
6. Solve for x:
   (a) $\log_x 100 = \frac{2}{3}$.   (b) $\log_{16} x = \frac{3}{4}$.   (c) $\log_e x^3 = 6$.
7. If $\log_b bx = 4$, find $\log_b x$.
8. Given $\log_{10} x + 2 \log_{10} (x + 1) - 2 = \log_{10} y$, find an expression for y.
9. If $R = \sqrt[3]{T} e^{u/v}$, find an expression for v.
10. Solve for x: $e^{2x} - 6e^x + 8 = 0$.
11. Find from tables (a) $\ln 81.210$, (b) $\ln (1.43 \times 9.25)$, and (c) $\ln (2.79)^2$.
12. If $\ln x = y - 5$ and $x = 1$, find y.
13. If $i = 6.5(1 - e^{-40t})$, find i when $t = 0.04$.
14. Solve for x: $12^{2x-3} = 24$.
15. Transform to exponential equations:
   (a) $\log_b (x^2 + 3) - 3 \log_b (2x + 1) = k$.   (b) $\log_b (y - 2) = \sqrt{2x + 5}$.
16. Simplify: (a) $e^{\log_e 3x}$, (b) $\log_e e^{5x-2}$.

17. Express any number a as (a) a natural log and (b) an exponential to base e.

18. If $\log (x^2 - 9) - \log (x - 3) = 0$, find x.

19. If $\log_e [\log_e (T/x)] - \log_e [\log_e (t/x)] = ML$, prove that $T = x(t/x)^{e^{ML}}$.

20. (a) Express a^x as a power of e, showing your derivation.

(b) Express $\log_a (x + 2)$ as a log to base e, showing your derivation.

21. A curve is given by $y = Ae^{mx}$. Describe and sketch the curve for the following conditions: (a) positive m, (b) negative m, and (c) $m = 0$.

What is the effect on the curve if (d) m is changed from 1 to 3, (e) m is changed from -1 to -3, and (f) A is changed from 1 to 3?

22. Assume that the present power load of a city is 1000 MW and that it doubles itself exponentially every ten years. Representing the present time as $t = 0$, find an equation to represent the power load as a function of t in years, and plot its curve on graph paper.

23. If q is an exponential function of t, and, at $t = 0$, $q = CE$ and at $t = RC$, $q = CE/e$, find q as a function of t.

24. A current i is given by $i = A(1 - e^{-mt})$. If, at $t = \infty$, $i = E/R$, and at $t = L/R$, $i = E/R - E/eR$, (a) solve for A and m, and thus find i as an exact function of t, and (b) if $E = 50$, $R = 12$, and $L = 0.25$, find i at $t = 0.020$.

25. If $\ln CE - \ln (CE - q) = t/RC$, (a) find q as a function of t and (b) sketch the graph of q against t.

9.4   HYPERBOLIC FUNCTIONS

The *hyperbolic functions* are based on the exponential functions and therefore they are not completely independent functions. It is often convenient in engineering work to express certain frequently occurring combinations of growth and decay as distinct functions. For instance, the mean value of a growth and decay function occurs frequently in engineering problems. This may be written as $(e^x + e^{-x})/2$. It is called the *hyperbolic cosine* of x, and abbreviated to cosh x. Its graph is developed from those of e^x and e^{-x} in Fig. 9-11.

There are six hyperbolic functions, named after the six trigonometric functions for reasons that will soon become apparent.

The two basic hyperbolic functions are defined as follows:

$$\textbf{sinh } x = \frac{e^x - e^{-x}}{2}$$

$$\textbf{cosh } x = \frac{e^x + e^{-x}}{2}$$

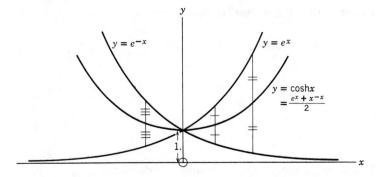

Fig. 9-11. Graph of cosh x developed from graphs of e^x and e^{-x}.

The other four hyperbolic functions depend on these two in the same way as the other trigonometric functions depend on the sine and cosine.

$$\tanh x = \frac{\sinh x}{\cosh x} = \frac{e^x - e^{-x}}{e^x + e^{-x}}$$

$$\operatorname{cosech} x = \frac{1}{\sinh x} = \frac{2}{e^x - e^{-x}}$$

$$\operatorname{sech} x = \frac{1}{\cosh x} = \frac{2}{e^x + e^{-x}}$$

$$\coth x = \frac{1}{\tanh x} = \frac{e^x + e^{-x}}{e^x - e^{-x}}$$

The reason why these functions are named after the trigonometric functions may now be explained in part. For every trigonometric formula or identity, there exists a corresponding hyperbolic formula, which either is identical in pattern or simply changes the sign of certain terms. The following tabulation shows this.

Trigonometric Formula	Hyperbolic Formula
$\cos^2 x + \sin^2 x = 1$	$\cosh^2 x - \sinh^2 x = 1$
$\sin 2x = 2 \sin x \cos x$	$\sinh 2x = 2 \sinh x \cosh x$
$\cos 2x = 1 - 2 \sin^2 x$	$\cosh 2x = 1 + 2 \sinh^2 x$
$\tan 2x = \dfrac{2 \tan x}{1 - \tan^2 x}$	$\tanh 2x = \dfrac{2 \tanh x}{1 + \tanh^2 x}$
$\sin (x + y) = \sin x \cos y$ $+ \cos x \sin y$	$\sinh (x + y) = \sinh x \cosh y$ $+ \cosh x \sinh y$
$\cos (x + y) = \cos x \cos y$ $- \sin x \sin y$	$\cosh (x + y) = \cosh x \cosh y$ $+ \sinh x \sinh y$

EXAMPLE 10.   Prove that $\cosh^2 x - \sinh^2 x = 1$.

Solution:

$$\cosh^2 x - \sinh^2 x = \left(\frac{e^x + e^{-x}}{2}\right)^2 - \left(\frac{e^x - e^{-x}}{2}\right)^2$$

$$= \frac{e^{2x} + 2 + e^{-2x}}{4} - \frac{e^{2x} - 2 + e^{-2x}}{4}$$

$$= \frac{4}{4} = 1.$$

This formula should become thoroughly familiar.

As stated previously, all the hyperbolic formulas are identical in pattern to the trigonometric formulas, but sometimes a sign changes. When do we change the sign of the trigonometric formula? A study of the tabulation shows that the sign changes preceding a two-dimensional sinh or tanh term, that is, before $\sinh^2 x$, $\tanh^2 x$, $\sinh x \sinh y$, or $\tanh x \tanh y$. This also implies that the sign will change preceding the reciprocals of this type of term, that is, $\operatorname{cosech}^2 x$, $\coth^2 x$, etc. Using this rule, many other hyperbolic identities can be constructed from the corresponding trigonometric identities, and it will only be necessary to know the trigonometric identities. Other similarities, or near-similarities, between the hyperbolic functions and the trigonometric functions are tabulated as follows:

Trigonometric Functions	Hyperbolic Functions
$\sin 0 = 0$	$\sinh 0 = 0$
$\cos 0 = 1$	$\cosh 0 = 1$
$\tan 0 = 0$	$\tanh 0 = 0$
$\cos x \le 1$	$\cosh x \ge 1$
The graph of $\cos x$ has even symmetry.	The graph of $\cosh x$ has even symmetry.
The graph of $\sin x$ has odd symmetry.	The graph of $\sinh x$ has odd symmetry.

The independent variable following hyperbolic functions is numerical only; it has nothing to do with an angle. Most tables and slide rules will have values of $\sinh x$ and $\cosh x$ for various values of x.

A brief, and only partial, explanation of the relationship of the hyperbolic functions to the hyperbola may be of interest. It was demonstrated in Section 9.1 that the trigonometric functions are related to circular rotation, and they are often referred to as the *circular functions*. In Fig. 9-12a, $x = \cos \theta$

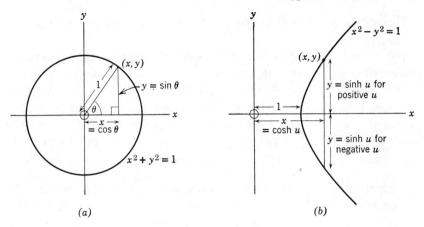

Fig. 9-12. (*a*) Circular and (*b*) hyperbolic functions related to the circle and hyperbola.

and $y = \sin \theta$, where x and y are the coordinates of any point on the graph of the circle $x^2 + y^2 = 1$. In Fig. 9-12*b* the graph of the rectangular hyperbola $x^2 - y^2 = 1$ is shown. If the cosh of any number is marked off along the x-axis, the sinh of this same number will be the y of the point on the hyperbola corresponding to this x, since $\cosh^2 u - \sinh^2 u = 1$. If u is negative, sinh u will be the negative y determined by the point on the lower branch.

Thus the hyperbolic functions are related to the hyperbola in much the same way that the trigonometric functions are related to the circle. In the pair of equations $x = \cos \theta$, $y = \sin \theta$, θ is a parameter, or an independent variable to which both x and y are related. The relationship between y and x may be established by eliminating y from the pair of equations; this may be done by squaring both equations and adding, so that

$$x^2 + y^2 = \cos^2 \theta + \sin^2 \theta = 1,$$

which is the equation of a circle. In the same way, the pair of equations $x = \cosh u$, $y = \sinh u$ use a parameter u, which we eliminate by using the fact that

$$x^2 - y^2 = \cosh^2 u - \sinh^2 u = 1,$$

and the xy equation is $x^2 - y^2 = 1$, the equation of a hyperbola. The parameter θ in the first pair of equations is well defined in Fig. 9-12*a* as the angle θ between the radius and the x-axis. The parameter u in the second pair of equations is not so easily defined. It, too, has a geometrical significance, but this is of little interest to us; however, this point is developed and explained in many mathematical texts.

The graphs of the three principal hyperbolic functions are shown in Fig. 9-13.

The *catenary*, the shape assumed by a cable of uniform weight distribution suspended between two supports of equal height, is a cosh function. It approximates a parabola near its central point but is not an exact parabola. The standard equation of a catenary is $y = a \cosh (x/a)$. The constant a takes care of all cases of *tight* or *loose* catenaries. a is the height of the lowest point above an origin which is fixed only after a has been determined to fit the tightness of the catenary. This origin is therefore very seldom at ground level.

EXAMPLE 11. A transmission line hangs in the form of a catenary with equation $y = 200 \cosh (x/200)$, x and y being measured in feet. If it is 30 ft above the ground at its central point, find its height above the ground 40 ft away from its central point.

Solution: With the y-axis through the central point, at a point 40 ft away from the central point, $x = 40$. Therefore $y = 200 \cosh \frac{40}{200} = 200 \times 1.0201 = 204.02$. However, at the central point, $y = 200$. Therefore the catenary is 4 ft higher at a point 40 ft from the central point. Thus this point is 34 ft above the ground.

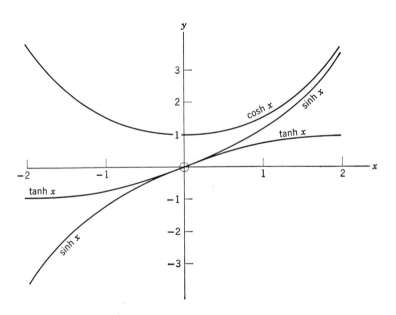

Fig. 9-13. Graphs of sinh x, cosh x, and tanh x.

9.5 INVERSE FUNCTIONS

If an angle is characterized by the value of one of its trigonometric functions, not by the number of degrees or radians in the angle, we use the symbol "arc sin," "arc tan," etc., to convey our meaning. The symbols "$\sin^{-1}$," "$\tan^{-1}$," etc., are also sometimes used.

arc sin x means "the angle whose sine is x."

$\sec^{-1} 1.2$ means "the angle whose secant is 1.2."

It is important here to distinguish between the angle and the value of its trigonometric function. It is recommended that the student read these symbols as *the angle whose sine is*, etc., to make this distinction.

Thus, arc sin $0.5 = 30°$, $150°$, $390°$, etc., and it is therefore multivalued. It is usual to take the *principal value* of this function as the one with lowest absolute value. Then arc tan (-1) has a principal value of $-45°$, and arc cos 0.707 has a principal value of $45°$, the positive angle being used in preference to $-45°$.

If $y = $ arc cos x, the statement may be written more directly by saying $x = \cos y$, y being an angle and x the value of its cosine. Therefore arc cos x is the inverse function of $\cos x$, and these types of functions are called *inverse functions*. They will be periodic along the y-axis, in the same way that the trigonometric functions are periodic along the x-axis. The graphs of $y = $ arc sin x, $y = $ arc cos x, and $y = $ arc tan x are shown in Fig. 9-14.

For the functions $y = $ arc sin x and $y = $ arc cos x, x cannot be numerically greater than 1.

Further clarification of the inverse trigonometric functions may often result from diagramming a right-angled triangle to illustrate the given function.

EXAMPLE 12. Find at least two other ways of expressing the function $y = $ arc cos x.

Solution: This implies that $x = \cos y$. The function may be illustrated by Fig. 9-15, where the value of $\cos y$, x or $x/1$, is used to indicate the relative lengths of the base and hypotenuse. Using the theorem of Pythagoras, the vertical side of the triangle $= \sqrt{1 - x^2}$. Therefore all six trigonometric functions of y may be listed; for example, $\sin y = \pm \sqrt{1 - x^2}$, $\tan y = \pm \sqrt{1 - x^2}/x$, etc., the double sign being necessary to take care of quadrants other than the first. Thus

$$y = \text{arc sin} \left(\pm \sqrt{1 - x^2} \right) = \text{arc tan} \left(\frac{\pm \sqrt{1 - x^2}}{x} \right), \quad \text{etc.}$$

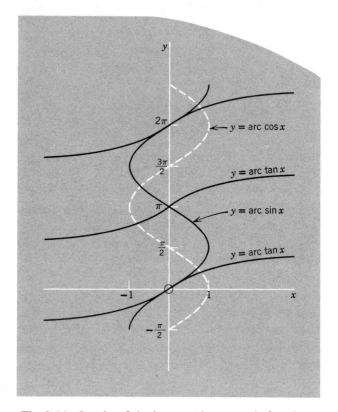

Fig. 9-14. Graphs of the inverse trigonometric functions.

EXAMPLE 13.   Simplify $\tan \arc \sin \dfrac{1}{\sqrt{1 + x^2}}.$

Solution: Let $\arc \sin \dfrac{1}{\sqrt{1 + x^2}} = \theta.$

$$\therefore \; \sin \theta = \frac{1}{\sqrt{1 + x^2}}.$$

The accompanying diagram illustrates this angle.

By the theorem of Pythagoras, the horizontal side is

$$\sqrt{(1 + x^2) - 1} = \pm x.$$

$$\therefore \; \tan \arc \sin \frac{1}{\sqrt{1 + x^2}} = \tan \theta = \pm \frac{1}{x}.$$

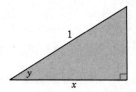

Fig. 9-15. Right-angled triangle to illustrate $y = \text{arc cos } x$.

The hyperbolic functions also have their corresponding inverse functions, and for the *inverse hyperbolic functions* the superscript symbol -1 is always used. Thus $y = \tanh^{-1} x$ means that $\tanh y = x$. There is no angle involved in the inverse hyperbolic functions, of course, and the example given could be read as "the number whose tanh is x."

EXAMPLE 14.   Prove that

$$\tanh^{-1} x + \tanh^{-1} y = \tanh^{-1} \frac{x + y}{1 + xy}.$$

Solution: Let $\tanh^{-1} x = m$, and $\tanh^{-1} y = n$. Therefore

$$x = \tanh m \text{ and } y = \tanh n.$$

$$\frac{x + y}{1 + xy} = \frac{\tanh m + \tanh n}{1 + \tanh m \tanh n} = \tanh (m + n).$$

(This has the same pattern as the trigonometric formula, but with the sign changed preceding a two-dimensional tanh). Thus

$$m + n = \tanh^{-1} \frac{x + y}{1 + xy}.$$

That is,

$$\tanh^{-1} x + \tanh^{-1} y = \tanh^{-1} \frac{x + y}{1 + xy}.$$

It may be difficult to get a clear conception of what is involved in inverse hyperbolic functions. However, remember that hyperbolic functions are actually combinations of exponential functions. We previously established that the inverse of an exponential function is a log function. Thus the inverse hyperbolic functions are types of log functions. To evaluate $\sinh^{-1} x$ as a log, let $\sinh^{-1} x = y$. Therefore

$$\sinh y = x \quad \text{and} \quad x = \frac{e^y - e^{-y}}{2}.$$

Multiplying both sides by e^y to eliminate negative exponents, we obtain

$$xe^y = \frac{e^{2y} - 1}{2}.$$

Arranging as a quadratic in e^y,

$$e^{2y} - 2xe^y - 1 = 0.$$

This may be solved for e^y by the quadratic formula

$$e^y = x \pm \sqrt{x^2 + 1}.$$
$$y = \ln (x \pm \sqrt{x^2 + 1}).$$

The value $\ln (x - \sqrt{x^2 + 1})$ must be rejected; it is the log of a negative number and does not exist. Thus

$$\sinh^{-1} x = \ln (x + \sqrt{x^2 + 1}).$$

EXERCISES

Establish the following formulas by using the exponential form of the hyperbolic functions involved:

1. $\operatorname{sech}^2 x = 1 - \tanh^2 x$.
2. $\cosh 2x = 1 + 2 \sinh^2 x$.
3. $\cosh 2x = 2 \cosh^2 x - 1$.
4. $\cosh 2x = \cosh^2 x + \sinh^2 x$.
5. $\sinh 2x = 2 \sinh x \cosh x$.
6. $\cosh (x - y) = \cosh x \cosh y - \sinh x \sinh y$.
7. Find from tables or with a slide rule (a) $\sinh (-1)$, (b) $\cosh (-2)$, and (c) $\tanh (-2.5)$.
8. Find the values of $\cosh x$ and $\sinh x$ for the following values of x by using tables or slide rule and hence verify that $\cosh^2 x - \sinh^2 x = 1$ for (a) $x = 0.5$, (b) $x = 1.22$, and (c) $x = -1.9$.

Simplify the following functions as far as possible:

9. $\cosh x - e^{-x}$.
10. $\cosh x + \sinh x$.

11. $(\cosh 2x + \sinh 2x)(\cosh 2x - \sinh 2x)$.

12. $\dfrac{\sqrt{(\cosh x + 1)(\cosh x - 1)}}{\sinh 2x}$.

13. $e^x \cosh x + e^x \sinh (-x)$.

14. A balanced T attenuator with attenuation N connecting circuits of impedance Z ohms has a resistance R ohms in its series arms, where $R = Z \tanh N/2$. Evaluate R if $Z = 220$ and $N = 1.4$.

15. The characteristic impedance of a transmission line is given by

$$Z_0 = 50 \ln \left(\frac{4w}{\pi d} \tanh \frac{\pi h}{w}\right).$$

Find Z_0 when $w = 3.0$, $h = 4.5$, and $d = 1.8$.

16. A transmission line hangs in the form of the catenary $y = 160 \cosh (x/160)$ ft, with the center point at $x = 0$. If its two points of suspension are 75 ft above the level ground and 250 ft apart, find the height of the center point above the ground.

17. Evaluate $\sinh^{-1} 2$ by using the logarithmic form of $\sinh^{-1} x$.

18. Evaluate $\sinh^{-1} (-2)$ by using the logarithmic form of $\sinh^{-1} x$.

19. Calculate $\cosh^{-1} x$ as a logarithmic function.

20. Calculate $\tanh^{-1} x$ as a logarithmic function.

Simplify the following functions as far as possible using principal values for the inverse functions.

21. $\tan^{-1} 1$.

22. $\cosh^{-1} 1 + \sinh^{-1} 1 + \tanh^{-1} 0$.

23. $\arcsin 0.5 + \arccos 0.5$.

24. $\arccos \sqrt{1 - x^2}$.

25. $\tan (\arcsin \sqrt{1 - 4x^2})$.

26. $\cos \left(\arcsin \dfrac{x}{\sqrt{1 - x^2}}\right)$.

27. $\arcsin \left(\cos \dfrac{\pi}{3}\right)$.

28. $\arccos (1 - 2 \sin^2 x)$.

29. $\arcsin (\sin x \cos y - \cos x \sin y)$.

30. $\sin \left(\arccos \dfrac{\sqrt{x^2 - 1}}{x}\right)$.

31. $\cos \left(\arcsin \dfrac{x^2 - 9}{x^2 + 9}\right)$.

32. $\arccos (\sin x)$.

Express as either an arc sin, arc cos, or arc tan:

33. $\operatorname{arc cosec} x$.

34. $\operatorname{arc sec} \dfrac{x + 2}{x}$.

35. $\operatorname{arc cot} \dfrac{3x - 1}{5}$.

CHAPTER 10

Calculus of the Trigonometric Functions

10.1 RELATIONSHIPS BETWEEN sin θ, θ, AND tan θ IN RADIAN MEASURE

So far, we have dealt with some of the fundamental principles of calculus, but rules for differentiation and integration have covered the power function only. We next discuss the trigonometric functions, to which the basic ideas of calculus still apply, of course, but for which rules for differentiation and integration must be developed.

The standard measurement of an angle may be in degrees or radians. The radian measure will be used almost exclusively in this chapter, since it facilitates the techniques of differentiation and integration. One important property of the radian for small angles that approach zero may be developed from Fig. 10-1.

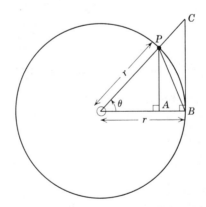

Fig. 10-1. Relationships between θ, sin θ, and tan θ.

One radian is the angle subtended at the center of any circle by an arc equal in length to the radius. If the arc length is twice the radius, the angle is 2 rad, etc. In general, the radian measure of an angle at the center of a circle is the ratio of the length of the subtending arc to the length of the radius. Since the radian measure is the ratio of two lengths, it is dimensionless; and if there is no indication of units following an angle, it is understood to be in radians.

In Fig. 10-1, BC is tangent to the circle at B, and PA is perpendicular to OB.

We see from the diagram that the area of $\triangle COB$ > area of sector OPB > area of $\triangle POB$.

Area of $\triangle COB$ is $\frac{1}{2}OB\,BC = \frac{1}{2}rr\tan\theta = \frac{1}{2}r^2\tan\theta$.

Area of sector OPB, treated as a fraction of the total circle, θ being in radians, is

$$\frac{\theta}{2\pi}\,\pi r^2 = \tfrac{1}{2}r^2\theta.$$

Area of $\triangle POB$ is

$$\tfrac{1}{2}OB\,PA = \tfrac{1}{2}rr\sin\theta = \tfrac{1}{2}r^2\sin\theta.$$

Thus

$$\tfrac{1}{2}r^2\tan\theta > \tfrac{1}{2}r^2\theta > \tfrac{1}{2}r^2\sin\theta \qquad \text{or} \quad \tan\theta > \theta > \sin\theta.$$

$$\therefore\ \frac{\tan\theta}{\sin\theta} > \frac{\theta\ (\text{in radians})}{\sin\theta} > 1.$$

$$\frac{\tan\theta}{\sin\theta} = \frac{\sin\theta}{\cos\theta\sin\theta} = \frac{1}{\cos\theta},$$

and, as θ approaches zero, this fraction approaches 1, the value of $1/\cos 0$. Thus the fraction $\theta/\sin\theta$, whose value lies between $\tan\theta/\sin\theta$ and 1, must also approach 1 as θ approaches zero. In other words,

$$\lim_{\theta\to 0}\frac{\theta}{\sin\theta} = \lim_{\theta\to 0}\frac{\sin\theta}{\theta} = \lim_{\theta\to 0}\frac{\sin\theta}{\tan\theta} = 1.$$

This fact may be corroborated from tables, where we may see that for angles less than about 6° the sine and tangent of the angle have almost the same values and the angle itself, expressed in radians, is also very close to this value. For smaller angles the three values are even closer. Slide rules use the same scale for the sine and tangent of angles less than 6°, indicating that their values, to three figures, are equal.

10.2 DERIVATIVES OF THE TRIGONOMETRIC FUNCTIONS

We now proceed to determine, from first principles, rules for differentiating the trigonometric functions using methods similar to those used for the power function. Fig. 10-2 may be used here.

Let $P(x,y)$ be any point on the graph of $y = \sin x$; let x take on an increase Δx, and y a corresponding increase Δy, so that $Q(x + \Delta x, y + \Delta y)$ lies on the graph. Then

$$y = \sin x. \qquad (1)$$

$$y + \Delta y = \sin (x + \Delta x). \qquad (2)$$

Subtracting (1) from (2),

$$\Delta y = \sin (x + \Delta x) - \sin x.$$

By using the trigonometric formula for changing a difference to a product, this may be written

$$\Delta y = 2 \cos \frac{2x + \Delta x}{2} \sin \frac{\Delta x}{2}.$$

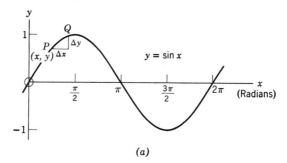

(a)

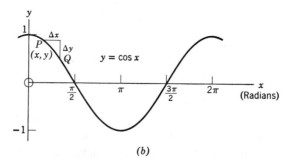

(b)

Fig. 10-2. Derivatives of $\sin x$ and $\cos x$.

Hence

$$\frac{\Delta y}{\Delta x} = 2 \cos \frac{2x + \Delta x}{2} \frac{\sin (\Delta x/2)}{\Delta x}$$

$$= \cos \frac{2x + \Delta x}{2} \frac{\sin (\Delta x/2)}{\Delta x/2}.$$

$$\frac{dy}{dx} = \lim_{\Delta x \to 0} \frac{\Delta y}{\Delta x}.$$

The second factor in the expression for $\Delta y/\Delta x$ is the ratio of the sine of an angle to the angle, and this angle approaches zero. Therefore, *provided x and Δx are the numerical values of the radians in the angles*, this ratio approaches 1. Therefore

$$\frac{dy}{dx} = \cos \frac{2x + 0}{2} \quad \text{or} \quad \frac{dy}{dx} = \cos x.$$

To make this more general, if y is a function of u, and u is a function of x, by the principles involved in a composite function (Section 6.3)

$$\frac{d}{dx} \sin u = \cos u \frac{du}{dx}.$$

We stress again that x and u must be in radians for this simple formula to hold true. It is because of the simplicity of this derivative formula that radians are used for the measurement of angles.

EXAMPLE 1.   Find the rate of change of y with respect to θ for the function $y = \sin \theta$ (a) at $\theta = 0$, (b) at $\theta = 60°$, (c) at $\theta = 90°$, and (d) at $\theta = 120°$.
Solution: $dy/d\theta = \cos \theta$. (a) At $\theta = 0$, $dy/d\theta = \cos 0 = 1$. Therefore y is changing at a rate of a numerical value of 1 per radian. (If 1 rad on the horizontal axis is the same length as 1 unit of numerical value on the vertical axis, the graphical slope of the curve at $\theta = 0$ is 1.)
(b) At $\theta = 60°$, $dy/d\theta = \cos 60° = 0.5$ per radian. (It makes no difference whether the angle is *expressed* in degrees or radians provided the rate is expressed per radian.)
(c) At $\theta = 90°$, $dy/d\theta = \cos 90° = 0$. This is easily verified on the graph of $y = \sin \theta$ (Fig. 10-2).
(d) At $120°$, $dy/d\theta = \cos 120° = -0.5$. The slope at $120°$ is the negative of the slope at $60°$, which also may be checked from the graph (Fig. 10-2).

EXAMPLE 2. Find (a) $(d/dx) \sin^3 x$, (b) $(d/dx)(1/\sin^2 3x)$, and (c) $(d/dx)(3 - \sin 5x)^2$.

Solution: These are all examples of composite functions, with both power functions and trigonometric functions involved.

(a) This means $(d/dx)(\sin x)^3$, and the outer structure is $(d/dx)(\quad)^3$, in which we use a power formula first and get $3(\quad)^2$. This is then multiplied by the derivative of the expression in the parentheses. Thus,

$$\frac{d}{dx} \sin^3 x = 3 \sin^2 x \cos x.$$

(b) This may be expressed as

$$\frac{d}{dx} (\sin 3x)^{-2} = -2(\sin 3x)^{-3} \cos 3x(3) = \frac{-6 \cos 3x}{\sin^3 3x}.$$

(c) There are three distinct steps in this calculation as we proceed from the outer structure to the details in the inner structure.

$$\frac{d}{dx} (3 - \sin 5x)^2 = 2(3 - \sin 5x)(- \cos 5x) (5)$$

$$= -10(3 - \sin 5x) \cos 5x.$$

To find $(d/dx) \cos x$, we may proceed from first principles using trigonometric formulas where required in much the same way as we have already done in differentiating $\sin x$. Alternatively, we may express $\cos x$ in terms of $\sin x$ and make use of the already established formula for the derivative of $\sin x$. Using the second method and making use of the relationships between $\sin x$ and $\cos x$ established in Section 9.1,

$$\frac{d}{dx} \cos x = \frac{d}{dx} \sin \left(x + \frac{\pi}{2}\right) = \cos \left(x + \frac{\pi}{2}\right) = \sin (x + \pi) = -\sin x.$$

$$\therefore \frac{d}{dx} \cos u = -\sin u \frac{du}{dx}.$$

EXAMPLE 3.   Find $(d/dx)(\cos^2 x - \sin^2 x)$.

Solution: Two methods are possible.

Method 1:

$$\frac{d}{dx} (\cos^2 x - \sin^2 x) = 2 \cos x(-\sin x) - 2 \sin x \cos x$$

$$= -4 \sin x \cos x.$$

Method 2:

$$\frac{d}{dx} (\cos^2 x - \sin^2 x) = \frac{d}{dx} \cos 2x = -2 \sin 2x.$$

An awareness of trigonometric formulas is important. We should also note that the answers are identical.

If $\sin \theta$ and $\cos \theta$ are plotted against θ in radians, with the length used for 1 rad on the horizontal axis equal to the length used for 1 (the maximum value of both functions) on the vertical axis (see Fig. 10-3), the slope of the sine curve at any point equals the vertical displacement of the cosine curve; and the slope of the cosine curve at any point is equal to the negative of the vertical displacement of the sine curve. This gives a graphical verification of the formulas established.

If degrees were used as the unit along the horizontal axis, with no distortion, the horizontal axis would be amplified by a factor of $180/\pi$, and each curve stretched out horizontally. This would decrease the numerical value of the slope at all points except at zero slope, and all slopes would be $\pi/180$ times their values in Fig. 10-3. In other words, rates of change per degree are $\pi/180$ times their corresponding rates per radian.

There are four other trigonometric functions besides sine and cosine, whose derivatives may be worked out by relating them to the sine and cosine. Thus $\tan x = \sin x/\cos x$; $\operatorname{cosec} x = (\sin x)^{-1}$; $\sec x = (\cos x)^{-1}$; $\cot x = \cos x/\sin x$. One convenient formula to know is that for $(d/dx) \tan x$. This is equivalent to

$$\frac{d}{dx}\frac{\sin x}{\cos x} = \frac{\cos x \cos x - \sin x(-\sin x)}{\cos^2 x} = \frac{\cos^2 x + \sin^2 x}{\cos^2 x} = \frac{1}{\cos^2 x}$$

$$= \sec^2 x.$$

$$\therefore \frac{d}{dx} \tan u = \sec^2 u \frac{du}{dx}.$$

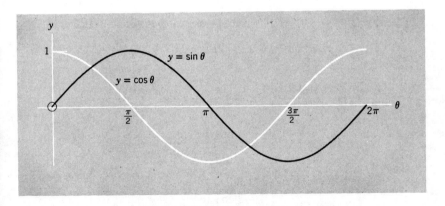

Fig. 10-3. Sin θ and cos θ plotted against θ in radians with undistorted scale.

Before we proceed to the integration of trigonometric functions and engineering problems, it is necessary to gain facility in working with derivatives, which now may involve power functions and trigonometric functions in many combinations. Most of these will be composite functions where the so-called chain rule must be applied.

One simple case that occurs frequently is where a trigonometric operator acts on a constant multiple of the independent variable, for example,

$$\frac{d}{d\theta} \sin 3\theta = \cos 3\theta(3) = 3 \cos 3\theta.$$

In the same way,

$$\frac{d}{d\theta} 4 \tan 2\theta = 8 \sec^2 2\theta.$$

The treatment of these multiples of the independent variable should become almost automatic.

Product and quotient formulas and implicit differentiation may also be involved. The basic techniques of calculus apply to all types of function.

EXAMPLE 4.   Find

$$\frac{d}{d\theta} \frac{\sqrt{2 - \cos^2 3\theta}}{\sin 3\theta}.$$

Solution: There is more than one way of solving this. An implicit differentiation is used here.

Let the expression to be differentiated be y. Therefore

$$y \sin 3\theta = \sqrt{2 - \cos^2 3\theta}.$$

$$y^2 \sin^2 3\theta = 2 - \cos^2 3\theta.$$

$$(\sin^2 3\theta)2y \, dy + y^2 2 \sin 3\theta(3 \cos 3\theta) \, d\theta = -2 \cos 3\theta(-3 \sin 3\theta) \, d\theta.$$

$$2y \sin^2 3\theta \, dy = 6 \sin 3\theta \cos 3\theta(1 - y^2) \, d\theta.$$

$$\frac{dy}{d\theta} = \frac{3 \cos 3\theta(1 - y^2)}{y \sin 3\theta} = \frac{3 \cos 3\theta\left(1 - \dfrac{2 - \cos^2 3\theta}{\sin^2 3\theta}\right)}{\sqrt{2 - \cos^2 3\theta}}$$

$$= \frac{3 \cos 3\theta(\sin^2 3\theta + \cos^2 3\theta - 2)}{\sin^2 3\theta\sqrt{2 - \cos^2 3\theta}}.$$

The expression in parentheses is $1 - 2 = -1$. Therefore the derivative is

$$\frac{-3 \cos 3\theta}{\sin^2 3\theta\sqrt{2 - \cos^2 3\theta}}.$$

1-15 odd, 29,31,33

EXERCISES

1. Find the following derivatives:

(a) $\dfrac{d}{dx}\sin 6x.$

(b) $\dfrac{d}{d\theta}\cos\left(2\theta - \dfrac{\pi}{3}\right).$

(c) $\dfrac{d}{dt}\tan\left(200\pi t - \dfrac{\pi}{4}\right).$

2. Evaluate the following:

(a) $\dfrac{d}{dt}\cos 120\pi t$ at $t = \dfrac{1}{360}.$

(b) $\dfrac{d}{dt}\sin\left(1000t + \dfrac{\pi}{6}\right)$ at $t = \dfrac{\pi}{3000}.$

(c) $\dfrac{d}{d\theta}\tan\left(\dfrac{\pi}{3} - 2\theta\right)$ at $\theta = \dfrac{\pi}{3}.$

Calculate the following derivatives:

3. $\dfrac{d}{d\theta}\cos^3 \theta.$

4. $\dfrac{d}{d\theta}\dfrac{1}{\sin^2 \theta}.$

5. $\dfrac{d}{dx}(4 + 3\sqrt{\tan 2x}).$

6. $\dfrac{d}{dt}\sin^2\left(\dfrac{\pi}{2} - 50t\right).$

7. $\dfrac{d}{dx}(\sin^2 x + \cos^2 x).$

8. $\dfrac{d}{d\theta}\sqrt{1 - 3\sin 5\theta}.$

9. $\dfrac{d}{dx}\dfrac{1}{\sqrt{\sin^2 3x + 2\cos^2 3x}}.$

10. Find $(d/dx)\cos x$ from first principles, using Δx and Δy.

Using the derivatives of $\sin x$ and $\cos x$, find the following derivatives:

11. $\dfrac{d}{dx}\operatorname{cosec} x.$

12. $\dfrac{d}{dx}\sec x.$

13. $\dfrac{d}{dx}\cot x.$

Evaluate the following:

14. $\dfrac{d}{d\theta}3\tan^2 \theta$ at $\theta = \dfrac{\pi}{4}.$

15. $\dfrac{d}{dt}(\sin 50\pi t - \cos 50\pi t)^2$ at $t = 0.001.$

16. $\dfrac{d}{dx}\cos(\sin x)$ at $x = \dfrac{\pi}{6}.$

17. $\dfrac{d}{dt}\cos^2\left(120\pi t - \dfrac{\pi}{3}\right)$ at $t = \dfrac{1}{720}.$

18. $\dfrac{d}{dt}(1 - \sin^2 100\pi t)^{3/2}$ at $t = \dfrac{1}{300}.$

19. Prove that $(d/dx)\cos x = -\sin x$ by expressing $\cos x$ as $\sqrt{1 - \sin^2 x}$ and using the formula for $(d/dx)\sin x$.

20. Use two methods to find $(d/dx)\,2 \sin x \cos x$.

Find the first, second, third, and fourth derivatives of the following functions:

21. $y = \sin 2x$.

22. $y = 2 \sin 2x$.

23. $y = A \cos x$.

24. $y = A \cos mx$.

Express y as a function of x to satisfy each of the following statements:

25. $\dfrac{d^2y}{dx^2} = -y$ (two answers).

26. $\dfrac{d^2y}{dx^2} = -9y$ (two answers).

27. $\dfrac{d^4y}{dx^4} = 16y$ (two answers).

28. $\dfrac{dy}{dx} = +\sqrt{1 - y^2}$.

Find the following derivatives:

29. $\dfrac{d}{d\theta} \sin 3\theta \tan 3\theta$.

30. $\dfrac{d}{d\theta} \dfrac{\sin 3\theta}{\tan 3\theta}$.

31. $\dfrac{d}{d\theta} \theta^3 \cos^2 \theta$.

32. $\dfrac{d}{dt} \dfrac{1 - \sin 377t}{\cos 377t}$.

33. $\dfrac{d}{dx} \dfrac{\cos x}{\sqrt{1 - 2 \sin^2 x}}$.

34. $\dfrac{d}{dx} \sin 2x \cos 2x \tan 2x$.

35. $\dfrac{d}{dx} \dfrac{x^3 \tan 2x}{\cos 2x}$.

36. $\dfrac{d}{dx} \sqrt{1 + \cos^2 \left(2x - \dfrac{\pi}{3}\right)}$.

37. $\dfrac{dy}{dx}$, where $y = \sin (x + y)$.

38. $\dfrac{dy}{dx}$, where $x \sin y + y \sin x = 1$.

39. Find the equation of the tangent to the graph of $y = \sin^2 5x$ at the point where $x = 0.05\pi$.

40. If $i_1 = 2 \cos \theta$ and $i_2 = 3 \sin \theta$, find d^2i_1/di_2^2 as a function of θ.

41. Find the maximum and minimum values of $i = \sin \theta + 2 \cos \theta$ by (a) the derivative method and (b) expressing i as a single sine function.

10.3   INTEGRATION OF TRIGONOMETRIC FUNCTIONS

Rules for integrating the simpler trigonometric functions are easily established from the rules for differentiation of these functions. Remembering that integration is the inverse of differentiation, the following rules are established:

$\because\ d \sin u = \cos u\, du,$ $\therefore\ \displaystyle\int \cos u\, du = \sin u + C.$

$\because\ d \cos u = -\sin u\, du,$ $\therefore\ \displaystyle\int \sin u\, du = -\cos u + C.$ where u is measured in radians.

$\because\ d \tan u = \sec^2 u\, du,$ $\therefore\ \displaystyle\int \sec^2 u\, du = \tan u + C.$

The *du* in each of these rules may be interpreted as "with respect to *u*"; however, more exactly, it means the differential of *u*, and, unless this differential appears as a factor of the expression to be integrated, these formulas do not apply. For example, $\int \cos 3x\, d(3x)$ is structurally identical to the left-hand side of the first formula, and the correct answer is $\sin 3x + C$; however, $\int \cos 3x\, dx$ has a different structure, since, considering $3x$ as the basic function of x, (the *u* of the formula) dx is not its correct differential.

A more fundamental way of dealing with this point is to examine the individual steps of a differentiation, and then to reverse these steps in an integration. As an example,

$$d \sin 3x = \cos 3x\, 3\, dx = 3 \cos 3x\, dx;$$

$$\therefore \int 3 \cos 3x\, dx = \sin 3x + C;$$

$$\therefore \int \cos 3x\, dx = \tfrac{1}{3} \sin 3x + k,$$

where k may have any constant value. The essential point here is that the differential of $3x$, $3\, dx$, is a factor in the answer to a problem in differentiation, but this same factor must occur as part of the expression to be integrated in an integration problem.

In every integration problem, the structure of the expression to be integrated must be broken into two factors. One factor contains a function of the independent variable, and the other is the differential of this function.

If the second factor only needs to be adjusted by a constant multiplier to become the differential of the main function of the independent variable, the solution is relatively simple. In our example,

$$\int \cos 3x\, dx = \tfrac{1}{3} \int \cos 3x\, d(3x) = \tfrac{1}{3} \sin 3x + C.$$

By using the same reasoning,

$$\int \sin \left(200t - \frac{\pi}{2} \right) dt = \frac{-1}{200} \cos \left(200t - \frac{\pi}{2} \right) + C,$$

$$\int \sec^2 (\omega t + \pi)\, dt = \frac{1}{\omega} \tan (\omega t + \pi) + C, \quad \text{etc.}$$

When the main function of the independent variable (here, $200t - \pi/2$ and $\omega t + \pi$, respectively) consists of a constant multiple of the independent variable with or without another constant term, the constant factor in the answer (1/200 and $1/\omega$ in these two examples) should be inserted almost automatically, although the reasoning behind this insertion is important.

Adjustments and substitutions may often be necessary in the expression to be integrated in order to make it fit the structure of one formula. For trigonometric functions we may often find a trigonometric formula that will result in an appropriate substitution.

EXAMPLE 5.   Find $\int \sin^2 \theta \, d\theta$.

Solution: The expression to be integrated is a power function controlling a trigonometric function, and we cannot use both a power formula and a trigonometric formula on the same single-termed function. We must find a substitution that eliminates the square power. This comes from one of the three formulas for $\cos 2\theta$ in terms of θ.

$$\cos 2\theta = \cos^2 \theta - \sin^2 \theta$$

$$= 2 \cos^2 \theta - 1$$

$$= 1 - 2 \sin^2 \theta.$$

From the last of these three formulas, we can express $\sin^2 \theta$ in terms of $\cos 2\theta$ and thus eliminate the square power.

$$2 \sin^2 \theta = 1 - \cos 2\theta. \qquad \therefore \ \sin^2 \theta = \tfrac{1}{2}(1 - \cos 2\theta).$$

Therefore $$\int \sin^2 \theta \, d\theta = \tfrac{1}{2} \int (1 - \cos 2\theta) \, d\theta.$$

Integrating term by term,

$$\int \sin^2 \theta \, d\theta = \tfrac{1}{2}(\theta - \tfrac{1}{2} \sin 2\theta) + C = \tfrac{1}{2}\theta - \tfrac{1}{4} \sin 2\theta + C.$$

It is worthwhile to verify answers to integration problems by taking the differential or derivative of the answer. Here,

$$d(\tfrac{1}{2}\theta - \tfrac{1}{4} \sin 2\theta) = (\tfrac{1}{2} - \tfrac{1}{2} \cos 2\theta) \, d\theta = \tfrac{1}{2}[1 - (1 - 2 \sin^2 \theta)] \, d\theta$$

$$= \tfrac{1}{2}(2 \sin^2 \theta \, d\theta) = \sin^2 \theta \, d\theta,$$

the original differential to be integrated.

$\int \cos^2 \theta \, d\theta$ is calculated in much the same way. The expression for $\cos^2 \theta$ in terms of $\cos 2\theta$ is derived from the second of the preceding three formulas for $\cos 2\theta$.

EXAMPLE 6.   Evaluate $\int_0^{1/800} 4 \cos^2 200\pi t \, dt$.

Solution: Since $\cos 2\theta = 2 \cos^2 \theta - 1$,

$$\cos^2 \theta = \tfrac{1}{2}(1 + \cos 2\theta).$$

$$\therefore \cos^2 200\pi t = \tfrac{1}{2}(1 + \cos 400\pi t).$$

$$\therefore \int_0^{1/800} 4 \cos^2 200\pi dt = (4)(\tfrac{1}{2}) \int_0^{1/800} (1 + \cos 400\pi t)\, dt$$

$$= 2\left(t + \frac{1}{400\pi} \sin 400\pi t\right)_0^{1/800}$$

$$= 2\left(\frac{1}{800} + \frac{1}{400\pi} \sin \frac{\pi}{2}\right)$$

$$= 2\left[\frac{1}{800} + \frac{1}{400\pi}(1)\right]$$

$$= \frac{\pi + 2}{400\pi} = 4.1 \times 10^{-3}.$$

EXERCISES

1. Evaluate the following definite integrals:

(a) $\displaystyle\int_0^{\pi/2} \cos \theta \, d\theta$.

(b) $\displaystyle\int_0^{\pi/2} \sin \theta \, d\theta$.

(c) $\displaystyle\int_0^{\pi/4} \sec^2 \theta \, d\theta$.

(d) $\displaystyle\int_{t=0}^{0.005\pi} \cos 50t \, d(50t)$.

2. Calculate the following integrals:

(a) $\int \sin 5x \, dx$.

(b) $\int \cos 50t \, dt$.

(c) $\int \sin 50t \, d(100t)$.

(d) $\int \sin 200\pi t \, dt$.

(e) $\displaystyle\int 3 \sin\left(1000t - \frac{\pi}{3}\right) dt$.

(f) $\displaystyle\int \frac{2 \, dt}{\cos^2 (2\pi f t + \pi)}$.

Evaluate the definite integrals in the following problems:

3. $\displaystyle\int_0^{0.2\pi} 2 \cos 5\theta \, d\theta$.

4. $\displaystyle\int_0^{1/240} 6 \sin 120\pi t \, dt$.

5. $\displaystyle\int_{1/2f}^{1/f} 0.4 \sin 2\pi f t \, dt$.

6. $\displaystyle\int_0^{\pi/360} 3 \sin\left(120t - \frac{\pi}{6}\right) dt$.

7. $\displaystyle\int_0^{0.005} 5 \cos\left(200\pi t + \frac{\pi}{6}\right) dt$.

8. Find the area enclosed between the first positive portion of $y = \sin x$ and the x-axis.

9. Find the area bounded by $y = \arcsin x$ and the y-axis between $y = 0$ and $y = \pi/3$.

10. Find the area enclosed by $y = \sin 2x$, $y = 1$, and $x = 0$.

11. Find the area enclosed by one positive section of $y = \sin^2 \theta$ and the θ-axis.

Calculate the following integrals:

12. $\int \sin^2 2\theta \, d\theta$.

13. $\int \left(\cos^2 \dfrac{\theta}{2} - \sin^2 \dfrac{\theta}{2} \right) d\theta$.

14. $\int (\cos^2 2\pi ft + \sin^2 2\pi ft) \, dt$.

15. $\int \sin^2 \left(2\theta - \dfrac{\pi}{2} \right) d\theta$.

16. $\int \cos^2 \left(2\pi ft + \dfrac{\pi}{6} \right) dt$.

17. $\int \sin x \cos x \, dx$.

18. $\int (1 - 3 \cos^2 377t) \, dt$.

19. $\int (\cos^4 50\pi t - \sin^4 50\pi t) \, dt$.

20. $\int \sin^2 \theta \cos^2 \theta \, d\theta$.

21. $\int \sin^4 \theta \, d\theta$.

Evaluate the following definite integrals:

22. $\int_0^{2\pi} \sin^2 \theta \, d\theta$.

23. $\int_0^{1/f} \sin^2 2\pi ft \, dt$.

24. $\int_0^{1/2f} \cos^2 2\pi ft \, dt$.

25. $\int_0^{0.005} \sec^2 50\pi t \, dt$.

26. $\int_0^{\pi/4} \sin^2 \left(2\theta - \dfrac{\pi}{4} \right) d\theta$.

27. $\int_0^{\pi/2} (\sin^2 \theta - \cos^2 \theta) \, d\theta$.

28. $\int_0^{\pi/2} (\cos \theta + \sin \theta)^2 \, d\theta$.

29. $\int_{0.005\pi}^{0.01\pi} \cos^2 \left(100t + \dfrac{\pi}{3} \right) dt$.

30. Prove that the area between the graph of $y = \sin^2 (500\pi t - \pi/6)$ and the t-axis between $t = 0$ and $t = 0.002$ is half its area between $t = 0$ and $t = 0.004$.

10.4 APPLICATIONS TO PROBLEMS

An examination of successive derivatives of the sine and cosine functions has considerable significance in the understanding of problems in the mechanical and electrical fields.

If $y = A \sin \omega x$:

$$\frac{dy}{dx} = \omega A \cos \omega x$$

$$\frac{d^2y}{dx^2} = -\omega^2 A \sin \omega x = -\omega^2 y.$$

If $y = B \cos \omega x$:

$$\frac{dy}{dx} = -\omega B \sin \omega x$$

$$\frac{d^2y}{dx^2} = -\omega^2 B \cos \omega x = -\omega^2 y.$$

The second derivative of a sine or cosine function is negatively proportional to the original function. These two are the only functions for which this is true. Therefore, if the second derivative of a function is negatively proportional to the original function, the function must be a sine or a cosine or a combination of the two.

If, for example, $d^2y/dx^2 = -16y$, we should instinctively recognize that y is a sine or cosine function or a combination of both, which means that y is sinusoidal. More exactly, $y = A \sin 4x$, $B \cos 4x$, or $A \sin 4x + B \cos 4x$, where A and B may have any arbitrary constant values. The multiple of y in the original equation here is the square of the multiple of x in the trigonometric functions, as seen from an examination of the derivatives previously listed.

In the mechanical field, the principles of simple harmonic motion may be derived from these laws. Figure 10-4 shows a spring suspended vertically with its free end compressed upwards a distance A from its neutral uncompressed position.

The internal force tending to restore the spring to its neutral position is proportional to the displacement of the free end from its neutral position at all times. The moment the spring is released, this force is translated to acceleration proportional to the force and therefore proportional to the displacement from N. Since the displacement is away from the neutral position and the acceleration is toward the neutral position, the acceleration is negatively proportional to the displacement. Hence $d^2s/dt^2 = -ks$. From the analysis at the beginning of this section it should be apparent that s is proportional to $\sin \sqrt{k}\,t$, $\cos \sqrt{k}\,t$, or a combination of both. Further analysis shows that s is at its maximum value of A at $t = 0$, the time of release. Since the cosine is maximum at $t = 0$, the function is given by $s = A \cos \sqrt{k}\,t$, with a maximum s of A at $t = 0$.

The free end will therefore oscillate sinusoidally with respect to time about its neutral position. The *s-t* graph is shown in Fig. 10-4. This oscillation is

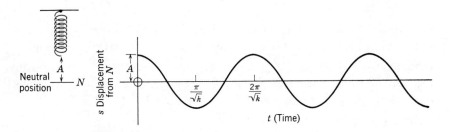

Fig. 10-4. A compressed spring. $\dfrac{d^2s}{dt^2} = -ks.$ $s = A \cos \sqrt{k}\,t.$

theoretically a perpetual motion. However, in our analysis the damping effect of spring and air resistance has been neglected. These forces would produce smaller and smaller oscillations until the spring finally comes to rest at N. The undamped motion is an example of *simple harmonic motion*, which may be defined as motion in which the acceleration towards a fixed point is proportional to and oppositely directed to its displacement from that point. The resulting motion, as we have seen, is sinusoidal with respect to time.

In the electrical field there is a condition similar to this mechanical simple harmonic motion. Suppose a capacitor is charged to a certain level of Q coulombs and then switched into series with an inductance in a closed loop as indicated in Fig. 10-5.

Assuming no resistance in the circuit, at all times following the closing

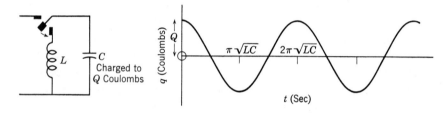

Fig. 10-5. A charged capacitor. $\dfrac{d^2q}{dt^2} = -\dfrac{1}{LC} q.$ $q = Q \cos \dfrac{t}{\sqrt{LC}}.$

of the switch ($t = 0$, say) the sum of the voltage drops across the capacitance and the inductance equals zero. Thus $q/C + L(di/dt) = 0$. Since $i = dq/dt$,

$$\frac{di}{dt} = \frac{d^2q}{dt^2} \quad \text{and} \quad \frac{q}{C} + L\frac{d^2q}{dt^2} = 0 \quad \text{or} \quad \frac{d^2q}{dt^2} = -\frac{1}{LC} q.$$

Once again the second derivative of a dependent variable is negatively proportional to the variable. Following the same reasoning as in the previous examples, $q = Q \cos (t/\sqrt{LC})$. The charge on the capacitor will oscillate between maximum positive and maximum negative charges Q and the oscillation will be sinusoidal with respect to time.

Since $i = dq/dt$ and $q = Q \cos (t/\sqrt{LC})$,

$$i = \frac{-Q}{\sqrt{LC}} \sin \frac{t}{\sqrt{LC}},$$

and the current is also sinusoidal with the same frequency as and 90° out of phase with the charge. The capacitor acts as a temporary a-c generator.

Here again, in actual practice, the oscillations of charge and current do not continue indefinitely. They are damped out by resistance, which will always be present even though it may be very small. The resistance, mechanical or electrical, constitutes the damping effect.

A more extensive study of these types of problems will be made in Chapters 16 and 17. For the moment it is interesting to note that the original application of mechanical and electrical laws often gives rise to equations involving derivatives, called *differential equations*. From these differential equations, by using knowledge of calculus plus a certain amount of intuition, we are able to express the dependent variable as a function of the independent variable, and thus see distinctly what type of motion, charge, or current results.

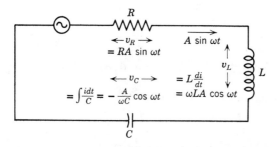

Fig. 10-6. The *RLC* series circuit.

We examine next a series circuit of resistance R ohms, inductance L henrys, and capacitance C farads through which an alternating current of $i = A \sin \omega t$ amperes flows, as depicted in Fig. 10-6.

The charge on the capacitor will be assumed to be sinusoidal without any extra constant amount.

$$q = \int i\, dt = \int A \sin \omega t\, dt = -\frac{A}{\omega} \cos \omega t + k, \qquad \text{where } k = 0.$$

The voltage across the capacitor is $q/C = (-A/\omega C) \cos \omega t$. Voltage across the inductor is

$$L\frac{di}{dt} = L\frac{d}{dt} A \sin \omega t = \omega LA \cos \omega t.$$

The voltages across the capacitor and the inductor are both proportional to $\cos \omega t$, and the opposite signs show that they oppose one another. The voltage across the resistor is $Ri = RA \sin \omega t$. The total voltage applied to the circuit is therefore $RA \sin \omega t + (\omega L - 1/\omega C)A \cos \omega t$.

We recall that the sum of a sine and a cosine function with the same frequency may be expressed as a single sine function with the same frequency and with a phase displacement, as analyzed in Section 9.1. Replacing $\omega L - 1/\omega C$ by X, the total applied voltage is $A(R \sin \omega t + X \cos \omega t)$, which may be expressed as a single sine in the form $\sqrt{R^2 + X^2} \, A \sin (\omega t + \phi)$, where $\tan \phi = X/R$. Thus the amplitude of the applied voltage is $\sqrt{R^2 + X^2}$ times the amplitude of the current, the voltage leads the current if X is positive, and lags the current if X is negative.

EXAMPLE 7.   Find the first maximum for the current

$$i = 12 \sin \theta + 2 \sin 3\theta.$$

Solution: For maximum i,

$$\frac{di}{d\theta} = 0.$$

$$12 \cos \theta + 6 \cos 3\theta = 0.$$

For solution of this equation, $\cos 3\theta$ must be expressed in terms of $\cos \theta$.

$$\cos 3\theta = 4 \cos^3 \theta - 3 \cos \theta.$$

$$\therefore \; 12 \cos \theta + 6(4 \cos^3 \theta - 3 \cos \theta) = 0.$$

$$24 \cos^3 \theta - 6 \cos \theta = 0.$$

$$6 \cos \theta(4 \cos^2 \theta - 1) = 0.$$

$$\cos \theta = 0 \quad \text{or} \quad \cos \theta = \pm \tfrac{1}{2}.$$

$$\theta = 90°, \; 60° \quad \text{or} \quad 120°.$$

$$\frac{d^2i}{d\theta^2} = \frac{d}{d\theta} (24 \cos^3 \theta - 6 \cos \theta) = 72 \cos^2 \theta(-\sin \theta) + 6 \sin \theta$$

$$= -72 \cos^2 \theta \sin \theta + 6 \sin \theta.$$

At $\theta = 90°$, $d^2i/d\theta^2 = +6$.

At $\theta = 60°$,

$$\frac{d^2i}{d\theta^2} = -72(\tfrac{1}{2})^2(0.866) + 6(0.866)$$

$$= 0.866(-18 + 6) = -0.866(12).$$

Thus the first maximum i occurs at $\theta = 60°$, and the maximum i is $12(0.866) + 0 = 10.4$.

EXAMPLE 8. A capacitor of 5000 microfarads is charged to 0.040 coulomb and then switched to discharge through a 20-millihenry inductor. Assuming no resistance, find as functions of time t in seconds after the switching: (a) the charge on the capacitor, (b) the current in the circuit, (c) the voltage across the capacitor, and (d) the voltage across the inductor. At the moment of switching, the current is zero, since no instantaneous change in current can occur through an inductance. *Solution:* The sum of the voltage drops across the inductor and capacitor must be zero at all times. Therefore $L(di/dt) + q/C = 0$. That is,

$$L\frac{d^2q}{dt^2} + \frac{q}{C} = 0.$$

$$\therefore \frac{d^2q}{dt^2} = \frac{-1}{LC}q = -\frac{1}{20 \times 10^{-3} \times 5000 \times 10^{-6}}q = -10{,}000q.$$

Since the second derivative of q is negatively proportional to q itself, q must be sinusoidal, of the form $A \sin 100t$, $B \cos 100t$, or a combination of both. To be perfectly sure of the right combination, we might let $q = A \sin 100t + B \cos 100t$, and use additional information to solve for A and B in much the same way as we have already solved for the integration constant in an indefinite integral. At $t = 0$, $q = 0.040$; substituting these values of the variables in the expression for q, $0.040 = A \times 0 + B \times 1 = B$. Therefore $B = 0.040$. Also, at $t = 0$, $i = 0$, since there can be no instantaneous change in current through an inductor. Since $i = dq/dt$, this means that at $t = 0$, $dq/dt = 0$.

$$\frac{dq}{dt} = 100A \cos 100t - 100B \sin 100t;$$

substituting $t = 0$ and $dq/dt = 0$,

$$0 = 100A \times 1 - 100B \times 0. \qquad 100A = 0. \qquad A = 0.$$

Hence

(a) $$q = 0.040 \cos 100t.$$

(b) $$i = \frac{dq}{dt} = \frac{d}{dt} 0.040 \cos 100t = -4.0 \sin 100t.$$

(c) Voltage across the capacitor is

$$\frac{q}{C} = \frac{0.040 \cos 100t}{5000 \times 10^{-6}} = 8.0 \cos 100t.$$

(d) Voltage across the inductor is

$$L \frac{di}{dt} = 20 \times 10^{-3} \frac{d}{dt}(-4.0 \sin 100t) = 20 \times 10^{-3}(-400 \cos 100t)$$

$$= -8.0 \cos 100t.$$

This, being the negative of the voltage across the capacitor, verifies that the sum of these two voltages is at all times equal to zero.

EXAMPLE 9.   A 4.0-ohm resistor, a 0.02-henry inductor, and a 0.005-farad capacitor are in series with an a-c voltage source, and the current is 2.0 sin 200t. Find (a) the voltage of the source as a function of t and (b) the maximum voltage of this source. Assume the voltage source to be sinusoidal with no constant term.

Solution: (a) Total voltage is

$$Ri + L\frac{di}{dt} + \frac{q}{C} = Ri + L\frac{di}{dt} + \frac{\int i\, dt}{C}$$

$$= 4.0 \times 2.0 \sin 200t + 0.02 \times 400 \cos 200t + \frac{-2.0 \cos 200t}{200 \times 0.005}$$

$$= 8.0 \sin 200t + 8.0 \cos 200t - 2.0 \cos 200t$$

$$= 8.0 \sin 200t + 6.0 \cos 200t.$$

The integration constant is zero, since the total voltage is sinusoidal. (b) The maximum voltage may be found by expressing it as a single sine function with a phase displacement, and verified by the methods of differential calculus. Let

$$8.0 \sin 200t + 6.0 \cos 200t = A \sin (200t + \phi)$$

$$= A \sin 200t \cos \phi + A \cos 200t \sin \phi.$$

A term-by-term comparison shows that

$$A \cos \phi = 8.0 \quad \text{and} \quad A \sin \phi = 6.0.$$

By squaring and adding these two equations,

$$A^2(\cos^2 \phi + \sin^2 \phi) = 64 + 36 = 100.$$

That is, $A^2 \times 1 = 100$. $A = 10$. By dividing the two equations,

$$\frac{A \sin \phi}{A \cos \phi} = \frac{6.0}{8.0} = 0.75 \qquad \tan \phi = 0.75 \qquad \phi = 36.9°.$$

Therefore the voltage of the source is $10 \sin (200t + 36.9°)$. Since the amplitude of a single sine function is its maximum value, the maximum voltage is 10.

To verify this, we use the original two-termed expression for the voltage. For a maximum, its derivative with respect to time must be 0. Hence

$$1600 \cos 200t - 1200 \sin 200t = 0.$$

$$\frac{\sin 200t}{\cos 200t} = \frac{1600}{1200}. \quad \tan 200t = 1.333.$$

$$200t = 53.1° \quad \text{or} \quad 233.1°.$$

The second derivative is $-320,000 \sin 200t - 240,000 \cos 200t$, which is negative for $200t = 53.1°$ and positive for $200t = 233.1°$. Therefore the maximum occurs at $200t = 53.1°$, and its value is

$$8.0 \sin 53.1° + 6.0 \cos 53.1° = 8.0 \times 0.80 + 6.0 \times 0.60$$
$$= 6.4 + 3.6 = 10.$$

EXERCISES

1. Find d^2y/dx^2 in terms of y for the following functions:

(a) $y = \sin 3x$. (b) $y = 2 \cos 3x$. (c) $y = A \sin kx$.

(d) $y = A \sin kx + B \cos kx$. (e) $y = A \cos (\sqrt{k}x + \phi)$.

2. Find i as a function of t in at least two ways, given:

(a) $\dfrac{d^2i}{dt^2} = -9i$. (b) $\dfrac{d^2i}{dt^2} = -ki$. (c) $\dfrac{d^2i}{dt^2} = -4\pi^2 f^2 i$.

(d) $\displaystyle\int i\, dt = -4\dfrac{di}{dt} + C$. (e) $\dfrac{d^4i}{dt^4} = 81i$.

3. The displacement of a point on a spring from its normal position is given by $s = 2.5 \cos 10\pi t$ in., t being in seconds.

(a) Find its maximum displacement from its normal position.

(b) Find expressions for its velocity and acceleration.

(c) Find its velocity and acceleration at the time of maximum displacement.

(d) Find its displacement and acceleration at the time of maximum velocity.

4. Given that the acceleration of a body is $-96 \sin (8t + \pi/3)$ in. per second per second, t being in seconds; at $t = 0$, velocity $v = 6.0$, and at $t = \pi/12$, displacement $s = 0$, find (a) velocity as a function of time, (b) displacement as a function of time, (c) maximum velocity, (d) maximum displacement, and (e) the velocity when the acceleration is zero.

5. Show that a body whose displacement from a fixed point is given by $s = 3 \sin 4t - 8 \cos 4t$ is moving in simple harmonic motion by proving that its acceleration is negatively proportional to its displacement.

6. The slope of a curve is given by the expression $\sin 2x + 3$ and the curve passes through the point (0,1). Find its equation.

7. If the voltage across an inductance of 2×10^{-4} henry is $1.2 \sin 120\pi t$, t being in seconds, find the current at $t = 1/360$, if the current is zero at $t = 1/240$.

8. Find the voltage across an inductance L henrys as a function of time t seconds when the current is given by $i = I_0 \cos^2 \omega t$ amperes.

9. A current, $i = 2.0 \sin 200\pi t$, flows through a resistance of 3.0 ohms.

(a) Find the power in the circuit as a function of time t.

(b) Sketch the power against time for one complete cycle of current.

(c) Find the time of the first maximum power, and find the value of this maximum, using calculus methods.

(d) Find by calculus the time at which the first point of inflection occurs in the graph of (b).

10. A resistance of 2.0 ohms, an inductance of 0.0040 henry, and a capacitance of 0.0010 farad are placed in series with an a-c voltage source. The charge on the capacitor is found to be sinusoidal, and equal to $0.006 \sin 500t$ coulombs, t being in seconds. Find the current in the circuit, the voltage across each element, and hence the source voltage as a function of time.

11. The *RLC* circuit of Exercise 10 is placed in series with a different a-c voltage source, and the current in the circuit is found to be $4.0 \sin 250t$ amperes, where t is in seconds. At $t = 0$, the voltage across the capacitor is -16 volts. Find the voltage of the source as a function of time.

12. In an *RLC* series circuit, the current is $A \sin 2\pi f t$.

(a) Find in terms of L and C the value of f for which the sum of the voltage drops across L and C is zero at all times, assuming sinusoidal voltages.

(b) For the value of f in (a), find the voltage of the source as a function of t. (This is called a resonant condition.)

13. In an *RLC* series circuit, $R = 5.0$ ohms, $L = 0.0010$ henry, and $C = 25$ microfarads. The voltage across the inductance L is found to be $8.0 \cos 4000t$ volts, t being in seconds. Assuming all voltages to be sinusoidal with no constant term, find (a) the current in the circuit, expressed as a function of t and (b) the voltage of the source, expressed as a single sine function.

14. The acceleration of a body is found to be negatively proportional to its displacement from a fixed point. When the displacement is 2.0 cm, the acceleration is -50 cm per second per second. At the time designated as zero, the displacement is 6.0 cm and the velocity is zero. Find as functions of time (a) the displacement and (b) the velocity. Verify that the acceleration is negatively proportional to the displacement.

15. Find the first maximum for the current $i = 2 \sin \theta + \cos 2\theta$.

16. Find the first maximum for the current $i = 7 \sin \theta + \sin 2\theta$.

17. Find the first maximum for the current $i = 12 \sin \theta + \sin 3\theta$.

18. It may be shown that the sending-end voltage v_1 and the receiving-end voltage v_2 of a transmission line of 50 miles or less are related by the equation

$$v_1{}^2 = (RI + v_2 \cos \phi)^2 + (XI + v_2 \sin \phi)^2.$$

If the values of R, I, v_2, and X remain fixed, find the value of ϕ for which v_1 is greatest.

19. The horizontal displacement x and the vertical displacement y of a missile from its firing point t seconds after firing are given by

$$x = v_0 t \cos \theta \quad \text{and} \quad y = v_0 t \sin \theta - \frac{gt^2}{2},$$

where g and v_0 are constants and θ is the angle of projection. Find (a) the maximum height for fixed θ and (b) the value of θ for maximum firing range.

20. A 0.0004-farad condenser is discharging through a 0.0016-henry inductor with no voltage applied to the circuit and no resistance in the circuit. At the time designated as zero, the voltage across the capacitor is 5.0 volts and the current is zero. Find as functions of time (a) the charge on the capacitor and (b) the current. Verify that the sum of the voltages around the closed path is always equal to zero. (c) What is the frequency of this oscillation?

21. A circuit consists of a resistance of 0.50 ohm in parallel with a 0.040-farad capacitance; a voltage of $2.0 \cos 80t$ volts is applied across the circuit, t being in seconds. Find as functions of t, the (a) charge on the capacitor, (b) current in the capacitive branch, (c) current in the resistive branch, (d) total current flowing to the circuit as a two-termed function and as a single-termed function, and (e) power in the circuit.

22. A voltage of $6.0 \sin 500t$ volts is applied to a circuit consisting of a resistance of 2.0 ohms in parallel with an inductance of 0.0025 henry, t being in seconds. Find the total current flowing to the circuit as a single-termed function of t, assuming all currents are sinusoidal.

23. An a-c voltage is applied to a circuit consisting of a 4.0-ohm resistance in series with a parallel arrangement of a 0.0025-henry inductor and a 0.0020-farad capacitor. At t seconds, the current in the inductive branch is $6.0 \sin 377t$ amperes. Find as functions of t the (a) voltage across the parallel branch, (b) charge on the capacitor, (c) current flowing to the capacitor, (d) current in the resistor, (e) source voltage, and (f) power in the circuit.

24. A resistance of R ohms is in series with a parallel arrangement of an inductance of L henrys and a capacitance of C farads, and at t seconds the charge on the capacitor is $A \cos 2\pi ft$ coulombs. Assuming all currents and voltages as sinusoidal, find:

(a) An expression for the current flowing to the capacitor.

(b) An expression for the current in the inductor.

(c) The value of f in terms of L and C necessary to produce zero current in the resistor.

(d) The power of the circuit with this value of f.

25. An a-c voltage is applied to an induction coil, of resistance 0.40 ohm and inductance 0.0020 henry, in parallel with a capacitance of 0.0010 farad. At t seconds, the current in the coil is 6.0 cos 400t amperes. Find, as a single-termed function of t, the (a) total current flowing to this parallel circuit and (b) power in the circuit.

26. The plate power in a rectifier circuit is

$$P = \frac{1}{2\pi} \int_0^\pi \frac{E_m}{R_L} \sin \theta (E_m \sin \theta - E_p) \, d\theta.$$

Evaluate P.

27. Evaluate the plate power input of a modulator, given by

$$P = \int_0^\pi I_0 E_0 (1 + m \cos \theta)^2 \, d\theta.$$

28. A man wishes to get from one point on the edge of a circular pond of 2-mile diameter to a point directly opposite in the shortest possible time. If he can swim at 2 miles per hour and walk at 4 miles per hour, how should he proceed?

29. Find the right-angled triangle of minimum area that may be circumscribed about a rectangle of dimensions 8 by 12 in., with the hypotenuse of the triangle falling along a 12-in. side. (*Hint:* Let one of the acute angles of the triangle = θ and use θ as the independent variable.)

10.5 MEAN AND ROOT MEAN SQUARE VALUES

In Section 8.5, the significance of the mean and root mean square of a variable current or voltage was discussed along with the methods of calculating these values. We now apply these ideas and calculations to the trigonometric functions.

The mean value of a sine or cosine function over a complete cycle is zero, since the positive half of the cycle is identical in numerical value to the negative half. The mean value of the positive half of the cycle, however, has some significance. If a cycle of alternating current or voltage is rectified and filtered in order to approach a constant value, the constant value it approaches, neglecting any losses, will be equal to the mean value of the original positive half of the wave.

EXAMPLE 10.   Find the mean value of the first half-cycle of $i = \sin \theta$.

Solution: The mean value will be the area under the graph divided by the width in units of θ from the lower to the upper limit. Hence, mean i is

$$\frac{\int_0^\pi i \, d\theta}{\pi} = \frac{\int_0^\pi \sin \theta \, d\theta}{\pi} = \frac{-[\cos \theta]_0^\pi}{\pi} = \frac{-\cos \pi - \cos 0}{\pi}$$

$$= \frac{-[-1 - 1]}{\pi} = \frac{2}{\pi} = 0.637.$$

Since the integration only holds true if θ is measured in radians, the width of the half-cycle must be in radians; hence we divide by the numerical value of π (3.142).

EXAMPLE 11.   A voltage consists of a fundamental cosine function of frequency 60 cps and amplitude 4.0 plus a third harmonic cosine function of amplitude 1.0. Find the mean value of one-fourth cycle of this voltage. *Solution:* A third harmonic has three times the frequency of the fundamental. Thus, since $\theta = 2\pi ft$, the fundamental function is $4.0 \cos 120\pi t$; the third harmonic is $1.0 \cos 360\pi t$; and

$$v = 4.0 \cos 120\pi t + 1.0 \cos 360\pi t.$$

In this form, t is the independent variable, and the upper and lower limits must be expressed in terms of t. We understand one-fourth cycle to mean one-fourth cycle of the fundamental. One cycle has a period of $\frac{1}{60}$ sec, and one-fourth cycle is completed in $\frac{1}{4} \times \frac{1}{60}$ or $\frac{1}{240}$ sec. Thus mean v is

$$\frac{\int_0^{1/240} (4.0 \cos 120\pi t + 1.0 \cos 360\pi t)\, dt}{1/240}$$

$$= 240\left[\frac{4}{120\pi} \sin 120\pi t + \frac{1}{360\pi} \sin 360\pi t\right]_0^{1/240}$$

$$= 240\left(\frac{1}{30\pi} \sin \frac{\pi}{2} + \frac{1}{360\pi} \sin \frac{3\pi}{2}\right) = \frac{240}{\pi}\left(\frac{1}{30} - \frac{1}{360}\right)$$

$$= \frac{240}{\pi}\left(\frac{11}{360}\right) = \frac{22}{3\pi} = 2.34 \text{ volts.}$$

The details of this calculation are tricky, and the student must be on the alert for possible errors. If we treat θ as the independent variable rather than t, we have a somewhat easier solution. Substituting $120\pi t = \theta$, and realizing that one-fourth cycle of $\theta = \pi/2$ radians, mean v is

$$\frac{\int_0^{\pi/2} (4.0 \cos \theta + 1.0 \cos 3\theta)\, d\theta}{\pi/2}$$

$$= \frac{2}{\pi}\left[4.0 \sin \theta + \frac{1}{3} \sin 3\theta\right]_0^{\pi/2} = \frac{2}{\pi}\left(4.0 \sin \frac{\pi}{2} + \frac{1}{3} \sin \frac{3\pi}{2}\right)$$

$$= \frac{2}{\pi}\left(4 - \frac{1}{3}\right) = \frac{2}{\pi}\left(\frac{11}{3}\right) = \frac{22}{3\pi} = 2.34 \text{ volts.}$$

The fact that the fundamental frequency is 60 cycles does not affect the answer here, since the interval over which the mean is taken is expressed as a fraction of a cycle.

In most problems involving the calculation of mean or rms values of trigonometric functions, the independent variable may be chosen as either t or θ. If t is chosen, the differential complementing the integration symbol is dt, the values of the upper and lower limits must be numerical values of t, and the divisor must be the width of the interval in units of t. If θ is chosen, the differential is $d\theta$, and the limits and divisor must be numerical values of θ in radians. Sometimes the calculation is easier using θ; but often it is easier to use t, especially in a problem where all information is given in terms of t. Both methods produce the same answer, since the numerator represents an area of which one dimension is width and we divide by a dimension of width, resulting in an answer that has a single dimension of height; provided the unit of width is identical in numerator and denominator, it makes no difference whether it is expressed in units of θ or of t.

The calculation of rms values is more involved than that of a simple mean. We recall that the rms value of a variable is that value which, when squared, gives the mean value of the square of the variable.

EXAMPLE 12.   Find the rms current over a complete cycle of the current $i = \sin \theta$.

Solution: We first find the mean of the square of the function, using θ as the independent variable. Mean square current is

$$\frac{\int_0^{2\pi} i^2\, d\theta}{2\pi} = \frac{\int_0^{2\pi} \sin^2 \theta\, d\theta}{2\pi} = \frac{\frac{1}{2}\int_0^{2\pi} (1 - \cos 2\theta)\, d\theta}{2\pi}$$

$$= \frac{1}{4\pi}\left[\theta - \frac{\sin 2\theta}{2}\right]_0^{2\pi} = \frac{1}{4\pi}(2\pi - 0 - 0) = \frac{1}{2}.$$

$$\therefore \text{ rms } i = \sqrt{\tfrac{1}{2}} = 0.707.$$

Figure 10-7 shows the graph of $i^2 = \sin^2 \theta$. The mean square i is the average height of this graph.

It has become customary in dealing with alternating currents and a-c voltages to use their rms values (sometimes called their *effective values*) to designate the current or voltage. Thus an alternating current of 2.0 amperes means a sinusoidal current whose rms value is 2.0 amperes. Since its rms value is 0.707 times its amplitude, its amplitude is $2.0/0.707 = 2.82$ amperes and this alternating current is expressed as $i = 2.82 \sin (\theta + \phi)$. Alter-

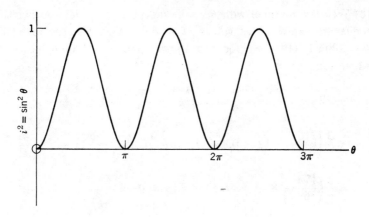

Fig. 10-7. Graph of i^2 against θ, where $i = \sin \theta$.

nating-current ammeters and voltmeters measure the rms values of a cycle of current and a cycle of voltage.

The designation of alternating currents and a-c voltages by their rms values enables us to adapt the basic electrical laws of d-c circuits for application to a-c circuits. In a d-c circuit, for instance, the constant power in a resistance of R ohms carrying a constant current of I amperes is I^2R watts. In an a-c circuit, the average power is I^2R watts, provided I^2 is the mean square current, which implies that I is the rms current.

EXAMPLE 13.   For the voltage $v = 2 + 3 \cos 100\pi t$ volts, find the rms voltage (a) over a half-cycle and (b) from $t = 1/100$ to $t = 4/300$.
Solution: (a) Replacing $100\pi t$ by θ, the upper limit for the half-cycle corresponds to $\theta = \pi$. Mean square value is

$$\frac{\int_0^{\pi} (2 + 3 \cos \theta)^2 \, d\theta}{\pi} = \frac{\int_0^{\pi} (4 + 12 \cos \theta + 9 \cos^2 \theta) \, d\theta}{\pi}$$

$$= \frac{\int_0^{\pi} [4 + 12 \cos \theta + \frac{9}{2}(1 + \cos 2\theta)] \, d\theta}{\pi}$$

$$= \frac{1}{\pi} \left[4\theta + 12 \sin \theta + \frac{9\theta}{2} + \frac{9}{4} \sin 2\theta \right]_0^{\pi}$$

$$= \frac{1}{\pi} \left(4\pi + 0 + \frac{9\pi}{2} + 0 - 0 \right) = \frac{1}{\pi} \left(\frac{17\pi}{2} \right) = \frac{17}{2} = 8.5.$$

Hence rms voltage is $\sqrt{8.5} = 2.92$ volts.

(b) We may continue with θ as the independent variable provided the limits are expressed as values of θ. $\theta = 100\pi t$. For the lower limit, $\theta = 100\pi(1/100) = \pi$; for the upper limit $\theta = 100\pi(4/300) = 4\pi/3$. Mean square value is

$$\left[4\theta + 12 \sin \theta + \frac{9\theta}{2} + \frac{9}{4} \sin 2\theta \right]_{\pi}^{4\pi/3} \div \left(\frac{4\pi}{3} - \pi \right)$$

$$= \frac{3}{\pi} \left[\left(\frac{16\pi}{3} - 12 \times 0.866 + 6\pi + \frac{9}{4} 0.866 \right) - \left(4\pi + 0 + \frac{9\pi}{2} + 0 \right) \right]$$

$$= \frac{3}{\pi} \left(\frac{17\pi}{6} - 8.44 \right) = 8.50 - 8.06 = 0.44.$$

Hence rms voltage is $\sqrt{0.44} = 0.66$ volt.
The reason for this low value may be seen if a graph of the voltage is sketched; the absolute values of v between $\theta = \pi$ and $\theta = 4\pi/3$ are all less than 1, except at the lower limit, where $|v| = |-1| = 1$.

EXAMPLE 14.   A periodic current with $i = 1 + \sin 120\pi t$ amperes for the first half-cycle and $i = 1$ ampere for the second half-cycle flows through a 2.0-ohm resistor. Find (a) the total energy consumed over one cycle and (b) the average power over one cycle.
Solution: The current function changes at the half-cycle point, a type of discontinuity. The graph is shown in Fig. 10-8.
(a)   Energy $= \int_a^b P \, dt$ and $P = i^2 R$. Thus energy $= R \int_a^b i^2 \, dt$. The integral $\int_a^b i^2 \, dt$ represents the area under the graph of $i^2 = f(t)$ for the interval of time from a to b. Here, the total area consists of two separate

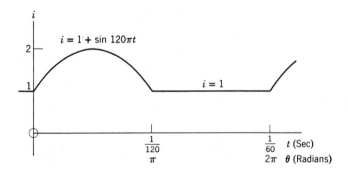

Fig. 10-8. A cycle of current with two separate wave forms.

portions and must be calculated by summing two separate integrals. Energy is then equal to

$$2.0\left[\int_0^{1/120}(1+\sin 120\pi t)^2\,dt + \int_{1/120}^{1/60}1^2\,dt\right]$$

$$= 2.0\left\{\int_0^{1/120}(1+2\sin 120\pi t+\sin^2 120\pi t)\,dt + [t]_{1/120}^{1/60}\right\}$$

$$= 2.0\left\{\int_0^{1/120}\left[1+2\sin 120\pi t+\frac{1}{2}(1-\cos 240\pi t)\right]dt + \frac{1}{120}\right\}$$

$$= 2.0\left\{\left[\frac{3}{2}t - \frac{2\cos 120\pi t}{120\pi} - \frac{\sin 240\pi t}{480\pi}\right]_0^{1/120} + \frac{1}{120}\right\}$$

$$= 2.0\left[\frac{3}{240} - \frac{2(-1)}{120\pi} - 0 + \frac{2(1)}{120\pi} + \frac{1}{120}\right]$$

$$= 2.0\left\{\frac{5}{240} + \frac{1}{30\pi}\right\} = 0.063 \text{ watt-sec or joule.}$$

The energy consumed for any number of complete cycles is the number of cycles times 0.063 joule.

(b) Average power equals $\dfrac{\text{energy per cycle}}{\text{period } T} = \dfrac{0.063}{\frac{1}{60}} = 3.78$ watts.

The calculation of average power does not depend on the previous calculation of energy. Average power for one cycle may be calculated as

$$R\frac{\int_a^b i^2\,dt}{b-a},$$

or often more simply as

$$R\frac{\int_0^{2\pi} i^2\,d\theta}{2\pi}$$

since average power, unlike energy, does not depend directly on time. This calculation should give 3.78 watts, as did the first one.

EXERCISES

1. For what type of current is the mean value equal to the rms value?

2. Find the mean value of $i = A\cos\omega t$ over (a) the first cycle, (b) the first $\frac{1}{2}$ cycle, (c) the first $\frac{1}{4}$ cycle, and (d) the first $\frac{1}{6}$ cycle.

3. Find the mean value of $v = 2.0 \sin (50\pi t + \pi/3)$, (a) over the first $\frac{1}{2}$ cycle, beginning at $t = 0$; (b) over the first $\frac{1}{4}$ cycle, beginning at $t = 0$; (c) from $t = -1/150$ to $t = 1/300$, and (d) from $t = 0$ to $t = 1/100$.

4. Find the mean value of $i = 2.0 \sin \theta + 2.0 \cos \theta$ over the first $\frac{1}{2}$ cycle, (a) by using the given two-termed expression, and (b) by expressing i as a single sine function with a phase displacement.

5. Find, neglecting losses, the amplitude of the direct current resulting from the rectification of the positive portion of the cycle of $i = \cos (\theta - \pi/3)$. (The rectified current is the mean value of the unrectified current.)

6. A current $i = A \sin \theta$ is rectified over the positive half of its cycle. Find (a) the rms value of the original current over its positive half-cycle, (b) the rms value of the rectified half-cycle, and (c) the percent power loss in the rectification. Neglect other losses.

7. Find the rms value of $v = 0.6 \sin 100\pi t$, (a) over the first $\frac{1}{4}$ cycle, (b) over the first $\frac{1}{12}$ cycle, and (c) from $t = 0.005$ to $t = 0.008$.

8. Find the rms i over a cycle of 4 sec if $i = 3 + t^2/2$ amperes for the first 2 sec and $i = 3$ amperes for the next 2 sec. What energy is used for one cycle of this current if it flows through a resistance of 2.0 ohms?

9. A current is given by $i = 5 - 2 \cos \theta$ amperes. Find (a) the rms current over a cycle and (b) the average power of the current through a 1.8-ohm resistor.

10. Find the rms voltage v for a cycle of the following voltage: $v = 3 \cos \theta$ from $\theta = 0$ to $\theta = \pi/2$, $v = 0$ from $\theta = \pi/2$ to $\theta = 3\pi/2$, and $v = 3 \cos \theta$ from $\theta = 3\pi/2$ to $\theta = 2\pi$.

11. Find the rms voltage v for a cycle of the following voltage: $v = 4 + 2 \sin \theta$ from $\theta = 0$ to $\theta = \pi$ and $v = 4$ from $\theta = \pi$ to $\theta = 2\pi$.

12. Find the rms i over one cycle of $i = 3.0 \cos (200\pi t - \pi/3)$.

13. Find the rms i over one cycle of $i = 3.0 \sin 100t + 5.0 \cos 100t$.

14. Find the rms value of $i = 8 \sin \theta + 2 \sin 3\theta$ from $\theta = 0$ to $\theta = 2\pi$. (*Note:* For $\int \sin \theta \sin 3\theta \, d\theta$, change the product to a difference by a trigonometric formula).

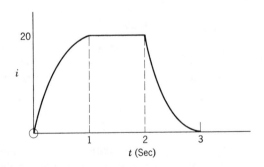

Fig. 10-9. One cycle of an electric arc.

15. The rms secondary voltage of a transformer is given by $E_s = 4Kf\phi_m n \times$ 10^{-8} volt, where K is the ratio rms voltage/mean voltage, called the *form factor*. If $f = 60$ cps, $\phi_m = 2 \times 10^6$ lines, and $n = 1000$ turns, calculate E_s for the following wave forms: (a) a sine wave, (b) a sawtooth wave where v is proportional to θ with a period of 2π, and (c) $v = A\cos\theta + 0.2A\cos 2\theta$. The mean of a sinusoidal wave is taken over the positive half-cycle only.

16. An electric arc is operated in a cycle illustrated in Fig. 10-9; $i = 20\sqrt{t}$ amperes between $t = 0$ and $t = 1$ sec., $i = 20$ amperes between $t = 1$ sec and $t = 2$ sec, and $i = 20(t-3)^2$ between $t = 2$ sec and $t = 3$ sec. If the resistance in the circuit is 4.5 ohms and the cost is 1.8 cents per kilowatt-hour, what does this operation cost if it is continued for 1 hr?

17. An a-c voltage is applied to a circuit consisting of a resistance of 1.2 ohms in series with a capacitor. The capacitor charge is $0.008\sin(377t - \pi/6)$ coulombs at t seconds. Find (a) the average power in the circuit and (b) the energy consumed during 1 min of operation.

18. An a-c voltage is applied to a circuit consisting of a resistance of 1.4 ohm in series with an inductance of 0.0020 henry. The voltage across the inductance at t seconds is $2.5\cos(100\pi t - \pi/2)$ volts, and the current is sinusoidal. Find (a) the average power in the circuit, and (b) the energy per cycle.

CHAPTER 11

Calculus of Exponential and Logarithmic Functions

11.1 DERIVATIVE OF AN EXPONENTIAL FUNCTION

The exponential function was examined in Chapter 9, and a thorough knowledge of this basic theory as well as the theory of logarithms is essential before we can proceed to the developments of this chapter. We are concerned now with the calculus of the exponential and logarithmic functions.

An exponential function has the same structure as a power function; in both functions, a base is raised to a power. Their distinguishing feature lies in the positions of variable and constant in this expression. In a power function, a variable is raised to a constant power; in an exponential function, a constant is raised to a variable power. The graph of a power function may appear on casual analysis to be somewhat similar to the graph of an exponential function, but they are basically different. The rules for differentiating and integrating them are also quite different.

Although any constant positive base other than 1 may be used in an exponential function, the base e lends itself to easiest manipulation in calculus. If the base does not happen to be e, it is fairly easy to convert the function to one using base e, and some of the examples in Chapter 9 indicate how this is done. We now proceed to find a rule for differentiating an exponential function whose base is e.

The symbol e is defined as $\lim_{n \to \infty} (1 + 1/n)^n$. Thus the exponential function

202

e^x means $\lim\limits_{n \to \infty} (1 + 1/n)^{nx}$. This may be expanded by the binomial theorem, from which

$$e^x = \lim_{n \to \infty} \left[1 + nx \frac{1}{n} + \frac{nx(nx - 1)}{2!} \frac{1}{n^2} + \frac{nx(nx - 1)(nx - 2)}{3!} \frac{1}{n^3} \cdots \right]$$

$$= \lim_{n \to \infty} \left[1 + x + \frac{1}{2!} (x)\left(x - \frac{1}{n}\right) + \frac{1}{3!} (x)\left(x - \frac{1}{n}\right)\left(x - \frac{2}{n}\right) \cdots \right].$$

n now only occurs as the denominator of fractions whose numerator is a finite number. When $n \to \infty$, all such fractions approach zero. Hence

$$e^x = 1 + x + \frac{x^2}{2!} + \frac{x^3}{3!} + \frac{x^4}{4!} \cdots \quad \text{to an infinite number of terms.}$$

This expression for e^x, which represents it as an infinite series of terms, each of which is a power function, holds true for any value of x. It may be used to find the value of any single exponential to any degree of accuracy required. However, the greater the numerical value of x, the greater the number of terms that must be used for the desired accuracy. If we wish to find $e^{0.5}$, for example, we substitute $x = 0.5$ in the series and find that the terms get progressively smaller quite rapidly so that they become negligible after the first few; this would not apply if we wish to find $e^{3.5}$.

The important consideration now is that, since this series is a correct representation of e^x for all values of x, we may differentiate e^x by differentiating a series of power functions. Hence,

$$e^x = 1 + x + \frac{x^2}{2!} + \frac{x^3}{3!} + \frac{x^4}{4!} \cdots \quad \text{to an infinite number of terms.}$$

Therefore

$$\frac{d}{dx} e^x = 0 + 1 + \frac{2x}{2!} + \frac{3x^2}{3!} + \frac{4x^3}{4!} \cdots \quad \text{to an infinite number of terms}$$

$$= 1 + x + \frac{x^2}{2!} + \frac{x^3}{3!} \cdots \quad \text{to an infinite number of terms}$$

$$= e^x.$$

We have here a function whose derivative with respect to x is equal to the original function. Graphically, this means that the slope of the graph of $y = e^x$ at any point is equal to the vertical coordinate of the point. We also see that

$$\frac{d}{dx} e^u = e^u \frac{du}{dx}.$$

The simplicity of this formula is the main reason for using e as a base. With any other base, an additional factor must be used in the derivative.

EXAMPLE 1. Find $(d/dt)(e^{2t} - e^{-5t})$.
Solution: Differentiating each term by following the formula and realizing that a chain rule applies in each term,

$$\frac{d}{dt}(e^{2t} - e^{-5t}) = e^{2t}(2) - e^{-5t}(-5) = 2e^{2t} + 5e^{-5t}.$$

EXAMPLE 2. Find $(d/dx)e^{-3x^2+4}$.
Solution: Here the principal structure of the function is exponential, with a substructure consisting of a power function.

$$\frac{d}{dx}e^{-3x^2+4} = e^{-3x^2+4}(-6x) = -6xe^{-3x^2+4}.$$

EXAMPLE 3. Find the maximum current i for (a) $i = 4.0te^{-200t}$ and (b) $i = 3.2e^{-80t} \sin 100t$.
Solution: For a maximum i, in either case, $di/dt = 0$ and d^2i/dt^2 is negative. Both (a) and (b) represent i as a product of two variables

(a) $\dfrac{di}{dt} = 4.0[t(e^{-200t})(-200) + e^{-200t}(1)] = 4.0e^{-200t}(1 - 200t)$

$= 0$ for maximum i.

Hence $e^{-200t} = 0$ or $1 - 200t = 0$; $t = \infty$ or $t = 1/200$.

$$\frac{d^2i}{dt^2} = 4.0[e^{-200t}(-200) + e^{-200t}(-200)(1 - 200t)]$$

$$= 800e^{-200t}(-1 - 1 + 200t) = 1600e^{-200t}(100t - 1).$$

At $t = 1/200$, $d^2i/dt^2 = 1600e^{-1}(\frac{1}{2} - 1)$, which is negative, since e^{-1} is positive and the factor in parentheses is negative. Therefore at $t = 1/200$, i has a maximum value of

$$4.0\left(\frac{1}{200}\right)e^{-1} = 0.02 \times 0.368 = 0.0074.$$

What happens at $t = \infty$ is best understood by drawing a graph.

(b) $\dfrac{di}{dt} = 3.2[e^{-80t} \cos 100t(100) + \sin 100te^{-80t}(-80)]$

$= 64e^{-80t}(5 \cos 100t - 4 \sin 100t) = 0$ for a maximum i.
$\therefore e^{-80t} = 0$ or $5 \cos 100t - 4 \sin 100t = 0$.

$t = \infty$ or $4 \sin 100t = 5 \cos 100t$.

Dividing by $\cos 100t$, $4 \tan 100t = 5$, $\tan 100t = 1.25$, and $100t = 51.3°$, which must be expressed in radians as 0.895 rad. There are other values of $100t$ as well, since arc tan 1.25 may be $0.895 + \pi$, $0.895 + 2\pi$, $0.895 + 3\pi$, etc., but for the moment, we will consider only 0.895. Hence $t = 0.00895$.

$$\frac{d^2i}{dt^2} = 64[e^{-80t}(-500 \sin 100t - 400 \cos 100t)$$

$$- 80e^{-80t}(5 \cos 100t - 4 \sin 100t)]$$

$$= 64e^{-80t}(-180 \sin 100t - 800 \cos 100t).$$

Since e^{-80t} is positive for all finite values of t, the sign of this expression depends on the sign of the function in parentheses; at $t = 0.00895$, this is negative. Thus $t = 0.00895$ gives a maximum i of $3.2e^{-0.72} \sin 51.3° = 3.2 \times 0.49 \times 0.78 = 1.22$. This is the first maximum. For the other values of t resulting from other values of $100t$ noted above, i is alternately minimum and maximum.

EXERCISES

By using the series for e^x, evaluate the following exponentials to three-figure accuracy:

1. $e^{0.4}$.

2. $e^{-0.4}$.

3. $e^{-0.6}$.

Find the following derivatives:

4. $\dfrac{d}{dx} e^{x+1}$.

5. $\dfrac{d}{dx} e^{3x}$.

6. $\dfrac{d}{dx} Ae^{-kx}$.

7. $\dfrac{d}{dx} 2e^{-x/3}$.

8. $\dfrac{d}{dt} Ae^{-kt+3}$.

9. $\dfrac{d}{dt} (Ae^{-kt} + 3)$.

10. $\dfrac{d}{dx} 2e^{3x^2}$.

11. $\dfrac{d}{dx} \dfrac{2e^{3x^2}}{e^{x+5}}$.

Find the following second derivatives:

12. $\dfrac{d^2}{dx^2} \dfrac{e^x - e^{-x}}{2}$.

13. $\dfrac{d^2}{dt^2} \dfrac{e^t + e^{-t}}{2}$.

14. $\dfrac{d^2}{dt^2} \dfrac{e^{10t} + e^{-10t}}{2}$.

Evaluate the following:

15. $\dfrac{d}{dx} e^{x-1}$ at $x = 1$.

16. $\dfrac{d}{dt} e^{-2t+1}$ at $t = 0.5$.

17. $\dfrac{d}{dx} e^{3x^2-9}$ at $x = 2$.

18. $\dfrac{d}{dt} 4.0(1 - e^{-12t})$ at $t = 0.20$.

Calculate the following derivatives:

19. $\dfrac{dy}{dx}$ if $e^{x+y} = 2x$.

20. $\dfrac{d}{dx} 3xe^{-x}$.

21. $\dfrac{d}{dt}(At + B)e^{-5t}$.

22. $\dfrac{d}{du} \dfrac{e^{3u} - e^{u}}{e^{2u}}$.

23. $\dfrac{d}{dy} \dfrac{e^{-2y} - e^{-5y}}{y}$.

24. $\dfrac{d}{dt} 3e^{-10t} \sin 20t$.

25. $\dfrac{dy}{dx}$ if $e^{-3x} \sin 5y = xy$.

26. $\dfrac{d}{dx} x^e e^x$.

27. $\dfrac{d}{dx} \sqrt{e^{-5x} + \cos 3x}$.

28. $\dfrac{d}{dz} 6e^{\tan 4z}$.

29. $\dfrac{d}{dx} e^{-5x} \cos 2x \operatorname{cosec} 2x$.

30. $\dfrac{d}{dx} e^{-5x} e^{3x^2} - 2e^{2x} e^{7 - x^2}$.

31. $\dfrac{d}{dx} \dfrac{e^{-3x} \sin 3x}{x^2}$.

32. A transmission line hangs in the form of a catenary whose equation is $y = 50(e^{0.01x} + e^{-0.01x})$. An approximation to this equation is the parabola $y = 100 + 0.005x^2$. From each of these equations:

(a) Find to four-figure accuracy the value of y at $x = 40$.

(b) Calculate to the nearest minute the angle that the line makes with the horizontal at $x = 40$.

33. The general exponential function is given by $y = Ae^{mx}$, where A and m are constant. Find the values of A and m if

(a) At $x = 0$, $y = 4$ and $dy/dx = 10$.

(b) At $x = 0$, $y = 10$ and $dy/dx = -50$.

(c) At $x = 0$, $y = 6$ and at $x = 2$, $y = 3$.

34. If i is an exponential function of t, and, at $t = 0$, $i = E/R$ and $di/dt = -E/L$, find i as a function of t.

35. The number of revolutions N after an electric motor is shut down is given by $N = 100(1 - e^{-0.05t})$, where t is measured in seconds after shutdown. Find

(a) The number of revolutions between $t = 70$ and $t = 80$.

(b) The machine velocity in revolutions per second as a function of t.

(c) The machine retardation in revolutions per second per second as a function of t.

(d) The initial velocity in revolutions per second.

(e) The total number of revolutions until it comes to rest.

36. A current i is given by $i = A(1 - e^{-mt})$, where A and m are constants. If, at $t = 0$, $di/dt = E/L$, and, at $t = \infty$, $i = E/R$, (a) find i as an exact function of t.

(b) If $E = 50$, $R = 12$, and $L = 0.25$, find i at $t = 0.020$.

(c) Sketch the graph of i against t

Find the first maximum current i and the time t at which it occurs for the following:

37. $i = 2.5e^{-50t} - 2.5e^{-100t}$.   **38.** $i = 2.0te^{-100t}$.

39. $i = 2.5e^{-200t} \sin 200t$.   **40.** $i = 1.4e^{-100t} \cos 50t$.

41. Given $v = E(1 - e^{-t/RC})$, find the value of $RC(dv/dt) + v$ in simplest form.

42. A current i is given by $i = 10(1 - e^{-40t})$. Prove that for this current the differential equation $0.005(di/dt) + 0.2i = 2$ holds true.

43. If $i = Ae^{-20t} + Be^{-50t}$, prove the validity of the differential equation $0.010(d^2i/dt^2) + 0.70(di/dt) + 10i = 0$.

44. If $i = 1.6e^{-250t} \sin 250t$, prove the validity of the differential equation $0.50(d^2i/dt^2) + 250(di/dt) + 10^6 i/16 = 0$.

45. A voltage is impressed on a circuit consisting of a 3.0-ohm resistor, a 0.0125-henry inductor, and a 0.0100-farad capacitor in series. At time t seconds, the charge q on the capacitor is $0.060 - 0.075e^{-40t} + 0.015e^{-200t}$ coulombs. Find: (a) the voltage across the capacitor as a function of t, (b) the voltage across the resistor as a function of t, (c) the voltage across the inductor as a function of t, and (d) the applied voltage.

11.2 NATURAL GROWTH AND DECAY

The relationship of exponential functions to natural (or continuous) growth and decay was studied in Section 9.3, in which it was indicated that the general growth function is $y = Ae^{mx}$ and the general decay function is $y = Ae^{-mx}$, where A and m are any positive constants. The constant A is the value of y at $x = 0$, and the constant m affects the rapidity of growth or decay.

The form of the derivative of exponential functions enables us to analyze natural growth and decay from a new standpoint. Examine the following exponential functions and their derivatives:

$y = Ae^x$	$y = Ae^{-x}$	$y = Ae^{mx}$	$y = Ae^{-mx}$
$y' = Ae^x$	$y' = -Ae^{-x}$	$y' = mAe^{mx}$	$y' = -mAe^{-mx}$
$= y$	$= -y$	$= my$	$= -my$
$y'' = Ae^x = y$	$y'' = Ae^{-x} = y$	$y'' = m^2Ae^{mx} = m^2y$	$y'' = m^2Ae^{-mx} = m^2y$

From the tabulation, we see that the first derivative of an exponential function is always directly proportional to the original function. For positive exponentials (growth functions), the proportionality factor is positive, and for negative exponentials (decay functions), the proportionality factor is negative. This proportionality factor is the value of the coefficient of x in the exponential.

The natural growth function may now be interpreted as a function in which, at any point, the rate of growth is positively proportional to the value of the function. Graphically, at any point, the slope is proportional to the height of the graph. When the height of the graph is y_1, the slope is my_1.

In the same way, the natural decay function is one in which the rate of decay is proportional to the function; that is, the rate of growth is negatively proportional to the value of the function. Graphically, at any point, the slope is negatively proportional to the value of the function. When the height of the graph is y_1, the slope is $-my_1$. Many cases of decaying current or voltage follow this law.

If we examine the graphs of $y = Ae^{mx}$ and $y = Ae^{-mx}$ in Fig. 9-9 from this viewpoint, we get a better understanding of the exponential functions. Both these functions are always positive (above the x-axis). In the graph of the growth function $y = Ae^{mx}$, for negative values of x, when y has a very small positive value, the slope of the curve also has a very small positive value. The closer the curve approaches the x-axis from right to left the smaller is its slope, so that the rate at which it approaches the x-axis is lessened. Hence the curve never reaches the x-axis. Thus the x-axis is an *asymptote* of the curve. Similarly in the right-hand portion of the decay curve $y = Ae^{-mx}$, as y becomes smaller, so does the numerical value of the negative slope. The smaller the value of the function, the less rapidly it decays. This principle governs all natural decay, whether it be cell structure or a continuously decaying electrical quantity such as current. The ultimate result of a decay is that there is nothing left but this occurs only at infinite values of the independent variable.

EXAMPLE 4.   Find the rate of change of i with respect to t for the function $i = 4e^{-20t}$ under the following conditions: (a) when $t = 0.025$, (b) when $i = 1.5$, and (c) when $i = 0$.
Solution: (a) When $t = 0.025$, $i = 4e^{-0.50} = 4 \times 0.606 = 2.43$.

$$\frac{di}{dt} = -80e^{-20t} = -20i.$$

Hence the rate of change of i with respect to t is $-20 \times 2.43 = -49$ ampere per second.
(b) $di/dt = -20 \times 1.5 = -30$ amperes per second.
(c) $i = 0$ at $t = \infty$ and di/dt also is zero at this time.

EXAMPLE 5.   Find v as a function of t for the following differential equations: (a) $dv/dt = -400v$; (b) $d^2v/dt^2 = 40{,}000v$.

Solution: (a) Since dv/dt is negatively proportional to v, v must be a decay function of t, and $v = Ae^{-400t}$. If further information about v were given, we could find a specific value of A to fit the conditions. For example, if, at $t = 0$, $v = 2.0$, then $A = 2.0$, and v would $= 2.0e^{-400t}$.

(b) When a second derivative of a function is positively proportional to the function, the function may be a growth function; but it may also be a decay function, since, for $y = Ae^{-mx}$, $d^2y/dx^2 = +m^2Ae^{-mx} = +m^2y$. In our example, v may equal Ae^{200t} or Be^{-200t}, the constant multiplier not necessarily being the same in the two functions. For the most general solution, v may contain both types of terms, and $v = Ae^{200t} + Be^{-200t}$. This may be verified by establishing that $d^2v/dt^2 = 40{,}000v$.

11.3   TRANSIENTS IN *RL* AND *RC* CIRCUITS

When the closing or opening of a switch changes the design of an electrical circuit, the currents, voltages, and charges within the circuit will usually change.

These changes in value are usually not instantaneous; instead, during a brief period of time, there is a continuous and smooth change in value from the sustained condition before switching to the final sustained condition after switching. The presence of inductance in a circuit opposes any instantaneous change in current in that circuit. The presence of capacitance in a circuit opposes any instantaneous change in voltage across the capacitance. The changing conditions in this interval of time between sustained conditions before and after switching are referred to as *transient conditions*; the changing current is called a *transient current*, the changing voltages are called *transient voltages*, etc.

In the circuit of Fig. 11-1*a*, the outside path constitutes a series *RL* circuit. With an open switch, the electrical quantities are all zero (zero current and zero voltage across all elements). After the switch has been closed for some time, a constant current $i = E/R$ is established through the *RL* circuit. However, the current cannot instantaneously change from 0 to E/R when the switch is closed. The transient current changes smoothly from one value to the other, and this change involves an exponential function. At all times t

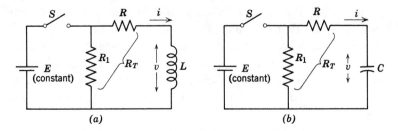

Fig. 11-1. Transients in RL and RC circuits.

after the switch is closed, the sum of the voltage drops must equal E. That is, $Ri + L(di/dt) = E$. Here we have an example of a differential equation from which i may be found as a function of t by methods to be discussed in Chapter 16. We are not yet at the stage where this equation can be solved directly.

The extra resistor paralleling the voltage E is necessary to provide a closed path for discharge when the switch S is open.

Appendix I indicates how the application of electrical laws produces a differential equation for each of the electrical quantities involved in these two circuits for both the charging and the discharging conditions. All the numbered differential equations have a term in one of the electrical quantities and another term in its derivative with respect to time. In several cases where the voltage E is constant, the differential equation indicates that the derivative is negatively proportional to the quantity itself, and we are now able to find the quantity as a function of t in these instances. For each situation, electrical laws impose certain conditions at the instant the switching is changed, the instant chosen as $t = 0$ for convenience. Mathematically these restrictions provide further information from which the quantity may be expressed as an exact function of t, without any indefinite constants.

The following differential equations and initial conditions may be determined from Appendix I, where E is constant. These are all of the problem types that may be solved by methods analyzed in Section 11.2.

For Fig. 11-1a: Constant voltage E

Switch closing. For voltage v across inductance L,

$$\frac{dv}{dt} = -\frac{R}{L}v.$$

At $t = 0$, $v = E$.

Switch opening. For current i,

$$L\frac{di}{dt} + R_T i = 0,$$

At $t = 0$, $i = E/R$.

For voltage v across inductance L,

$$\frac{dv}{dt} = -\frac{R_T}{L} v.$$

At $t = 0$, $v = -(E/R)R_T$.

For Fig. 11-1b: Constant voltage E

Switch closing with discharged capacitor. For current i,

$$R\frac{di}{dt} + \frac{i}{C} = 0.$$

At $t = 0$, $i = E/R$.

Switch opening for discharge of a fully-charged capacitor. For current i,

$$R_T\frac{di}{dt} + \frac{i}{C} = 0.$$

At $t = 0$, $i = -E/R_T$.

For charge q on capacitor,

$$R_T\frac{dq}{dt} + \frac{q}{C} = 0.$$

At $t = 0$, $q = CE$.

For voltage v across capacitor,

$$R_T C\frac{dv}{dt} + v = 0.$$

At $t = 0$, $v = E$.

EXAMPLE 6.   On closing the switch in the circuit of Fig. 11-1a, (a) find v as a function of t, assuming $t = 0$ at the instant the switch closes; (b) find the voltage across the resistance R as a function of t; and (c) hence find the current i in the outside RL circuit in terms of t.

Solution: (a) $dv/dt = -(R/L)v$, by applying the laws previously given. Since dv/dt is negatively proportional to v itself, v must be a negative exponential function of t, and $v = Ae^{-Rt/L}$.

Referring to the foregoing analysis of the RL circuit on closing the switch, at $t = 0$, $v = E$. Substituting these values for the two variables,

$$E = Ae^0 = A \quad \text{and} \quad A = E.$$
$$\therefore\ v = Ee^{-Rt/L}, \quad \text{a decay function.}$$

(b) Voltage across $R = E - v = E - Ee^{-Rt/L} = E(1 - e^{-Rt/L})$.

(c) $$i = \frac{\text{voltage across } R}{R} = \frac{E(1 - e^{-Rt/L})}{R}.$$

EXAMPLE 7.   (a) Verify that $i = (E/R)(1 - e^{-Rt/L})$ is a correct solution of the differential equation $Ri + L(di/dt) = E$, where R, L, and E are constants.

(b) Verify that $i = E/R + Ae^{-Rt/L}$ is also a correct solution of this differential equation, independent of the value of A.

(c) Show that $i = (E/R)(1 - e^{-Rt/L})$ satisfies the electrical condition at $t = 0$ of the RL circuit in Fig. 11-1a.

(d) Find the value of i at $t = \infty$ for this function.

(e) Sketch the graph of this function.

Solution: (a) If di/dt is calculated, substitutions may be made for i and di/dt in the left-hand side of the differential equation.

$$\frac{di}{dt} = \frac{E}{R}\left(\frac{R}{L}e^{-Rt/L}\right) = \frac{E}{L}e^{-Rt/L}.$$

$$\therefore \; Ri + L\frac{di}{dt} = \frac{RE}{R}(1 - e^{-Rt/L}) + \frac{LER}{RL}e^{-Rt/L}$$

$$= E(1 - e^{-Rt/L} + e^{-Rt/L}) = E,$$

which verifies the differential equation.

(b) $di/dt = -(AR/L)e^{-Rt/L}$ and

$$Ri + L\frac{di}{dt} = \frac{ER}{R} + RAe^{-Rt/L} - \frac{LAR}{L}e^{-Rt/L} = E,$$

which also verifies the differential equation.

(c) For $i = (E/R)(1 - e^{-Rt/L})$, at $t = 0$, $i = (E/R)(1 - 1) = 0$. The current should be 0 at $t = 0$; therefore the electrical condition is satisfied. For the equation $i = E/R + Ae^{-Rt/L}$, at $t = 0$, $i = E/R + A$. Thus $i = 0$ only if $A = -E/R$, which results in the equations (a) and (c).

(d) At $t = \infty$, $i = (E/R)(1 - e^{-\infty}) = (E/R)(1 - 0) = E/R$. This is the sustained current after the transient conditions are over. Theoretically this does not occur in any finite time, but usually after a short fraction of a second the transient conditions are negligible. $e^{-5}(=0.0067)$ is usually taken as 0 for practical purposes.

(e) The graph may be sketched in steps by sketching $i = (E/R)e^{-Rt/L}$, a decay curve, then $i = -(E/R)e^{-Rt/L}$, its negative, and finally $i = E/R - (E/R)e^{-Rt/L}$ by translating the entire second curve up E/R units. This is demonstrated in Fig. 11-2.

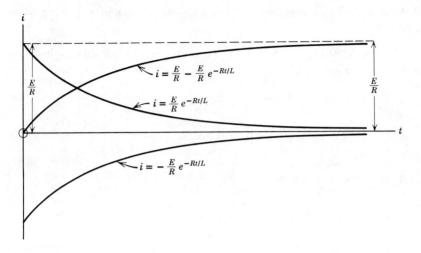

Fig. 11-2. Graph of $i = \dfrac{E}{R}(1 - e^{-Rt/L})$ by steps.

EXERCISES

1. If the slope of an exponential graph is -3 when its height is 2, what is the slope when the height is 0.5?

2. Without finding any values of x, find the value of dy/dx at the specified values of y for the following functions:

(a) $y = e^{-20x}$ at $y = 0.5$.   (b) $y = Ae^{200x}$ at $y = 0.02$.

(c) $y = 3e^{-50x}$ at $y = 0.03$.   (d) $y = 10e^{-mx}$ at $y = h$.

3. Using an indefinite constant A in the answer, express the dependent variable as a function of the independent variable for the following differential equations:

(a) $\dfrac{di}{dt} = -40i$.

(b) $R\dfrac{di}{dt} = -\dfrac{i}{C}$.

(c) $RC\dfrac{dv}{dt} + v = 0$.

(d) $a\dfrac{dy}{dx} + by = c\dfrac{dy}{dx} + dy$.

(e) $\dfrac{dy}{dx} = -40x$.

4. Using indefinite constants A and B and two functions in the answer, solve the following differential equations:

(a) $\dfrac{d^2y}{dx^2} = 225y$.

(b) $\dfrac{d^2i}{dt^2} = \dfrac{R^2}{L^2}i$.

(c) $R^2\dfrac{d^2q}{dt^2} - \dfrac{q}{C^2} = 0$.

(d) $\dfrac{d^2v}{dt^2} = -\omega^2 v$.

5. If $d^4y/dx^4 = y$, express y as a single-termed function of x in four possible ways.

Solve the following differential equations exactly (that is, with no indefinite constant in the function):

6. $\dfrac{dy}{dx} = 12y$ and, at $x = 0$, $y = 2.5$.

7. $\dfrac{di}{dt} = -80i$ and, at $t = 0.010$, $i = 2.0$.

8. $R\dfrac{di}{dt} + \dfrac{i}{C} = 0$ and, at $t = 0$, $i = -I$.

9. $L\dfrac{di}{dt} + i(R + R_1) = 0$ and, at $t = 0$, $i = \dfrac{E}{R}$.

10. For the graph of $y = 5.0e^{-10x}$, find: (a) the slope where y is $\frac{1}{2}$ its value at $x = 0$ and (b) the value of y when the slope is -20.

11. A current i is decaying exponentially. At $t = 0$, the current is 2.5 amperes and its rate of change is -125 amperes per second. Find i as an exact function of t.

In each of Exercises 12–16, use the appropriate differential equation and initial conditions listed in Section 11.3 as a means of calculating the answer.

12. For the RC circuit of Fig. 11-1b on closing the switch S, find the current i as a function of t, using the constants E, R, and C.

13. For the RL circuit of Fig. 11-1a on opening the switch S, find the current i as a function of t, using the constants E, R, R_T, and L.

14. For the RC circuit of Fig. 11-1b on opening the switch S, find the current i as a function of t, using constants E, R_T, and C.

15. For the RC circuit of Fig. 11-1b on opening the switch S, find the charge q on the capacitor as a function of t, using constants E, R_T, and C.

16. For the RL circuit of Fig. 11-1a on circuit closing, if $E = 3.0$, $R = 2.0$, and $L = 4.0 \times 10^{-3}$, find v, the voltage drop across the inductance L, as a function of t.

17. Newton's Law of Cooling states that when a hot body is surrounded by cooler air, its rate of cooling is directly proportional to the temperature difference between the body and the surrounding air. Interpreting this statement as an equation by using T to represent the temperature difference between the body and the air and t to represent time, we see that $dT/dt = -kT$, the negative sign being necessary to indicate a cooling or drop in temperature. If a machine at a temperature of 400°F stands in a room at 75°F, and after 20 min it has cooled to 200°F, assuming room temperature remains constant, find: (a) T as an exact function of t in minutes, (b) machine temperature after 40 min of cooling, and (c) the total time required for the machine to cool to 80°F.

18. In an *RL* series circuit with an a-c voltage applied, the differential equation for current *i* is $L(di/dt) + Ri = E_m \sin \omega t$.

(a) Prove that the solution

$$i = Ae^{-Rt/L} + \frac{E_m}{R^2 + \omega^2 L^2} (R \sin \omega t - \omega L \cos \omega t)$$

satisfies the differential equation.

(b) Solve for *A* in terms of E_m, *R*, *L*, and ω in this solution if, at $t = 0$, $i = 0$.

(c) What part of this current is transient, and what is the expression for *i* after the transient current disappears?

19. For the *RC* circuit of Fig. 11-1*b*, on closing the switch when time $t = 0$, make use of the current *i* as a function of *t* which has been established as the answer to Exercise 12 to find (a) the voltage drop across the resistance *R* as a function of *t*, (b) the voltage drop across the capacitance *C* as a function of *t*, (c) the charge *q* on the capacitor as a function of *t*, and (d) the charge *q* on the capacitor at $t = 2RC$. Sketch the graph of *q* against *t*.

20. The differential equation involving the charge *q* and time *t* for the *RC* circuit of Fig. 11-1*b* on closing the switch is $R(dq/dt) + q/C = E$, and, at $t = 0$, $q = 0$.

(a) Verify that $q = CE - Ae^{-t/RC}$ satisfies the differential equation.

(b) Find the value of *A* in equation (a) that satisfies the initial condition at $t = 0$, and hence find *q* as an exact function of *t*.

21. In a certain electrical circuit, the differential equation involving the voltage *v* across a capacitive element is $I - C(dv/dt) = v/R$, and at $t = 0$, $v = 0$. If $I = 3.0$, $C = 5.0 \times 10^{-3}$, and $R = 4.0$, (a) verify that $v = 12 + Ae^{-50t}$ satisfies the differential equation, and (b) find *v* as an exact function of *t* to satisfy the condition at $t = 0$.

11.4   DIFFERENTIATION OF THE LOG FUNCTION

We may use the fact that the log function is the inverse of the exponential function to find the derivative of ln *x* with respect to *x*. If $y = \ln x$, $x = e^y$. By use of differentials, $dx = e^y \, dy$, and $dy/dx = 1/e^y = 1/x$. Therefore $(d/dx) \ln x = 1/x$, and

$$\frac{d}{dx} \ln u = \frac{1}{u} \frac{du}{dx}.$$

EXAMPLE 8.   Find $(d/dx) \ln 7x$.

Solution: $(d/dx) \ln 7x = (1/7x) \, 7 = 1/x$. It may seem strange that the derivative of the log of a multiple of *x* is the same as the derivative of the log of x, but this may be verified from first principles. $\ln 7x = \ln 7 + \ln x$, and its graph will be the graph of $\ln x$ shifted upwards a constant amount of $\ln 7$, which does not affect the slope.

EXAMPLE 9.   Find $\dfrac{d}{dt} \ln \dfrac{2t^3 \cos 5te^{3t-1}}{t + 2}$.

Solution: The $\ln$ function controls a complicated expression involving four functions in product and quotient, and to complete the work needed for taking the derivative, long extensions of the product and quotient formulas may have to be used. However, this may be neatly avoided by the simplification resulting from the laws of logs. The multiplications and divisions may be changed to additions and subtractions, and hence derivatives of separate terms may be taken one at a time. The derivative is

$$\frac{d}{dt} [\ln 2 + 3 \ln t + \ln \cos 5t + (3t - 1) \ln e - \ln (t + 2)]$$

$$= \frac{d}{dt} [\ln 2 + 3 \ln t + \ln \cos 5t + 3t - 1 - \ln (t + 2)]$$

$$= 0 + \frac{3}{t} + \frac{1}{\cos 5t} (-5 \sin 5t) + 3 - 0 - \frac{1}{t + 2}$$

$$= \frac{3}{t} - 5 \tan 5t + 3 - \frac{1}{t + 2}.$$

The simplification resulting from the use of logs proves effective in taking derivatives of certain complicated expressions even when there is no log function involved.

EXAMPLE 10.   Find $\dfrac{d}{dx} \sin^2 3xe^{2x}\sqrt{x^2 + 5}$.

Solution: Let the expression to be differentiated be y. Therefore
$$\ln y = 2 \ln \sin 3x + 2x + \tfrac{1}{2} \ln (x^2 + 5).$$
Taking differentials,

$$\frac{1}{y} dy = \left[\frac{2}{\sin 3x} 3 \cos 3x + 2 + \frac{1}{2} \left(\frac{1}{x^2 + 5} \right) 2x \right] dx$$

$$\frac{dy}{dx} = y \left(6 \cot 3x + 2 + \frac{x}{x^2 + 5} \right)$$

$$= \sin^2 3xe^{2x}\sqrt{x^2 + 5} \left(6 \cot 3x + 2 + \frac{x}{x^2 + 5} \right).$$

This method is effective in taking derivatives of functions that are not standard functions or for which no standard differentiation formula applies. We have defined the power function and the exponential function, the first being a variable base raised to a constant power, and the second a constant base raised to a variable power. Moreover, we have developed a differentiation formula for both of these functions. However, if both base and power are variable, we have a nonstandard function. Another example of a nonstandard function is where any base other than e is used in the logarithmic and exponential functions. In all of these nonstandard functions, the derivative may readily be found by using a logarithmic method similar to that of Example 10.

EXAMPLE 11.   Find $(d/dv)(\tan 3v)^{v^2-1}$.

Solution: We have here a variable base raised to a variable power; the function is neither an exponential nor a power function, and is therefore not one of the standard functions. Logarithmic methods are used to get the derivative. Let $y = (\tan 3v)^{v^2-1}$. Hence $\ln y = (v^2 - 1) \ln \tan 3v$. By taking differentials,

$$\frac{1}{y} dy = \left[(v^2 - 1) \frac{1}{\tan 3v} 3 \sec^2 3v + \ln \tan 3v(2v)\right] dv.$$

$$\frac{dy}{dv} = y\left[3(v^2 - 1) \frac{\sec^2 3v}{\tan 3v} + 2v \ln \tan 3v\right]$$

$$= (\tan 3v)^{v^2-1}\left[\frac{3(v^2 - 1)}{\sin 3v \cos 3v} + 2v \ln \tan 3v\right].$$

EXAMPLE 12.   Find $(d/dx)a^{5x-2}$.

Solution: This is an exponential with a nonstandard base. Let $y = a^{5x-2}$. $\ln y = (5x - 2) \ln a$. Note that $\ln a$ is a constant; then $(1/y) dy = 5 \ln a \, dx$.

$$\frac{dy}{dx} = 5y \ln a = 5 \ln a(a^{5x-2}).$$

EXAMPLE 13.   Find $(d/dy) \log_{10} (3y^2 + \sin 2y)$.

Solution: This example is a log function with a nonstandard base. Let $z = \log_{10} (3y^2 + \sin 2y)$. Taking the ln of both sides introduces a double log and this complicates the calculation. By changing the log equation to an exponential equation, we avoid this difficulty. Thus $10^z = 3y^2 + \sin 2y$. By taking a natural log of both sides, $z \ln 10 = \ln (3y^2 + \sin 2y)$.

Therefore

$$\ln 10 \, dz = \frac{1}{3y^2 + \sin 2y} (6y + 2 \cos 2y) \, dy$$

$$\frac{dz}{dy} = \frac{6y + 2 \cos 2y}{\ln 10(3y^2 + \sin 2y)}.$$

EXERCISES

1. Find the following derivatives

(a) $\dfrac{d}{dx} \ln (x - 2)$.

(b) $\dfrac{d}{dx} \ln (3x - 2)$.

(c) $\dfrac{d}{dx} \ln (2 - 3x)$.

(d) $\dfrac{d^2}{dx^2} \ln x$.

(e) $\dfrac{d}{dx} \ln (6x^2 - 4)$.

(f) $\dfrac{d}{dx} \ln e^{2x}$.

(g) $\dfrac{d}{d\theta} \ln (\sin \theta - \cos \theta)$.

(h) $\dfrac{d}{dt} \ln \dfrac{1}{t}$.

(i) $\dfrac{d}{dx} \ln (2x + 5)^2$.

(j) $\dfrac{d}{dx} [\ln (2x + 5)]^2$.

2. Using the fact that logs of negative numbers do not exist, what limitations are placed on the value of x in Exercises: (a) 1a, (b) 1b, and (c) 1c?

3. Sketch the graphs of $y = \ln (3x - 2)$ and $y = \ln (2 - 3x)$.

Find the following derivatives:

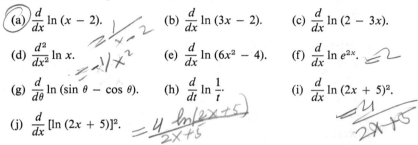

4. $\dfrac{d}{dx} \ln \dfrac{x^2}{x - 3}$.

5. $\dfrac{d}{dx} \dfrac{\ln x^2}{\ln x}$.

6. $\dfrac{d}{dx} \dfrac{\ln x^2}{x}$.

7. $\dfrac{d}{dx} \dfrac{\ln x^2}{\ln (x - 3)}$.

8. $\dfrac{d}{dx} \ln \dfrac{x^2 + x + 1}{x^2 - x + 1}$.

9. $\dfrac{d}{dx} \ln \left(\dfrac{x^2 + x + 1}{x^2 - x + 1} \right)^{1/2}$.

Find the following derivatives:

10. $\dfrac{d}{dx} x \ln x$.

11. $\dfrac{d}{dx} \ln 2x (\tan 2x)$.

12. $\dfrac{d}{dx} \ln (2x \tan 2x)$.

13. $\dfrac{d}{dx} e^{2x} \ln (x^2 e^{2x})$.

14. If $iR/E = e^{-Rt/L}$, find di/dt as a function of i, (a) by an exponential differentiation, and (b) by expressing t as a log function of i and using a logarithmic differentiation.

15. If $v = e^{-100t} \sin 50t$, (a) express $100t$ as a log function of v and t, (b) use the function of (a) to find dv/dt, and (c) verify the answer to (b) by taking a direct derivative of the original.

16. For the function $y = (\ln x)/x$ find (a) the equation of the tangent at $x = 1$, and (b) the maximum y and the value of x for maximum y.

17. Evaluate d^2y/dx^2 at $x = e$ for the function $x = y \ln x$.

Evaluate the following derivatives at the specified value of the independent variable:

18. $\dfrac{d}{dx} 3 \cos 2xe^{x-1}(x + 1)$ at $x = 0$.

19. $\dfrac{d}{dt} t^5 \ln \dfrac{1}{t}$ at $t = 1$.

20. $\dfrac{d}{dt} \dfrac{2(t + 3)^3 e^{2t-1}}{\ln (t + 2)}$ at $t = 1$.

21. $\dfrac{d}{dv} \sqrt{\dfrac{(2v - 1)e^{-3v}}{\tan^3 v}}$ at $v = \dfrac{\pi}{4}$.

22. $\dfrac{d}{dx} \left(\dfrac{-4x^2 \ln x}{e^{2x}} \right)$ at $x = 1$.

Find the following derivatives:

23. $\dfrac{d}{dx} k^x$.

24. $\dfrac{d}{dx} x^k$.

25. $\dfrac{d}{dx} x^x$.

26. $\dfrac{d}{dt} 10^{\cos 20t}$.

27. $\dfrac{d}{dt} \cos^{10} 20t$.

28. $\dfrac{d}{dx} (\sin 20x)^{10x}$.

29. $\dfrac{d}{dx} \log_{10} (x + 2)$.

30. $\dfrac{d}{dx} (x^{x^x})$.

31. $\dfrac{d}{dx} (x^x)^x$.

32. $\dfrac{d}{dt} \dfrac{4t^{2t+1} \sin^3 4t}{2^{3t-1}}$.

33. Find the maximum current i if $i = 6t \times 10^{-100t}$. Show distinctly that it is maximum.

11.5   INTEGRATION OF EXPONENTIALS AND 1/u

By using the basic principle that an integral is the inverse of a differential, since $de^u = e^u \, du$,

$$\int e^u \, du = e^u + C$$

and, since $d \ln u = (1/u) \, du$,

$$\int \frac{1}{u} \, du = \ln u + C.$$

The second formula recalls the fact that, for the integration of a power function,

$$\int u^n \, du = \frac{u^{n+1}}{n + 1} + C,$$

and this formula applies for all values of n except $n = -1$. Where $n = -1$,

$$\int u^{-1}\, du = \int \frac{1}{u}\, du = \ln u + C.*$$

EXAMPLE 14. Find the following integrals: (a) $\int e^{5x}\, dx$, (b) $\int a^{-3x}\, dx$, (c) $\int e^{-50t+1}\, dt$, (d) $\int e^{-20t}\, e^{5t-1}\, dt$, and $\int e^{6x^2-2}\, x\, dx$.

Solution: Use the formula $\int e^u\, du = e^u + C$; it is important to make the structure of the expression to be integrated fit the formula. In particular, du must be the differential of u.

(a) Here $5x = u$ in the formula; therefore $du = d(5x) = 5\, dx$. Thus the integral is

$$\frac{1}{5} \int e^{5x}\, d(5x) = \frac{1}{5} e^{5x} + C.$$

(b) In order to use the standard formula, a^{-3x} must be expressed as an exponential with base e. Let $a^{-3x} = y$; thus $\ln y = -3x \ln a$, and $y = e^{-3x \ln a}$. The required integral is

$$\int e^{-3x \ln a}\, dx = \frac{-1}{3 \ln a} \int e^{-3x \ln a}\, d(-3x \ln a)$$

$$= \frac{-1}{3 \ln a} e^{-3x \ln a} + C = \frac{-1}{3 \ln a} a^{-3x} + C.$$

---

* The fact that $\int x^n\, dx$ is a power function for every value of n except the one isolated case where $n = -1$ may not appear logical. If the value of n approaches -1 as closely as we wish, provided n is not equal to -1 exactly, the integral is still a power function.

$$\lim_{n \to -1} \int x^n\, dx = \lim_{n \to -1} \left(\frac{x^{n+1}}{n+1} \right) + C,$$

and this value approaches $x^0/0 + C = 1/0 + C$, which appears to be infinite. Why, then, does the integral approach infinity and then suddenly change to $\ln x + C$, which is not infinite for a finite x, at the point where $n = -1$ exactly? The fallacy in this argument is that the integration constant may be infinitely different for the two cases, and it is impossible to form a conclusion using an indefinite integral. The integration constant may be eliminated in the analysis if we investigate a definite integral.

$$\lim_{n \to -1} \int_a^b x^n\, dx = \lim_{n \to -1} \left[\frac{x^{n+1}}{n+1} \right]_a^b = \lim_{n \to -1} \left[\frac{b^{n+1} - a^{n+1}}{n+1} \right].$$

This expression approaches the indeterminate form $0/0$, but it may be determined (by l'Hospital's rule, which is not developed in this text) as the limit of the derivative of the numerator over the derivative of the denominator, treating n as the independent variable.

$$\lim_{n \to -1} \int_a^b x^n\, dx = \lim_{n \to -1} \left(\frac{b^{n+1} \ln b - a^{n+1} \ln a}{1} \right) = b^0 \ln b - a^0 \ln a = \ln b - \ln a = [\ln x]_a^b.$$

This is the answer developed more directly by treating $n = -1$ as an exceptional case. Thus, as $n \to -1$, the value of any definite integral of this form approaches the value resulting from the special formula for the case where $n = -1$. There is therefore no sudden change nor any discontinuity in the integral at $n = -1$.

(c) The integral is

$$\frac{-1}{50}\int e^{-50t+1}\, d(-50t+1) = \frac{-1}{50}\, e^{-50t+1} + C.$$

By this time, division by the constant multiple of the independent variable should be a matter of course.

(d) Using a law of indices, the integral is

$$\int e^{-20t+5t-1}\, dt = \int e^{-15t-1}\, dt = \left(\frac{-1}{15}\right)e^{-15t-1} + C.$$

(e) $6x^2 - 2$ replaces u in the formula, and $du = 12x\, dx$. Hence the integral is

$$\frac{1}{12}\int e^{6x^2-2}\, d(6x^2-2) = \left(\frac{1}{12}\right)e^{6x^2-2} + C.$$

If the factor x had not appeared in this integral, the integration would not be so simple.

EXAMPLE 15.   Find the following integrals: (a) $\int dx/3x$, (b) $\int dx/(3-2x)$, (c) $\int_0^Q dq/(EC-q)$, and (d) $\int x\, dx/(1-3x^2)$.

Solution: (a)
$$\int \frac{dx}{3x} = \frac{1}{3}\int \frac{dx}{x} = \frac{1}{3}\ln x + C$$

or
$$\int \frac{dx}{3x} = \frac{1}{3}\int \frac{d(3x)}{3x} = \frac{1}{3}\ln 3x + c = \frac{1}{3}\ln 3 + \frac{1}{3}\ln x + c$$

$$= \frac{1}{3}\ln x + C.$$

This is obtained by grouping the two constant terms of the previous form.

(b)
$$\int \frac{dx}{3-2x} = -\frac{1}{2}\int \frac{d(3-2x)}{3-2x} = -\frac{1}{2}\ln(3-2x) + C.$$

(c)
$$\int_0^Q \frac{dq}{EC-q} = -\int_0^Q \frac{d(EC-q)}{EC-q} = -[\ln(EC-q)]_0^Q$$

$$= -\ln(EC-Q) + \ln EC = \ln\frac{EC}{EC-Q}.$$

(d)
$$\int \frac{x\, dx}{1-3x^2} = -\frac{1}{6}\int \frac{d(1-3x^2)}{1-3x^2} = -\frac{1}{6}\ln(1-3x^2) + C,$$

or
$$\ln(1-3x^2)^{-1/6} + C.$$

EXAMPLE 16.   A current of $i = 3e^{-100t} - 3e^{-200t}$ amperes flows in an *RLC* series circuit at a time t seconds. If the capacitor is uncharged at $t = 0$ and the resistance $R = 2$ ohms, find:
(a) Two separate values of t at which $i = 0$.
(b) The charge on the capacitor at $t = \infty$.
(c) The total energy stored in the circuit at $t = \infty$.
Solution: (a) $i = 0$ when $3e^{-100t} - 3e^{-200t} = 0$. $3e^{-100t}(1 - e^{-100t}) = 0$.

$$e^{-100t} = 0 \quad \text{or} \quad 1 - e^{-100t} = 0.$$
$$t = \infty \quad \text{or} \quad e^{-100t} = 1$$
$$-100t = 0$$
$$t = 0.$$

(b) $q = \int_0^\infty i\, dt$, which is the graphical area under the curve of i against t. Therefore

$$q = \int_0^\infty (3e^{-100t} - 3e^{-200t})\, dt = 3\left[-\frac{1}{100} e^{-100t} + \frac{1}{200} e^{-200t} \right]_0^\infty$$

$$= 3\left(-0 + 0 + \frac{1}{100} - \frac{1}{200} \right) = \frac{3}{200} = 0.015 \text{ coulomb.}$$

(c) Energy is $\int_0^\infty P\, dt$, where power is
$$P = i^2 R = (3e^{-100t} - 3e^{-200t})^2\, 2$$
$$= 18(e^{-200t} - 2e^{-300t} + e^{-400t}).$$
Therefore energy is

$$18 \int_0^\infty (e^{-200t} - 2e^{-300t} + e^{-400t})\, dt$$

$$= 18 \left[-\frac{e^{-200t}}{200} + \frac{2e^{-300t}}{300} - \frac{e^{-400t}}{400} \right]_0^\infty$$

$$= 18 \left(-0 + 0 - 0 + \frac{1}{200} - \frac{2}{300} + \frac{1}{400} \right) = \frac{18}{1200} = 0.015 \text{ joule.}$$

EXERCISES

1. Find the following integrals and verify all answers:
(a) $\int e^{x-2}\, dx$.
(b) $\int e^{-2x}\, dx$.
(c) $\int e^{mt+6}\, dt$.
(d) $\int e^{kx}e^{mx}\, dx$.
(e) $\int e^{-x/3}\, dx$.
(f) $\int e^{-3}\, dt$.
(g) $\int 10^{2x}\, dx$.
(h) $\int (2x)^{10}\, dx$.
(i) $\int \sqrt{e^{-x}}\, dx$.
(j) $\int \frac{dx}{e^{3x-2}}$
(k) $\int e^{\ln t}\, dt$.

2. Find the following integrals and verify all answers:

(a) $\int \dfrac{dx}{4x}$.

(b) $\int \dfrac{dx}{x-3}$.

(c) $\int (4x-3)^{-1}\,dx$.

(d) $\int (3-4x)^{-1}\,dx$.

(e) $\int (3-4x)^{-2}\,dx$.

(f) $\int \dfrac{di}{E-iR}$.

(g) $\int \dfrac{dv}{I-v/R}$.

(h) $\int y\,dx$ if $xy=4$.

3. In Exercise 2, there is no limitation on the value of the independent variable, and yet an examination of the answers to the integrals shows that there are limitations to the values of the independent variables in the answers. What limitations on the value of x are implied in the answers to Exercises 2a, 2b, 2c, and 2d?

4. Since the answers to Exercises 2a, 2b, 2c, and 2d are correct only for certain values of x, they do not represent complete answers. Complete the answer to Exercise 2c for the case where x is less than $\tfrac{3}{4}$, and complete the answer to 2d for the case where x is greater than $\tfrac{3}{4}$. Sketch the graph of $y=(4x-3)^{-1}$ from $x=0$ to $x=2$. (*Hint:* Express $1/(4x-3)$ as $-1/(3-4x)$.)

Evaluate the following definite integrals:

5. $\int_1^e \dfrac{dx}{2x}$.

6. $\int_0^\infty 4e^{-4t}\,dt$.

7. $\int_{-\infty}^0 e^{2t}\,dt$.

8. $\int_0^{L/R} e^{-Rt/L}\,dt$.

9. $\int_0^\infty 2e^{-Rt/L}\,dt$.

10. $\int_0^v \dfrac{R\,dv}{IR-v}$.

11. $\int_0^{(E-1)/R} \dfrac{di}{E-iR}$.

12. The emitter current I_e and emitter voltage E_e of a transistor used as a class A amplifier are related by the differential equation $dE_e/dI_e = k_2/(k_1+I_e)$. Find E_e as a function of I_e.

13. The power supplied to the magnetic field of an inductance in an RL circuit is given by $P = E^2(e^{-Rt/L}-e^{-2Rt/L})/R$. Find the total energy in the field when this power is consumed for an infinite time beginning at $t=0$.

14. If $PV=500$, find $\int_{20}^{50} P\,dV$.

15. If $\int_0^v dv/(E-v) = t/RC$, find v as a function of t, stating the limitation on the value of v in the answer.

16. Find the area between $y=1/(2x-1)$ and the x-axis (a) between $x=2$ and $x=4$, and (b) between $x=-1$ and $x=0$.

17. If $\int_t^{t+1} dt/2t = \int_{t+1}^{t+2} dt/t$, solve for t.

18. For the current $i=1-e^{-0.50t}$, (a) find the mean current between $t=0$ and $t=4.0$, and (b) find the rms current for the interval of (a).

19. A current of $i = 5e^{-20t} - 5e^{-50t}$ amperes flows in an RLC series circuit t seconds after a source voltage is switched in.

(a) Sketch the graph of i against t.

(b) Find the charge on the capacitor at $t = \infty$ assuming it was uncharged at $t = 0$.

(c) If $R = 4.0$ ohms, find the energy stored in the circuit at $t = \infty$.

20. A voltage is applied to a circuit consisting of a 2.0-ohm resistor in series with a 0.0025-henry inductor. After t seconds the voltage across the inductor is $v = -4.0e^{-800t} + 4.0 \cos 400t + 2.0 \sin 400t$ volts. If the current is zero at $t = 0$, find (a) the current as a function of t, and (b) the applied voltage as a function of t.

Find the following integrals:

21. $\int e^{5x^2}(10x\, dx)$.

22. $\int e^{5x^2}x\, dx$.

23. $\int e^{-x^3}x^2\, dx$.

24. $\int \dfrac{t^3\, dt}{(e)^{t^4}}$.

25. $\int e^{x^2}e^{4x-x^2}\, dx$.

26. $\int e^{x^2}e^{-4x}x\, dx$.

27. $\int \dfrac{e^{3x^2}x\, dx}{e^{x^2}}$.

28. $\int \sqrt{\dfrac{e^{x^2}}{e^{2x^2}}}\, x\, dx$.

29. $\int a^{2x^3-5}x^2\, dx$.

30. $\int \dfrac{6x\, dx}{2 + 3x^2}$.

31. $\int \dfrac{6\, dx}{(2 + 3x)^2}$.

32. $\int \dfrac{x\, dx}{2 - 3x^2}$.

33. $\int \dfrac{6x\, dx}{(2 + 3x^2)^2}$.

34. $\int \dfrac{t\, dt}{(t - 2)(t + 2)}$.

35. $\int \left(\dfrac{-1}{t - 2} + \dfrac{2t}{t^2 - 4} \right) dt$ (by two separate methods).

36. Find the area bounded by the curve $y = 9/(4 - x)$, the two axes, and the line $x = 1$, (a) by using x as the independent variable, and (b) by using y as the independent variable and combining two areas.

37. In an RLC series circuit, the voltage drop across the inductance L is $v = 10e^{-500t} - 4.0e^{-200t}$ volts at a time t seconds. If $L = 0.0050$ henry, $R = 3.5$ ohms, and $C = 0.0020$ farad, and the capacitor is uncharged with no current flowing at $t = 0$, find (a) the voltage drops across the resistor and capacitor as functions of t, (b) the source voltage, (c) the separate voltage drops across the three elements at $t = 0$, and (d) the separate voltage drops across the three elements at $t = \infty$.

38. The equation of the path in which a transmission line hangs, assuming the y-axis passes through its central point, is $y = 40(e^{-x/80} + e^{x/80})$, x and y being expressed in feet. If the distance between supporting towers is 100 ft, find the length of the line, given that the length of a curve between a and b is equal to $\int_a^b \sqrt{1 + (dy/dx)^2}\, dx$.

CHAPTER 12

Calculus of Hyperbolic and Inverse Functions

12.1 THE CALCULUS OF HYPERBOLIC FUNCTIONS

The hyperbolic functions were discussed in Section 9.4. We recall that these functions are special combinations of the exponential function. More specifically,

$$\sinh x = \frac{e^x - e^{-x}}{2} \qquad \cosh x = \frac{e^x + e^{-x}}{2} \qquad \tanh x = \frac{e^x - e^{-x}}{e^x + e^{-x}};$$

cosech x, sech x, and coth x are the reciprocals of sinh x, cosh x, and tanh x, respectively.

To find the derivatives of the hyperbolic functions, we may use the established derivatives of the exponential functions. Thus,

$$\frac{d}{dx} \sinh u = \frac{d}{dx} \frac{e^u - e^{-u}}{2} = \frac{e^u + e^{-u}}{2} \frac{du}{dx} = \cosh u \frac{du}{dx}.$$

In the same manner, we may readily see that

$$\frac{d}{dx} \cosh u = \sinh u \frac{du}{dx}.$$

$$\frac{d}{dx} \tanh u = \frac{d}{dx} \frac{\sinh u}{\cosh u} = \frac{\cosh u \cosh u - \sinh u \sinh u}{\cosh^2 u} \frac{du}{dx}$$

$$= \frac{\cosh^2 u - \sinh^2 u}{\cosh^2 u} \frac{du}{dx} = \frac{1}{\cosh^2 u} \frac{du}{dx} = \text{sech}^2 u \frac{du}{dx}.$$

In summary,

$$\frac{d}{dx}\sinh u = \cosh u \frac{du}{dx}. \qquad \frac{d}{dx}\cosh u = \sinh u \frac{du}{dx}. \qquad \frac{d}{dx}\tanh u = \operatorname{sech}^2 u \frac{du}{dx}.$$

Since the integral is the inverse of the differential,

$$\int \cosh u \; du = \sinh u + C. \qquad \int \sinh u \; du = \cosh u + C.$$

$$\int \operatorname{sech}^2 u \; du = \tanh u + C.$$

EXAMPLE 1.   Find $(d/dx)\operatorname{sech}^2(4x - 2)$.
Solution:

$$\frac{d}{dx}\operatorname{sech}^2(4x - 2) = \frac{d}{dx}\cosh^{-2}(4x - 2)$$

$$= -2[\cosh(4x - 2)]^{-3}\sinh(4x - 2)4$$

$$= \frac{-8\sinh(4x - 2)}{\cosh^3(4x - 2)}.$$

EXAMPLE 2.   Evaluate: (a) $\int_0^{0.5}\sinh 3x \, dx$ and (b) $\int_0^{0.5}\sinh^2 3x \, dx$.
Solution: (a) In order that this example may fit the second of the foregoing formulas exactly, the differential factor must be the exact differential of $3x$, which is $3 \, dx$. Hence the integral is

$$\tfrac{1}{3}\int_0^{0.5}\sinh 3x \, d(3x) = \tfrac{1}{3}[\cosh 3x]_0^{0.5} = \tfrac{1}{3}(\cosh 1.5 - \cosh 0)$$

$$= \tfrac{1}{3}[2.35 - 1] = 0.45.$$

(b) Here we have a power function of a hyperbolic function, and the structure of a composite function cannot be used for integration. The power function can be eliminated, however, in much the same way as we used for $\int \sin^2 \theta \, d\theta$. Thus, $\sin^2 3x = \tfrac{1}{2}(1 - \cos 6x)$. The corresponding hyperbolic formula is $-\sinh^2 3x = \tfrac{1}{2}(1 - \cosh 6x)$; thus $\sinh^2 3x = \tfrac{1}{2}(\cosh 6x - 1)$. Hence

$$\int_0^{0.5}\sinh^2 3x \, dx = \tfrac{1}{2}\int_0^{0.5}(\cosh 6x - 1) \, dx = \tfrac{1}{2}[\tfrac{1}{6}\sinh 6x - x]_0^{0.5}$$

$$= \tfrac{1}{2}(\tfrac{1}{6}\sinh 3 - 0.5 - \tfrac{1}{6}\sinh 0) = \tfrac{1}{2}(\tfrac{1}{6} \, 10.02 - 0.5 - 0)$$

$$= \tfrac{1}{2}(1.17) = 0.58.$$

EXERCISES

Find the following derivatives:

1. $\dfrac{d}{dt}\sinh 3t.$

2. $\dfrac{d}{dz}\cosh(5z^2 + 2).$

3. $\dfrac{d}{dx}\tanh^3 4x.$

4. $\dfrac{d}{dp}\operatorname{cosech}(2p - 1).$

5. $\dfrac{d}{dx}\coth(4x^2 - x).$

6. $\dfrac{d}{dt}\sqrt{\cosh^2 t + 1}.$

7. $\dfrac{d}{dx}\sqrt{\cosh^2 3x - 1}.$

8. $\dfrac{d}{dx}\ln\cosh 2x.$

Find the following integrals:

9. $\int \cosh(2x + 3)\, dx.$

10. $\int \operatorname{sech}^2 4t\, dt.$

11. $\int (\sinh 3x^2)6x\, dx.$

12. $\int (\cosh 3x^2)x\, dx.$

13. $\int \cosh^2 x\, dx.$

14. $\int \sinh^2(2x - 1)\, dx.$

15. $\int \sinh^2 \dfrac{x}{4}\, dx.$

16. $\int (\sinh^2 2x - \cosh^2 2x)\, dx.$

17. $\int \dfrac{(\cosh^3 5u - 2)\, du}{\cosh^2 5u}.$

18. Find $(d/dx)\tanh u$ by using exponential equivalents throughout and express the result in terms of a hyperbolic function.

Find the area enclosed by each of the following curves, the x-axis, and the lines $x = 0$ and $x = 2$:

19. $y = \sinh\dfrac{x}{2}.$

20. $y = \cosh^2\dfrac{x}{2}.$

21. $y = \cosh 2x - \sinh 2x.$
22. $y = \cosh 2x - e^{-2x}.$ (*Hint:* Change to simplest exponential form.)
23. Express $\tanh^2 x$ in terms of $\operatorname{sech}^2 x$, and hence find $\int \tanh^2 x\, dx.$

Find the following derivatives:

24. $\dfrac{d}{dt}\dfrac{\sinh t}{1 + \cosh t}.$

25. $\dfrac{d}{dx}\sinh 2x \cos 3x.$

26. $\dfrac{d}{dx}\sinh 2x \cosh 2x.$

27. $\dfrac{d}{dp}e^{-3p}(\cosh 3p + \sinh 3p).$

28. $\dfrac{d}{dp}e^{-3p}(\cosh 3p - \sinh 3p).$

29. $\dfrac{d}{dx}e^x \ln\sinh x.$

30. $\dfrac{d}{dx}e^x \ln(\cosh x + \sinh x).$

31. $\dfrac{d}{dx}(\cosh 5x)^{\cosh 5x}.$

32. A transmission line suspended from towers 320 ft apart hangs in the form of a catenary whose equation is $y = 400 \cosh (x/400)$, x being measured in feet horizontally from the center point of the span.

(a) Find the angle of inclination of the line at the point of suspension.

(b) The total length of the line may be calculated as $\int_a^b \sqrt{1 + (dy/dx)^2}\, dx$, where a and b are the values of x at the two end points. Find the length of the transmission line.

33. Express y as a single-termed function of x in four different ways, each satisfying the given differential equation:

(a) $\dfrac{d^2y}{dx^2} = y.$ (b) $\dfrac{d^2y}{dx^2} = 9y.$ (c) $\dfrac{d^2y}{dx^2} = ky.$

34. If $y = \tanh x$, show that $d^2y/dx^2 + 2y(dy/dx) = 0$.

35. Find the minimum value of $y = 4 \sinh x + 5 \cosh x$.

36. The characteristic impedance of a transmission line is given by

$$Z_0 = 90 \ln \left(\frac{4w \tanh (\pi h/w)}{\pi d} \right).$$

Find the rate at which Z_0 changes with respect to h.

12.2   DERIVATIVES OF INVERSE FUNCTIONS

The inverse functions were discussed in Section 9.5. The derivatives of these functions will now be investigated.

If $y = \text{arc sin } u$, alternatively expressed as $\sin^{-1} u$, then $u = \sin y$, and $du = \cos y\, dy$. Thus

$$dy = \frac{du}{\cos y} \quad \text{and} \quad \frac{dy}{dx} = \frac{1}{\cos y} \frac{du}{dx}.$$

In order to express the result as a function of u, we see that $\cos y = \pm \sqrt{1 - \sin^2 y} = \pm \sqrt{1 - u^2}$. Therefore

$$\frac{d}{dx} \text{arc sin } u = \frac{1}{\pm \sqrt{1 - u^2}} \frac{du}{dx}.$$

Since $u = \sin y$, u is limited in value to the range from 1 to -1, the same values that produce a real value in the expression for the derivative. Referring to the graph of $y = \text{arc sin } x$ (Fig. 9-14), we may see that, for any value of x from 1 to -1, there is an infinite number of values of y as we move farther and farther upward. But the absolute values of the slopes at all of these points are identical, the sign of the slope alternating between plus and minus, except where $x = \pm 1$; then the slope is always infinite. This substantiates the fact that the expression for the slope should have the alternative plus or minus signs. If it is specified that $-(\pi/2) \leq$ arc sin $u \leq \pi/2$, the positive sign only would be used, and this is usual practice.

With the same type of calculation as we used for (d/dx) arc sin u, we may establish that (d/dx) arc cos $u = (1/\pm \sqrt{1 - u^2})(du/dx)$. We may also see from a comparison of the graphs of arc sin x and arc cos x that the numerical values of the slope for any given value of x are the same for the two graphs, and that the slopes for the graph of $y =$ arc cos x also alternate between minus and plus, the slopes for the first positive values of y being negative. Thus it is usual to take the minus sign here.

If $y =$ arc tan u, then $u = \tan y$, and $du = \sec^2 y \, dy = (1 + \tan^2 y) \, dy = (1 + u^2) \, dy$. Therefore

$$dy = \frac{1}{1 + u^2} \, du \quad \text{and} \quad \frac{d}{dx} \text{ arc tan } u = \frac{1}{1 + u^2} \frac{du}{dx}.$$

EXAMPLE 3.   Find $(d/dx)x$ arc sin (x/a), assuming arc sin (x/a) lies between $-\pi/2$ and $\pi/2$, inclusive. What are the restrictions on the value of x?

Solution: Using the product formula,

$$\frac{d}{dx} x \text{ arc sin } \frac{x}{a} = \frac{x}{\sqrt{1 - x^2/a^2}} \frac{1}{a} + \text{arc sin} \frac{x}{a}$$

$$= \frac{x}{a\sqrt{(a^2 - x^2)/a^2}} + \text{arc sin} \frac{x}{a}$$

$$= \frac{x}{\sqrt{a^2 + x^2}} + \text{arc sin} \frac{x}{a}.$$

Since the sine function has extreme values of 1 and -1, x^2 cannot be greater than a^2.

The inverse hyperbolic functions may be differentiated in a somewhat similar manner.

If $y = \sinh^{-1} u$, $u = \sinh y$.

$$du = \cosh y \, dy$$

$$= +\sqrt{1 + \sinh^2 y} \, dy \quad (\cosh y \text{ is always positive})$$

$$= +\sqrt{1 + u^2} \, dy.$$

$$\therefore \ dy = \frac{1}{+\sqrt{1 + u^2}} \, du.$$

and

$$\frac{d}{dx}\sinh^{-1}u = \frac{1}{+\sqrt{1+u^2}}\frac{du}{dx}.$$

If $y = \cosh^{-1}u$, $u = \cosh y$.

$$du = \sinh y\, dy = \pm\sqrt{\cosh^2 y - 1}\, dy = \pm\sqrt{u^2 - 1}\, dy.$$

$$\therefore\ dy = \frac{1}{\pm\sqrt{u^2 - 1}}\, du.$$

and

$$\frac{d}{dx}\cosh^{-1}u = \frac{1}{+\sqrt{u^2 - 1}}\frac{du}{dx}\ \text{for principal values.}$$

In a similar manner we may prove that

$$\frac{d}{dx}\tanh^{-1}u = \frac{1}{1-u^2}\frac{du}{dx}.$$

Since the tanh function lies between 1 and -1 inclusive, u^2 cannot be greater than 1.

In summary, for principal values,

$$\frac{d}{dx}\text{arc sin }u = \frac{1}{+\sqrt{1-u^2}}\frac{du}{dx}. \qquad \frac{d}{dx}\sinh^{-1}u = \frac{1}{+\sqrt{1+u^2}}\frac{du}{dx}.$$

$$u^2 \leq 1.$$

$$\frac{d}{dx}\text{arc cos }u = \frac{1}{-\sqrt{1-u^2}}\frac{du}{dx}. \qquad \frac{d}{dx}\cosh^{-1}u = \frac{1}{+\sqrt{u^2-1}}\frac{du}{dx}.$$

$$u^2 \leq 1. \qquad\qquad\qquad\qquad u \geq 1.$$

$$\frac{d}{dx}\text{arc tan }u = \frac{1}{1+u^2}\frac{du}{dx}. \qquad \frac{d}{dx}\tanh^{-1}u = \frac{1}{1-u^2}\frac{du}{dx}.$$

$$u^2 \leq 1.$$

As discussed in Section 9.5, inverse hyperbolic functions may be expressed as natural logs. In summary,

$$\sinh^{-1}x = \ln\left(x + \sqrt{x^2 + 1}\right).$$

$$\cosh^{-1}x = \ln\left(x \pm \sqrt{x^2 - 1}\right) \qquad \text{where}\quad x \geq 1.$$

$$\tanh^{-1}x = \tfrac{1}{2}\ln\frac{1+x}{1-x} \qquad \text{where}\quad x^2 \leq 1.$$

The derivatives of the inverse hyperbolic functions may therefore be calculated by using these logarithmic forms.

EXAMPLE 4.   Find $(d/dx)\cosh^{-1}u$ by using the logarithmic equivalent of $\cosh^{-1}u$.

Solution: $\dfrac{d}{dx}\cosh^{-1}u = \dfrac{d}{dx}\ln(u \pm \sqrt{u^2-1})$

$$= \frac{1}{u \pm \sqrt{u^2-1}}\left(1 \pm \frac{u}{\sqrt{u^2-1}}\right)\frac{du}{dx}$$

$$= \frac{1}{u \pm \sqrt{u^2-1}}\,\frac{\sqrt{u^2-1}\pm u}{\sqrt{u^2-1}}\frac{du}{dx}$$

$$= \frac{\pm 1}{\sqrt{u^2-1}}\frac{du}{dx}.$$

EXERCISES

1. Find the following derivatives:

(a) $\dfrac{d}{dx}\arcsin 2x$.   (b) $\dfrac{d}{dx}\arctan x^2$.   (c) $\dfrac{d}{dx}2\arcsin(x^2-1)$.

(d) $\dfrac{d}{dx}\sinh^{-1}\dfrac{x}{2}$.   (e) $\dfrac{d}{dx}\tanh^{-1}\dfrac{x}{3}$.   (f) $\dfrac{d}{dx}\cosh^{-1}\dfrac{\sqrt{3}x}{2}$.

2. Find the slope of the curve $y = \arcsin(x/a)$, (a) as a function of x, and (b) as a function of y, by each of the following methods:

(i) Use the formula for the derivative of $\arcsin u$.

(ii) Express the equation as a direct function and differentiate a sine.

3. Establish the formula for $(d/dx)\arctan u$ by expressing $\arctan u$ as an arc cos and using $(d/dx)\arccos u$.

4. Establish the formula for $(d/dx)\tanh^{-1}u$ by the method used for $(d/dx)\sinh^{-1}u$.

5. Establish the formula for $(d/dx)\tanh^{-1}u$ by first expressing it as a logarithmic function and using the derivative of a natural log.

6. Establish the formula for (d/dx) arc cot u by using a direct function.

7. Find (d/dx) arc cosec x by expressing x in terms of $\sin y$.

8. Find $(d/dx)\sin(\arccos x)$ by (a) the method of a composite function, and (b) simplifying $\sin(\arccos x)$ originally.

Evaluate the following derivatives at the specified value of the variable:

9. $\dfrac{d}{dx}\tfrac{1}{2}\arctan 4x$ at $x = 2$.   10. $\dfrac{d}{dt}$ arc sec t when arc sec $t = \dfrac{\pi}{3}$.

11. $\dfrac{d}{dx}\arcsin\dfrac{x^2}{5}$ at $x = 2$.   12. $\dfrac{d}{dx}\arccos\dfrac{1}{\sqrt{x^2+1}}$ at $x = \sqrt{3}$.

13. $\dfrac{d}{dx} \operatorname{cosech}^{-1} x$ at $x = \sqrt{2}$.

14. $\dfrac{d}{dx} \cosh^{-1} \dfrac{1}{\sqrt{x}}$ when $\cosh^{-1} \dfrac{1}{\sqrt{x}} = 0$.

15. Find the maximum value of $2 \operatorname{arc} \tan x - x$, assuming $\operatorname{arc} \tan x < \pi/2$.

16. Evaluate $(d^2/dx^2) \operatorname{arc} \sin (x/2)$ at $x = \sqrt{3}$.

Find the following derivatives:

17. $\dfrac{d}{dx} \sin 2x \operatorname{arc} \sin 2x$.

18. $\dfrac{d}{dx} \cosh 2x \cosh^{-1} 2x$.

19. $\dfrac{d}{dx} \dfrac{\operatorname{arc} \sin x}{\sqrt{1 - x^2}}$.

20. $\dfrac{d}{dx} (a^2 + x^2) \operatorname{arc} \tan \dfrac{x}{a}$.

21. $\dfrac{d}{dx} \sqrt{a^2 - x^2} \operatorname{arc} \cos \dfrac{x}{a}$.

22. $\dfrac{d}{dx} \dfrac{\tanh^{-1} ax}{1 - a^2 x^2}$.

23. $\dfrac{d}{dx} \left(2 \sinh^{-1} \dfrac{x}{2} + \dfrac{x\sqrt{4 + x^2}}{2} \right)$.

24. Find d^2y/dx^2 for the following functions:

(a) $y = x \operatorname{arc} \sin x$.

(b) $y = (a^2 - x^2) \tanh^{-1} \dfrac{x}{a}$.

12.3   INTEGRALS OF EXPRESSIONS RESULTING IN INVERSE FUNCTIONS

It is noteworthy that the derivatives of inverse functions yield expressions that are very similar in pattern. We are now prepared to integrate expressions of this pattern.

In the following calculations and formulas, a represents any positive number.

Since $d \operatorname{arc} \sin \dfrac{u}{a} = \dfrac{1}{\sqrt{1 - u^2/a^2}} \dfrac{1}{a} du = \dfrac{du}{a\sqrt{(a^2 - u^2)/a^2}} = \dfrac{du}{\sqrt{a^2 - u^2}}$,

it follows that

$$\int \dfrac{du}{\sqrt{a^2 - u^2}} = \operatorname{arc} \sin \dfrac{u}{a} + C. \ u^2 \leq a^2 \text{ on both sides of this formula.}$$

Similarly, since $d \operatorname{arc} \tan \dfrac{u}{a} = \dfrac{1}{1 + u^2/a^2} \dfrac{1}{a} du = \dfrac{a \, du}{a^2 + u^2}$,

it follows that

$$\int \dfrac{du}{a^2 + u^2} = \dfrac{1}{a} \operatorname{arc} \tan \dfrac{u}{a} + C.$$

In a similar manner, by using derivatives of inverse hyperbolic functions, three more integration formulas are developed. A complete list of these formulas follows.

$$\int \frac{du}{\sqrt{a^2 - u^2}} = \text{arc sin } \frac{u}{a} + C.$$

$$(u^2 \leq a^2)$$

$$\int \frac{du}{\sqrt{a^2 + u^2}} = \sinh^{-1} \frac{u}{a} + C$$

$$= \ln \frac{u + \sqrt{a^2 + u^2}}{a} + C.$$

$$\int \frac{du}{\sqrt{u^2 - a^2}} = \cosh^{-1} \frac{u}{a} + C$$

$$= \ln \frac{u + \sqrt{u^2 - a^2}}{a} + C.$$

$$(u \geq a)$$

$$\int \frac{du}{a^2 + u^2} = \frac{1}{a} \text{ arc tan } \frac{u}{a} + C.$$

$$\int \frac{du}{a^2 - u^2} = \frac{1}{a} \tanh^{-1} \frac{u}{a} + C$$

$$= \frac{1}{2a} \ln \frac{a + u}{a - u} + C.$$

$$(u^2 \leq a^2)$$

EXAMPLE 5.   Evaluate

$$\int_0^{0.75} \frac{dx}{\sqrt{9 - 4x^2}}.$$

Solution: When a coefficient other than 1 occurs with the x^2 term, some care must be taken. Possibly the easiest method is to remove a common factor $\sqrt{4} = 2$ from the denominator, thus producing a coefficient of 1 with the x^2 term.

$$\int_0^{0.75} \frac{dx}{\sqrt{9 - 4x^2}} = \frac{1}{2} \int_0^{0.75} \frac{dx}{\sqrt{(3/2)^2 - x^2}} = \frac{1}{2} \left[\text{arc sin } \frac{x}{3/2} \right]_0^{0.75}$$

$$= \frac{1}{2} (\text{arc sin } 0.5 - \text{arc sin } 0) = \frac{1}{2} \frac{\pi}{6} = \frac{\pi}{12}.$$

EXAMPLE 6.   Evaluate $\int_0^1 \frac{dx}{9 - 4x^2}.$

Solution: The structure of the expression is $\int_0^1 \frac{dx}{3^2 - (2x)^2}.$

The integral is

$$\frac{1}{2} \int_0^1 \frac{d(2x)}{3^2 - (2x)^2} = \frac{1}{2} \left[\frac{1}{2 \times 3} \ln \frac{3 + 2x}{3 - 2x} \right]_0^1 = \frac{1}{12} (\ln 5 - \ln 1)$$

$$= \frac{\ln 5}{12} = 0.134.$$

For definite integrals, the logarithmic form of answer is more adaptable than the inverse hyperbolic form. The integration might also have been calculated by removing 4 as a common factor in the denominator.

EXAMPLE 7.   Find $\int \frac{(3x - 2) \, dx}{1 + 4x^2}$.

Solution: In order to express this in terms of standard integrals, it is necessary to split the expression into two fractions, and proceed as follows:

$$\int \frac{(3x - 2) \, dx}{1 + 4x^2} = \int \frac{3x}{1 + 4x^2} \, dx - \int \frac{2 \, dx}{1 + 4x^2}.$$

These two integrals may be made to fit two different standard integrals. In the first integral, the whole denominator may be considered as a basic function of x whose differential $= 8x \, dx$; hence the given numerator is $\frac{3}{8}$ times the differential of the denominator, and this integral fits the general form $\int (du/u)$. In the second integral, the numerator is the differential of a term in x to the first power, and therefore the basic function must involve only a first power of x. Thus the integral is

$$\frac{3}{8} \int \frac{d(1 + 4x^2)}{1 + 4x^2} - \int \frac{d(2x)}{1 + (2x)^2} = \frac{3}{8} \ln (1 + 4x^2) - \arctan 2x + C.$$

EXAMPLE 8.   Find

$$\int \frac{dx}{4 + 3x - x^2}.$$

Solution:

By the method of completing the square, this may be put in the form

$$\int \frac{du}{a^2 - u^2}.$$

$$4 + 3x - x^2 = 4 - (x^2 - 3x + \tfrac{9}{4}) + \tfrac{9}{4} = \tfrac{25}{4} - (x - \tfrac{3}{2})^2.$$

Therefore the integral is

$$\int \frac{dx}{(\frac{5}{2})^2 - (x - \frac{3}{2})^2} = \int \frac{d(x - \frac{3}{2})}{(\frac{5}{2})^2 - (x - \frac{3}{2})^2}$$

$$= \frac{1}{5} \ln \frac{\frac{5}{2} + (x - \frac{3}{2})}{\frac{5}{2} - (x - \frac{3}{2})} + C$$

$$= \frac{1}{5} \ln \frac{1 + x}{4 - x} + C.$$

This answer is correct if the logarithm is real, which implies that x lies between -1 and 4, inclusive. If x does not lie between -1 and 4, the solution involves a technique to be analyzed in Chapter 13.

EXERCISES

Find the following integrals:

1. $\int \dfrac{dx}{4 + x^2}.$

2. $\int \dfrac{dx}{4 + x}.$

3. $\int \dfrac{dx}{\sqrt{4 + x^2}}.$

4. $\int \dfrac{dx}{\sqrt{4 + x}}.$

5. $\int \dfrac{dx}{\sqrt{4 - 4x^2}}.$

6. $\int \dfrac{dx}{\sqrt{1 + 4x^2}}.$

7. $\int \dfrac{dx}{3 - 4x^2}$ (if $4x^2 \le 3$).

8. $\int \dfrac{dt}{9t^2 - 1}$ (if $9t^2 \le 1$).

9. $\int \dfrac{1 + 2x}{1 - 4x^2}\, dx$ (if $2x \le 1$).

Evaluate the following definite integrals:

10. $\displaystyle\int_0^2 \dfrac{dx}{4 + x^2}.$

11. $\displaystyle\int_0^1 \dfrac{dx}{\sqrt{4 - x^2}}.$

12. $\displaystyle\int_0^2 \dfrac{dx}{4 + x}.$

13. $\displaystyle\int_0^2 \dfrac{x\, dx}{4 + x^2}.$

14. $\displaystyle\int_0^3 \dfrac{dx}{\sqrt{9 + x^2}}.$

15. $\displaystyle\int_1^2 \dfrac{dx}{\sqrt{9x^2 - 4}}.$

16. $\displaystyle\int_0^{0.5} \dfrac{dx}{\sqrt{1 - 3x^2}}.$

17. $\displaystyle\int_{0.1}^{0.2} \dfrac{dx}{4x^2 - 1}.$

18. (a) Find the area bounded by the positive branch of $y^2 = 9/(4 - x^2)$, the x-axis, the y-axis, and the line $x = 1$.

(b) Find the volume of revolution generated by rotating the area of (a) about the x-axis.

19. A current i is given by $i^2 = 9/(4 + t^2)$ amperes squared at t seconds. Calculate: (a) the mean current from $t = 0$ to $t = 2$, and (b) the rms current from $t = 0$ to $t = 2$.

20. Find the average power in a resistor of 2.0 ohms carrying a current i from $t = 0$ to $t = 0.40$ sec, where $i^2 = 1/(1 - 4t^2)$ amperes squared.

Find the following integrals:

21. $\int \dfrac{2 - 4x}{4x^2 + 1} \, dx.$
　　　　22. $\int \dfrac{2x^2 - 1}{4x^4 - 1} \, dx.$
　　　　23. $\int \dfrac{x - 2}{x^2 + 4} \, dx.$

24. $\int \dfrac{2 - x}{1 - 4x^2} \, dx$ (if $4x^2 < 1$).

25. $\int \dfrac{1 - 2x}{\sqrt{4 - x^2}} \, dx.$
　　26. $\int \dfrac{x^3 - x}{4 + x^4} \, dx.$
　　27. $\int \dfrac{dx}{4 + (x - 3)^2}.$

28. $\int \dfrac{dx}{\sqrt{1 + 4(x - 2)^2}}.$
　29. $\int \dfrac{dx}{x^2 - 4x + 8}.$
　30. $\int \dfrac{dx}{\sqrt{5 + 4x - x^2}}.$

31. $\int \dfrac{dx}{\sqrt{x^2 - 4x + 3}}.$
　32. $\int \dfrac{dx}{\sqrt{8 - 4x - 4x^2}}.$

33. (a) What limitation is necessary on the value of x in the answer to Exercise 30?

(b) Does this limitation appear in the form of the original problem?

CHAPTER 13

Special Integration Techniques

13.1 THE STRUCTURE OF AN INTEGRAL

The basic problems of integration arise from the fact that it is the inverse of differentiating, and the direct calculations involved in finding a differential must be reversed in all respects when an integral is to be determined. This "backwards" calculation is fundamentally a problem in structure which becomes more apparent when we consider how the differential of a composite function is calculated and then reverse the calculation.

EXAMPLE 1.   Find the differential of $\sin^3 2x$, and use the answer to calculate $\int \sin^2 2x \cos 2x \, dx$.
Solution:
   $d \sin^3 2x = d(\sin 2x)^3 = 3 \sin^2 2x \, 2 \cos 2x \, dx = 6 \sin^2 2x \cos 2x \, dx.$
By the original definition of an integral, the integral of this answer must be $\sin^3 2x$. Therefore

$$\int \sin^2 2x \cos 2x \, dx = \tfrac{1}{6} \int 6 \sin^2 2x \cos 2x \, dx = \tfrac{1}{6} \sin^3 2x + C.$$

The most important thing to note in Example 1 is that, in calculating the differential, the expression $\cos 2x \, dx$ appears as an extra factor in the answer, (by using a *chain rule*). However, in calculating the integral, the factor $\cos 2x \, dx$ forms part of the original expression to be integrated and does not appear in the answer. To perform this integration it is necessary to see that $2 \cos 2x \, dx$ is the differential of $\sin 2x$, that is, $\cos 2x \, dx = \tfrac{1}{2} d \sin 2x$, and the structure of the integral is $\tfrac{1}{2} \int (\quad)^2 \, d(\quad)$. The structure of the answer will be $\tfrac{1}{2}(\quad)^3/3 = (\quad)^3/6$ plus the integration constant.

The structure of any integral must be shown to correspond to the structure of the integral in one formula only. In Example 1, the structure is $\int u^n \, du$.

EXAMPLE 2.   Find

$$\int \frac{\sin 2x \, dx}{1 - 3 \cos 2x}.$$

Solution: A preliminary analysis yields the fact that the expression $1 - 3 \cos 2x$ will appear somewhere in the answer, and that the factor $\sin 2x \, dx$ is simply some multiple of the differential of this principal expression which is present as part of the structure of the integral. Thus, if we represent $1 - 3 \cos 2x$ as (   ), $d($ $) = +6 \sin 2x \, dx$; the integral is therefore $\frac{1}{6} \int d($ $)/($ $)$. Using the appropriate formula, the answer is $\frac{1}{6} \ln ($ $) + C$, and the detailed answer is $\frac{1}{6} \ln (1 - 3 \cos 2x) + C$. Another technique would be to actually substitute a single letter (u, say) for the basic expression which preliminary analysis indicates will appear in the answer. Thus, let $1 - 3 \cos 2x = u$.

$$6 \sin 2x \, dx = du. \qquad \sin 2x \, dx = \tfrac{1}{6} \, du.$$

The integral may now be completely translated into the u system as

$$\frac{1}{6} \int \frac{du}{u} = \frac{1}{6} \ln u + C = \frac{1}{6} \ln (1 - 3 \cos 2x) + C.$$

Example 2 might have been calculated by considering only $\cos 2x$ as the basic function. However, the inclusion of the constant multiplier -3 and the constant term 1 does not change the structure of the differential, and it produces a less complicated result. It is best, therefore, to include as much as possible in the basic expression.

Following is a tabulation of various important functions of x that may appear as basic functions u in an integral, along with their differentials, which must appear to complement the basic function in the structure of the integral:

u	du	u	du
$k + k_1 x^n$	$k_2 x^{n-1} \, dx$	$k + k_1 \ln x$	$\dfrac{k_1}{x} \, dx$
$k + k_1 \sin mx$	$k_2 \cos mx \, dx$	$k + k_1 e^{mx}$	$k_2 e^{mx} \, dx$
$k + k_1 \cos mx$	$k_2 \sin mx \, dx$	$k + k_1 \sinh mx$	$k_2 \cosh mx \, dx$
$k + k_1 \tan mx$	$k_2 \sec^2 mx \, dx$	$k + k_1 \arcsin mx$	$\dfrac{k_2 \, dx}{\sqrt{1 - m^2 x^2}}$

The expression referred to as u in the foregoing tabulation may, of course, be contained within another function for which a rule of integration has been established.

This tabulation is by no means complete, but it should be sufficient for most situations. The student should be alert to note the occurrence of these types of expressions in pairs in the structure of an integral.

EXAMPLE 3. Calculate $\int \dfrac{e^{2x}\,dx}{\sqrt{1-e^{2x}}}$.

Solution: $e^{2x}\,dx$ is a multiple of the differential of $1-e^{2x}$, which means that $1-e^{2x}$ should be taken as the basic function u. Let

$$1-e^{2x}=u. \qquad -2e^{2x}\,dx=du. \qquad e^{2x}\,dx=-\tfrac{1}{2}\,du.$$

Therefore the integral is

$$-\frac{1}{2}\int\frac{du}{\sqrt{u}} = -\frac{1}{2}\int u^{-\frac{1}{2}}\,du = \frac{-\frac{1}{2}u^{\frac{1}{2}}}{\frac{1}{2}}+C = -u+C$$

$$= -\sqrt{1-e^{2x}}+C.$$

Verification may further clarify the steps in solving the problem. The student should, therefore, take the differential of this answer to show that this results in the original expression that was integrated.

EXAMPLE 4. Calculate $\int \dfrac{t^2\,dt}{4-t^6}$.

Solution: Here the numerator is the differential of a multiple of t^3. Hence, let

$$t^3=u. \qquad 3t^2\,dt=du. \qquad t^2\,dt=\tfrac{1}{3}\,du.$$

Therefore the integral is

$$\frac{1}{3}\int\frac{du}{4-u^2}=\frac{1}{3}\frac{1}{2}\tanh^{-1}\frac{u}{2}+C=\frac{1}{6}\tanh^{-1}\frac{t^3}{2}+C \quad \text{assuming } t^6\le 4.$$

There is, of course, an alternative logarithmic answer.

EXAMPLE 5. Calculate $\int_1^e \dfrac{\cos(\ln x)\,dx}{x}$.

Solution: The combination of $\ln x$ and $(1/x)\,dx$ should be recognized here, and the principal function, $\ln x$, has a cosine function controlling it. Let

$$\ln x=u. \qquad \therefore \frac{dx}{x}=du.$$

Therefore the integral is

$$\int_{x=1}^{e} \cos u \; du = \int_{u=0}^{1} \cos u \; du = [\sin u]_0^1 = \sin 1 - \sin 0$$

$$= \sin 57.3° = 0.842.$$

The translation of the limits to values of u avoids switching back to the x system; when $x = 1$, $u = \ln 1 = 0$, and when $x = e$, $u = \ln e = 1$.

EXERCISES

Find the following integrals:

1. $\displaystyle\int \frac{3x^2 \, dx}{x^3 + 5}.$

2. $\displaystyle\int \frac{x^3 \, dx}{x^4 + 5}.$

3. $\displaystyle\int \frac{x^3 + 5}{x^2} \, dx.$

4. $\displaystyle\int \frac{x^2 \, dx}{\sqrt{x^3 + 5}}.$

5. $\displaystyle\int \frac{dx}{(3 - x)^2}.$

6. $\displaystyle\int \frac{dx}{x + 1/x}.$

7. $\displaystyle\int \frac{x - 4}{x^2 + 4} \, dx.$

8. $\displaystyle\int \frac{x^2 \, dx}{(3 - 2x^3)^2}.$

9. $\displaystyle\int \frac{x + 2}{x^2 - 4} \, dx.$

10. $\displaystyle\int \frac{x^3 + 2x}{x^4 + 4} \, dx.$

11. $\displaystyle\int \frac{x^3 + 1}{x^4 + 4x} \, dx.$

12. $\displaystyle\int \frac{e^{3x} \, dx}{\sqrt{e^{3x} - \pi}}.$

13. $\displaystyle\int \cos 2x \sqrt{3 - \sin 2x} \; dx.$

14. $\displaystyle\int \cot 2x \; dx.$

15. $\displaystyle\int \frac{e^x \, dx}{e^x - 1}.$

16. $\displaystyle\int \frac{e^x - 1}{e^x} \, dx.$

17. $\displaystyle\int \frac{e^{-x}}{4 + e^{-2x}} \, dx.$

18. $\displaystyle\int \frac{4 \, dx}{e^x + e^{-x}}.$ (*Hint:* Multiply numerator and denominator by e^x.)

19. $\displaystyle\int \operatorname{sech} x \; dx.$

20. $\displaystyle\int \frac{e^x + e^{2x}}{4 + e^{2x}} \, dx.$

21. $\displaystyle\int \frac{e^x + e^{2x}}{\sqrt{1 + e^{2x}}} \, dx.$

22. $\displaystyle\int \frac{\ln x}{x} \, dx.$

23. $\displaystyle\int \frac{1}{x} \, d(\ln x).$

24. $\displaystyle\int \frac{\sin x \, dx}{1 + \cos^2 x}.$

25. $\displaystyle\int \frac{dx}{x[1 + (\ln x)^2]}.$

26. $\displaystyle\int \ln x^{1/x} \; dx.$

27. $\displaystyle\int \frac{\cosh \ln x}{x} \, dx.$

28. $\displaystyle\int \frac{\ln (1/x)}{x} \, dx.$

29. $\displaystyle\int \frac{x^3 \, dx}{\sqrt{9 - x^8}}.$

30. $\displaystyle\int \frac{x^2 \, dx}{\sqrt{x^2 + x^6}}.$

31. $\displaystyle\int \frac{\cosh^2 x \, dx}{\sqrt{1 + \sinh^2 x}}.$

32. $\displaystyle\int \frac{\arctan 2x}{1 + 4x^2} \, dx.$

33. $\displaystyle\int \frac{\sinh^{-1} (x/2)}{\sqrt{4 - x^2}} \, dx.$

34. $\displaystyle\int \frac{\cos x \, dx}{\sqrt{e^{\sin x}}}.$

35. $\displaystyle\int (2\theta - \sin 2\theta) \sin^2 \theta \; d\theta.$

36. $\displaystyle\int \frac{x^{-1/3}}{1 + x^{2/3}} \, dx.$

37. $\displaystyle\int \frac{x^{-2/3}}{1 + x^{2/3}} \, dx.$

38. $\int \cos \text{ arc} \sin \dfrac{x^2 - 4}{x^2 + 4}\, dx.$

39. $\int \left(1 - \dfrac{2}{e^x + 2}\right) dx.$

40. $\int \dfrac{(x - 1)\, dx}{\sqrt{x^2 - 2x + 2}}.$

41. $\int \dfrac{dx}{\sqrt{x^2 - 2x + 2}}.$

42. $\int \dfrac{(x - 3)\, dx}{\sqrt{x^2 - 2x + 2}}.$

43. Calculate $\int \ln \sin (3x^2 - 2e^{-5x}) \dfrac{(6x + 10e^{-5x}) \cos (3x^2 - 2e^{-5x})\, dx}{\sin (3x^2 - 2e^{-5x})}.$

Evaluate the following definite integrals:

44. $\int_0^{\pi/4} e^{\tan \theta} \sec^2 \theta\, d\theta.$

45. $\int_0^3 x\sqrt{25 - x^2}\, dx.$

46. $\int_{\pi/6}^{\pi/2} \cot \theta \ln (2 \sin \theta)\, d\theta.$

47. $\int_0^1 \dfrac{e^{-x}\, dx}{1 - e^{-2x}}.$

48. $\int_0^\infty v e^{-mv^2/2eE}\, dv.$

49. Find the rms value of $y = e^{x/2}/\sqrt{1 + e^{2x}}$ from $x = 0$ to $x = 1$.

13.2 DIVIDING OUT

Many answers to derivative problems are simplified algebraically before the final answer is achieved. The problem in integration, therefore, is often resolved into our ability to see what these answers were before simplification. The constant numerical factor of an integrand, for example, may have to be expressed as the product of two appropriate numerical factors before the answer to the integration is apparent. Example 2 illustrates this point; the original numerical factor of 1 was split into $\frac{1}{6} \times 6$.

Here and in Section 13.3, the essential method of integration requires us to examine a single fraction which has resulted from combining two fractions or more over a common denominator, to see how it can be split again into these two fractions or more. The preliminary work here and in most of the integration methods in this chapter consists of reversing standard techniques of simplification.

Let us consider the integration of fractions in which the numerator and denominator consists of terms in whole positive powers of the independent variable.

Many of these types of expression have already been analyzed for integration in such problems as Examples 6 and 7 of Chapter 12 and Example 4 of this chapter. In all expressions considered up to this point, the highest power in the numerator has been smaller than the highest power in the denominator. If this is not true, it is necessary to divide the denominator into the numerator and thus express the fraction as single terms plus a fraction in which the highest power in the numerator *is* smaller than the highest power in the denominator. All standard methods of integrating fractional expressions involving powers apply only to fractions in which the highest power in the numerator is smaller than the highest power in the denominator.

EXAMPLE 6.   Find $\int \dfrac{x^2 - 2x + 2}{3x^2 + 4}\, dx$.

Solution: Division is necessary to reduce the power of the final numerator.

$$3x^2 + 4 \overline{)x^2 - 2x + 2}\,|\tfrac{1}{3}$$
$$\underline{\quad x^2 \qquad\quad + \tfrac{4}{3}}$$
$$-2x + \tfrac{2}{3}$$

Hence the integral is

$$\int \left(\frac{1}{3} + \frac{-2x + \frac{2}{3}}{3x^2 + 4}\right) dx = \int \frac{dx}{3} - \int \frac{2x}{3x^2 + 4}\, dx + \int \frac{\frac{2}{3}\, dx}{3x^2 + 4}.$$

The final calculations are left to the student. The final answer is $x/3 - \tfrac{1}{3} \ln (3x^2 + 4) + (\sqrt{3}/9)$ arc tan $(\sqrt{3}\,x/2) + C$.
This type of division may also work in integrating fractions in which the terms are powers of e^x.

EXAMPLE 7.   Find $\int \dfrac{e^x - 1}{e^x + 1}\, dx$.

Solution: Since the differential of the denominator is $e^x\, dx$, it would be convenient to have merely a multiple of $e^x\, dx$ in the numerator without the extra $-1\, dx$. The -1 can be eliminated in the same way that we used for eliminating the x^2 term from the numerator in Example 6, that is, by a division in which both divisor and dividend begin with the numerical term.

$$1 + e^x \overline{)-1 + e^x}\,|-1$$
$$\underline{\quad -1 - e^x}$$
$$2e^x$$

Therefore the integral is

$$\int \left(-1 + \frac{2e^x}{1 + e^x}\right) dx = -x + 2 \ln (1 + e^x) + C.$$

Exercises 10, 11, and 12 following Section 13.3 are also cases where division will produce expressions that may be integrated by fairly standard methods, but no rule can be given regarding the order of the terms of the divisor and dividend in this division. The method used may be varied to suit the particular problem, and no hard and fast rule will replace the student's ingenuity.

13.3 PARTIAL FRACTIONS

The method of combining two or more fractions into a single fraction with a common denominator is well known. For example, the expression $3/(x - 2) - 2/(x + 3)$ may easily be transformed to

$$\frac{3(x + 3) - 2(x - 2)}{(x - 2)(x + 3)} = \frac{x + 13}{(x - 2)(x + 3)}.$$

The last expression is preferable to the original in most algebraic problems. However, if the problem involves an integration, the original expression with two separate fractions may be integrated quite simply, whereas the last expression could be integrated as it stands, but with considerable difficulty. Thus, when a fraction, in which the denominator may be expressed as the product of two or more factors is to be integrated, it would be very helpful to know how to split the fraction into two or more separate fractions.

This procedure is known as changing a fraction to *partial fractions*, and, as with most reverse calculations, it is not as straightforward as the direct method of changing the sum of two or more fractions to a single fraction. In this section we will examine methods of finding partial fractions. These methods can only be applied to single fractions in which the highest power of the numerator is smaller than the highest power of the denominator; if this is not true, the procedure of division will produce single terms plus a fraction in which it is true.

EXAMPLE 8.   Find $\int \dfrac{x + 13}{x^2 + x - 6}\, dx$.

Solution: Since the denominator can be factored, it should be possible to split the single fraction into two partial fractions. If the highest power of the original numerator is smaller than the highest power of the original denominator, the highest power in the numerator of each partial fraction will always be smaller than the highest power in its denominator. This is the guiding principle in determining the separate numerators. The fraction, when factored, becomes $(x + 13)/(x + 3)(x - 2)$, and this may therefore be considered identical to $A/(x + 3) + B/(x - 2)$, where

A and B are unknown constants whose value is to be determined. We may now directly express the sum of the two fractions as one fraction whose numerator may be compared with the original numerator in order to evaluate A and B.

$$\frac{A}{x+3} + \frac{B}{x-2} = \frac{A(x-2) + B(x+3)}{(x+3)(x-2)} = \frac{(A+B)x - 2A + 3B}{(x+3)(x-2)}.$$

The numerator here must be identical to the original numerator. Thus $A + B = 1$, and $-2A + 3B = 13$. Solving these equations, $A = -2$ and $B = 3$. Hence the fraction equals $3/(x-2) - 2/(x+3)$. This result should be verified at this point by the direct method of changing the sum of the two fractions to a single fraction. Therefore the integral is

$$\int \frac{3}{x-2}\,dx - \int \frac{2}{x+3}\,dx = 3\ln(x-2) - 2\ln(x+3) + C.$$

From log theory this is

$$\ln \frac{K(x-2)^3}{(x+3)^2} \qquad \text{where } C = \ln K.$$

Some shortcuts may be taken in this calculation. It is not necessary to regroup the numerator $A(x-2) + B(x+3)$ if it is realized that this expression must be equal to $x + 13$ for all values of x. If we let $x = -3$, the term involving B will disappear. Thus $A(x-2) = x + 13$ when $x = -3$, A equals the value of $(x+13)/(x-2)$ when $x = -3$, and

$$A = \frac{-3+13}{-3-2} = \frac{10}{-5} = -2.$$

Similarly B equals the value of $(x+13)/(x+3)$ when $x = 2$, and

$$B = \frac{2+13}{2+3} = 3.$$

This calculation may be quickly performed by the following method, which is often referred to as the *cover-up rule*:

For a fraction of the form

$$\frac{\text{numerator}}{(x+a)(x+b)(x+c)\cdots}$$

where all factors in the denominator are linear and the numerator consists only of terms in whole positive powers of the variable numerically smaller than the number of factors in the denominator, the numerator for the partial fraction with denominator of $(x + a)$ may be evaluated by covering up the factor $(x + a)$ in the original and evaluating the uncovered expression after replacing x by $-a$. Numerators for additional partial fractions may be similarly calculated. Thus, for the fraction

$$\frac{x^2 - 2}{x(x + 1)(x - 3)},$$

the numerator of the partial fraction with x as denominator is the value of

$$\frac{x^2 - 2}{(x + 1)(x - 3)}$$

where $x = 0$, which is

$$\frac{0 - 2}{(0 + 1)(0 - 3)} = +\frac{2}{3}.$$

The numerator of the partial fraction with $x + 1$ as denominator is the value of $(x^2 - 2)/x(x - 3)$ where $x = -1$, which is

$$\frac{+1 - 2}{-1(-1 - 3)} = -\frac{1}{4}.$$

The numerator of the partial fraction with $x - 3$ as denominator is the value of $(x^2 - 2)/x(x + 1)$ where $x = 3$, which is

$$\frac{9 - 2}{3(3 + 1)} = \frac{7}{12}.$$

Hence the original fraction equals

$$\frac{\frac{2}{3}}{x} - \frac{\frac{1}{4}}{x + 1} + \frac{\frac{7}{12}}{x - 3}.$$

This quick calculation may be used only if the numerator of the partial fraction concerned is a single constant term. This may not be equivalent to saying that the denominator of the partial fraction concerned is linear. For example, the fraction

$$\frac{x^2 + 5}{(x^2 + 2)(x^2 - 3)}$$

may be split into partial fractions by using the cover-up rule, since all factors of the denominator are linear in x^2 (which could be replaced by z, say), and thus all numerators of the partial fractions will be constant.

EXERCISES

Calculate the following integrals:

1. $\int \dfrac{x+1}{x-1}\,dx.$

2. $\int \dfrac{x+4}{2-x}\,dx.$

3. $\int \dfrac{1+4x}{1-2x}\,dx.$

4. $\int \dfrac{1+x}{2-3x}\,dx.$

5. $\int \dfrac{x^3-3x}{x^2+4}\,dx.$

6. $\int \dfrac{3x^3+12x-2}{x^2+4}\,dx.$

7. $\int \dfrac{x^3-2x^2+3x-4}{x^2+9}\,dx.$

8. $\int \dfrac{(x-1)(x^2+2x+1)}{x^2-1}\,dx.$

9. $\int \dfrac{ax^2+b}{c-kx}\,dx.$

10. $\int \dfrac{2\,dx}{e^x+1}.$

11. $\int \dfrac{e^{2x}\,dx}{e^x+1}.$

12. $\int \dfrac{e^{2x}-e^x}{e^x+1}\,dx.$

Calculate the following integrals:

13. $\int \dfrac{x+3}{x^2-4}\,dx.$

14. $\int \dfrac{x-4}{x^2-4x}\,dx.$

15. $\int \dfrac{x-2}{x^2-4x}\,dx.$

16. $\int \dfrac{x^3}{x^2-9}\,dx.$

17. $\int \dfrac{x^3-3x}{x^2-5}\,dx.$

18. $\int \dfrac{3\,dx}{x^2-5}.$

19. $\int \dfrac{dx}{2x^2+7x+6}.$

20. $\int \dfrac{x^2\,dx}{2x^2-x-6}.$

21. $\int \dfrac{x^2\,dx}{2x^4+7x^2+6}.$

22. Find $\int \dfrac{dx}{x^2-a^2}$ assuming $x^2 > a^2$. Explain the reason for this limitation in the answer.

23. Find $\int \dfrac{dx}{a^2-x^2}$ assuming $x^2 < a^2$ (two separate methods of solution).

24. Find $\int \dfrac{dx}{x^2-a^2}$ assuming $x^2 < a^2$.

25. Find $\int \dfrac{dx}{a^2-x^2}$ assuming $x^2 > a^2$.

Evaluate the following definite integrals:

26. $\int_3^6 \dfrac{dx}{x^2-4}.$

27. $\int_0^1 \dfrac{dx}{x^2-4}.$

28. $\int_3^7 \dfrac{dx}{4-x^2}.$

29. $\int_0^1 \dfrac{dx}{x^2-5x+6}.$

30. $\int_1^2 \dfrac{dx}{x^2-x-6}.$

31. $\int_4^5 \dfrac{dx}{x^2-x-6}.$

32. Sketch the graphs of $y = 1/(x^2-9)$ and $y = 1/(9-x^2)$, and calculate the area bounded by these two graphs and the lines $x = 0$ and $x = 2$.

33. For a short interval of time t seconds, the voltage drop across an inductance of 80 millihenries is $(3t^2-1)/(t^2+7t+6)$ volts. Find the current as a function of t if the current is zero at $t = 0$.

Partial Fractions with Nonlinear Denominators. If the numerator of the partial fraction is not simply one constant term, the cover-up rule cannot be used. However, the more general method of introducing unknown constants for the coefficients in the numerators of the partial fractions may always be used. Assuming that all powers of the variable involved in the original numerator are numerically smaller than the highest power in the denominator, this same relationship holds true in the partial fractions, and thus the form of each numerator can be determined.

EXAMPLE 9. Find $\int \dfrac{x^2 + 3x + 5}{x^3 + 4x}\, dx$.

Solution: Since the denominator can be factored, a solution by partial fractions is indicated. The fraction equals

$$\frac{x^2 + 3x + 5}{x(x^2 + 4)} = \frac{\frac{5}{4}}{x} + \frac{Ax + B}{x^2 + 4} = \frac{\frac{5}{4}(x^2 + 4) + x(Ax + B)}{x(x^2 + 4)}$$

$$= \frac{(\frac{5}{4} + A)x^2 + Bx + 5}{x(x^2 + 4)}.$$

Notice that the numerator of the first partial fraction may be obtained by cover-up rule, and the numerator of the second partial fraction may contain a term in all positive whole powers of x lower than the second power. Equating coefficients of x^2 and x in the numerators, $\frac{5}{4} + A = 1$, and $A = -\frac{1}{4}$; $B = 3$. The fact that the $+5$ occurs as the constant term in both forms is a check on the accuracy of the numerator of the first partial fraction. Hence the fraction equals

$$\frac{5}{4x} + \frac{-\frac{1}{4}x + 3}{x^2 + 4}, \text{ and the integral is}$$

$$\int \left(\frac{5}{4x} - \frac{1}{4}\frac{x}{x^2 + 4} + \frac{3}{x^2 + 4} \right) dx$$

$$= \frac{5}{4} \ln x - \frac{1}{8} \ln (x^2 + 4) + \frac{3}{2} \text{ arc tan} \frac{x}{2} + C.$$

This answer should be verified.

EXAMPLE 10. Find $\int \dfrac{(s^2 + 1)(s^2 + 4)}{s(s^2 + 2)(s^2 + 3)}\, ds$.

Solution: The fraction may be put equal to

$$\frac{\frac{2}{3}}{s} + \frac{As + B}{s^2 + 2} + \frac{Cs + D}{s^2 + 3},$$

where the first partial fraction is determined by the cover-up rule and all powers of each numerator are smaller than the highest power of the corresponding denominator. Noting that the original numerator when multiplied would contain only even powers of s, and that, when the partial fractions are combined over a common denominator, B and D will only appear with s^3 and s terms, it might be seen before proceeding further that B and D must be zero, and the solution will be simpler if B and D are left out of the calculation immediately. Thus

$$\frac{(s^2 + 1)(s^2 + 4)}{s(s^2 + 2)(s^2 + 3)} = \frac{\frac{2}{3}(s^2 + 2)(s^2 + 3) + As^2(s^2 + 3) + Cs^2(s^2 + 2)}{s(s^2 + 2)(s^2 + 3)}.$$

The numerators of the two sides must be identical for all values of s. If s^2 is made equal to -2 in both numerators, two of the three terms in the numerator of the right-hand side will be zero, and A is evaluated quickly. Thus, $A(-2)(-2 + 3) = (-2 + 1)(-2 + 4)$, $-2A = -2$, and $A = 1$. Similarly, by putting $s^2 = -3$ in both numerators, $C(-3)(-3 + 2) = (-3 + 1)(-3 + 4)$, and $C = -\frac{2}{3}$. Therefore the fraction equals

$$\frac{2}{3s} + \frac{s}{s^2 + 2} - \frac{2}{3}\frac{s}{s^2 + 3}.$$

This result should be verified before the integration is worked out. Therefore the integral equals

$$\frac{2}{3}\ln s + \frac{1}{2}\ln (s^2 + 2) - \frac{1}{3}\ln (s^2 + 3) + \ln C = \ln \frac{Cs^{\frac{2}{3}}(s^2 + 2)^{\frac{1}{2}}}{(s^2 + 3)^{\frac{1}{3}}}.$$

A fraction in which the denominator has a repeating factor (that is, this factor occurs as a square or higher whole power in the denominator) may usually be split into partial fractions as an aid to integration. A fraction, such as $3/(2x + 5)^2$, where all powers of the variable in the numerator are numerically less than the highest power inside the parentheses in the denominator, cannot be split up, but in most cases such fractions can be integrated directly. A fraction of the form $(ax + b)/(cx + d)^2$ has partial fractions of the form $A/(cx + d)^2 + B/(cx + d)$. Similarly, the form $(ax^2 + bx + k)/(cx + d)^3$ has partial fractions of the form $A/(cx + d)^3 + B/(cx + d)^2 + C/(cx + d)$. A fraction of the form

$$\frac{ax^{2n-1} + bx^{2n-2} \cdots + k}{(cx^2 + dx + e)^n}$$

has partial fractions of the form

$$\frac{Ax + B}{(cx^2 + dx + e)^n} + \frac{Cx + D}{(cx^2 + dx + e)^{n-1}} \cdots + \frac{Mx + N}{cx^2 + dx + e}.$$

Whenever there is a repeating factor in the denominator, all powers of the variable in each separate numerator are smaller than the highest power inside the parentheses of the denominator.

EXAMPLE 11.   Find $\displaystyle\int \frac{3x^2 - 5}{(x - 2)^2(x + 3)}\, dx$.

Solution: The fraction may be considered equal to

$$\frac{22/25}{x + 3} + \frac{A}{(x - 2)^2} + \frac{B}{x - 2},$$

the first numerator being calculated by the cover-up rule, and the forms of the next two fractions being determined by the preceding rules. Therefore

$$\frac{3x^2 - 5}{(x - 2)^2(x + 3)} \equiv \frac{22/25(x - 2)^2 + A(x + 3) + B(x + 3)(x - 2)}{(x - 2)^2(x + 3)}.$$

Substituting $x = 2$ in each numerator produces a solution for A directly. $3(2)^2 - 5 = A(5)$. $A = \tfrac{7}{5}$. B may be found fairly rapidly by substituting $x = 0$ in each numerator. $-5 = (4)22/25 + \tfrac{7}{5}(3) - 6B$. $6B = 318/25$; $B = 53/25$. Therefore the fraction equals

$$\frac{22/25}{x + 3} + \frac{7/5}{(x - 2)^2} + \frac{53/25}{x - 2},$$

and the integral equals

$$\frac{22}{25}\ln (x + 3) - \frac{7}{5(x - 2)} + \frac{53}{25}\ln (x - 2) + C.$$

EXAMPLE 12.   Calculate $\displaystyle\int_0^1 \frac{x^3 - 2x^2 + x - 6}{(x^2 + 3)^2}\, dx$.

Solution: Since there is a repeating factor in the denominator, it should be realized that the fraction may be split into two partial fractions, each of which should be easier to integrate than the original single fraction. According to the rules, let

$$\begin{aligned}
\frac{x^3 - 2x^2 + x - 6}{(x^2 + 3)^2} &= \frac{Ax + B}{(x^2 + 3)^2} + \frac{Cx + D}{x^2 + 3} \\
&= \frac{Ax + B + (Cx + D)(x^2 + 3)}{(x^2 + 3)^2} \\
&= \frac{Cx^3 + Dx^2 + (A + 3C)x + (B + 3D)}{(x^2 + 3)^2}.
\end{aligned}$$

By equating corresponding coefficients in the numerators,

$$C = 1, \quad D = -2, \quad A + 3C = 1, \quad B + 3D = -6,$$
$$A + 3 = 1, \quad B - 6 = -6,$$
$$A = -2. \quad B = 0.$$

Therefore the integral is

$$\int_0^1 \left[\frac{-2x}{(x^2 + 3)^2} + \frac{x - 2}{x^2 + 3} \right] dx$$

$$= \int_0^1 \left[\frac{-2x}{(x^2 + 3)^2} + \frac{x}{x^2 + 3} - \frac{2}{x^2 + 3} \right] dx$$

$$= \left[\frac{1}{x^2 + 3} + \frac{1}{2} \ln (x^2 + 3) - \frac{2}{\sqrt{3}} \arctan \frac{x}{\sqrt{3}} \right]_0^1$$

$$= \frac{1}{4} + \frac{1}{2} \ln 4 - \frac{2}{\sqrt{3}} \arctan \frac{1}{\sqrt{3}} - \frac{1}{3} - \frac{1}{2} \ln 3 + \frac{2}{\sqrt{3}} \arctan 0$$

$$= \frac{1}{4} + \frac{1}{2} \ln \frac{4}{3} - \frac{2}{\sqrt{3}} \frac{\pi}{6} - \frac{1}{3} + 0 = -0.54.$$

EXAMPLE 13.   Find $\displaystyle\int \frac{x^2 + 4}{x^2(x^2 - 4x + 6)} \, dx.$

Solution: By using the given rules for forming the numerators, let

$$\frac{x^2 + 4}{x^2(x^2 - 4x + 6)}$$

$$= \frac{A}{x^2} + \frac{B}{x} + \frac{Cx + D}{x^2 - 4x + 6}$$

$$= \frac{(A + Bx)(x^2 - 4x + 6) + x^2(Cx + D)}{x^2(x^2 - 4x + 6)}$$

$$= \frac{(B + C)x^3 + (A - 4B + D)x^2 + (6B - 4A)x + 6A}{x^2(x^2 - 4x + 6)}.$$

Comparing numerators, from the right,

$$6A = 4 \quad 6B - 4A = 0 \quad A - 4B + D = 1 \quad B + C = 0$$
$$A = \tfrac{2}{3} \quad 6B - \tfrac{8}{3} = 0 \quad \tfrac{2}{3} - \tfrac{16}{9} + D = 1 \quad \tfrac{4}{9} + C = 0$$
$$B = \tfrac{4}{9} \quad D = \tfrac{19}{9} \quad C = -\tfrac{4}{9}.$$

Therefore the integral is $\displaystyle\int \left[\frac{2}{3x^2} + \frac{4}{9x} + \frac{-4x + 19}{9(x^2 - 4x + 6)} \right] dx.$

In order to integrate the last of these three fractions, it must be further split into two fractions, the numerator of the first fraction being a multiple of the derivative of the denominator in which the first term is $-4x$, and the numerator of the second fraction being numerical.

$$\frac{-4x + 19}{9(x^2 - 4x + 6)} = \frac{-4x + 8}{9(x^2 - 4x + 6)} + \frac{11}{9(x^2 - 4x + 6)}$$

$$= -\frac{2}{9}\frac{2x - 4}{x^2 - 4x + 6} + \frac{11}{9[(x - 2)^2 + 2]}.$$

The original integral may now be worked out completely as

$$-\frac{2}{3x} + \frac{4 \ln x}{9} - \frac{2 \ln (x^2 - 4x + 6)}{9} + \frac{11}{9\sqrt{2}} \arctan \frac{x - 2}{\sqrt{2}} + C.$$

It must be emphasized that all rules for forming partial fractions apply only if the highest power in the numerator is smaller than the highest power in the denominator. If this is not true, the fraction must be divided out.

EXERCISES

Find the following integrals:

1. $\int \dfrac{3\,dx}{x^3 + 3x}.$

2. $\int \dfrac{3x\,dx}{x^3 + 3x}.$

3. $\int \dfrac{3x + 1}{x^2(x - 1)}\,dx.$

4. $\int \dfrac{dx}{x^2(4 - x)}.$

5. $\int \dfrac{3x + 1}{x(x - 1)^2}\,dx.$

6. $\int \dfrac{3\,dx}{x^2(x - 1)^2}.$

7. $\int \dfrac{3x^3 + 1}{x(x - 1)^2}\,dx.$

8. $\int \dfrac{3x^3 + 1}{x(x^2 - 1)}\,dx.$

9. $\int \dfrac{x^2 + 1}{x^3 + 3x}\,dx.$

10. $\int_0^1 \dfrac{(x^4 - x^2 + 1)\,dx}{(x^2 + 1)(x^2 + 3)}.$

11. $\int_0^1 \dfrac{(x^3 - x)\,dx}{(x^2 + 1)(x^2 + 3)}.$

12. $\int \dfrac{-3x^4 - 1}{x(x^2 + 1)^2}\,dx.$

Find the following integrals using the method indicated for the integration of the last partial fraction in the solution of Example 13:

13. $\int \dfrac{(2x - 2)\,dx}{x^2 - 2x + 2}.$

14. $\int \dfrac{(2x - 1)\,dx}{x^2 - 2x + 2}.$

15. $\int \dfrac{(4x - 2)\,dx}{x^2 - 2x + 2}.$

16. $\int \dfrac{4x\,dx}{\sqrt{x^2 - 4x + 5}}.$

17. $\int \dfrac{4x\,dx}{\sqrt{4x - x^2}}.$

18. $\int \dfrac{2x\,dx}{\sqrt{x}\sqrt{x - 4}}.$

Find the following integrals:

19. $\int \dfrac{6x - 12}{x^3 + 8}\,dx.$

20. $\int \dfrac{(2x - 3)\,dx}{x(x^2 + 6x + 10)}.$

21. $\int \dfrac{(2x - 3)\,dx}{x(x^2 + 6x + 9)}.$

22. $\int \dfrac{(3x^2 + 2)\,dx}{x(x^2 + 1)(x^2 + 4)}.$

23. $\int \dfrac{6 - 4x^4}{x(1 - x^2)(3 - x^2)}\,dx,\ \text{if } x^2 < 1.$

24. $\int_1^2 \dfrac{(4x + 1)\,dx}{x^3 + 4x^2 + 5x}.$

25. $\int \dfrac{(e^{2x} + 2)\,dx}{e^{3x} + 4e^x}.$

26. Sketch the graph of $i = 1/(9 - t^2)$ from $t = 0$ to $t = 3$, and find the rms value of i from $t = 0$ to $t = 2$.

27. Sketch the graph of $y = 1/(x^2 - 9)$ from $x = 0$ to $x = 3$ and find the volume of revolution when the area bounded by this graph, the x-axis, and the lines $x = 0$ and $x = 2$ is rotated about the x-axis.

13.4   SUBSTITUTIONS FOR IRRATIONAL EXPRESSIONS

When a square root, cube root, or root of any order is a factor in the expression to be integrated, a method of solution involving no fractional indices is often possible by substituting for the root. This is especially true if the expression under the radical consists only of a term in the first power of the variable and a constant term.

EXAMPLE 14.   Find $\int x^2 \sqrt{3 - x}\, dx$.

Solution: Let $\sqrt{3 - x} = u$.

Then $3 - x = u^2$.

$$x = 3 - u^2. \qquad dx = -2u\, du.$$

The integral is

$$-2 \int (3 - u^2)^2 uu\, du = -2 \int u^2 (9 - 6u^2 + u^4)\, du$$

$$= -2 \int (9u^2 - 6u^4 + u^6)\, du$$

$$= -2\left(3u^3 - \frac{6}{5} u^5 + \frac{u^7}{7}\right) + C$$

$$= -2\left[3(3 - x)^{3/2} - \frac{6}{5}(3 - x)^{5/2} + \frac{(3 - x)^{7/2}}{7}\right] + C.$$

EXAMPLE 15.   Find

$$\int \frac{x\, dx}{(2x - 1)^{2/3}}.$$

Solution: To avoid any fractional indices in the solution, it is best to substitute for the portion of the irrational expression that represents a root. Let $(2x - 1)^{1/3} = u$.

Then $(2x - 1)^{2/3} = u^2$. $2x - 1 = u^3$.

$$x = \frac{u^3 + 1}{2}. \qquad dx = \tfrac{3}{2} u^2\, du.$$

The integral is

$$\frac{3}{4} \int \frac{u^2(u^3 + 1)}{u^2} \, du = \frac{3}{4} \int (u^3 + 1) \, du$$

$$= \frac{3}{4} \left(\frac{u^4}{4} + u \right) + C$$

$$= \frac{3}{4} \left[\frac{(2x - 1)^{\frac{4}{3}}}{4} + (2x - 1)^{\frac{1}{3}} \right] + C$$

$$= \frac{3}{16} (2x - 1)^{\frac{1}{3}}(2x - 1 + 4) + C$$

$$= \frac{3}{16} (2x - 1)^{\frac{1}{3}}(2x + 3) + C.$$

This method may sometimes be used when the expression under the radical contains a higher power of the variable and a constant term.

EXAMPLE 16.   Evaluate

$$\int_1^2 \frac{\sqrt{9 - x^3}}{x} \, dx.$$

Solution: Let $\sqrt{9 - x^3} = u.$ Then $9 - x^3 = u^2,$ $x^3 = 9 - u^2,$ $x = (9 - u^2)^{\frac{1}{3}},$ and $dx = (-2u/3)(9 - u^2)^{-\frac{2}{3}} \, du.$ When $x = 1,$ $u = 2\sqrt{2},$ and when $x = 2,$ $u = 1.$ Therefore the integral is

$$\int_{2\sqrt{2}}^1 \left(-\frac{2}{3} \right) \frac{u^2(9 - u^2)^{-\frac{2}{3}}}{(9 - u^2)^{\frac{1}{3}}} \, du = -\frac{2}{3} \int_{2\sqrt{2}}^1 \frac{u^2}{9 - u^2} \, du$$

$$= -\frac{2}{3} \int_{2\sqrt{2}}^1 \left(-1 + \frac{9}{9 - u^2} \right) du$$

(by dividing)

$$= -\frac{2}{3} \left[-u + \frac{9}{6} \ln \frac{3 + u}{3 - u} \right]_{2\sqrt{2}}^1$$

$$= -\frac{2}{3} \left[-1 + 2\sqrt{2} + \frac{3}{2} \left(\ln \frac{4}{2} - \ln \frac{5.83}{0.17} \right) \right]$$

$$= 1.61.$$

The method of evaluating the limits as values of u avoids the necessity of replacing u by $\sqrt{9 - x^3}.$

In other cases, a different kind of substitution is necessary, but too many rules regarding the type of substitution become unwieldy. In many cases the type of substitution can only be determined by trial and error. Some type of substitution will usually work in integrating an expression containing a root.

EXAMPLE 17.   Find $\int \sqrt{3x^2 - 1}\ x^3\ dx$.

Solution: This may be calculated by letting $\sqrt{3x^2 - 1} = u$. However, another method here is to substitute $3x^2 - 1 = u$. Then $6x\ dx = du$, and $x^2 = (u + 1)/3$.

The integral is

$$\frac{1}{6} \int \sqrt{3x^2 - 1}\ x^2 6x\ dx = \frac{1}{18} \int u^{\frac{1}{2}}(u + 1)\ du$$

$$= \frac{1}{18} \int (u^{\frac{3}{2}} + u^{\frac{1}{2}})\ du$$

$$= \frac{1}{18} \left(\frac{2}{5} u^{\frac{5}{2}} + \frac{2}{3} u^{\frac{3}{2}} \right) + C$$

$$= \frac{(3x^2 - 1)^{\frac{5}{2}}}{45} + \frac{(3x^2 - 1)^{\frac{3}{2}}}{27} + C.$$

EXERCISES

Find the following integrals and verify the answers:

1. $\int x\sqrt{2x - 1}\ dx.$ 2. $\int x\sqrt{2x^2 - 1}\ dx.$ 3. $\int \dfrac{x\ dx}{\sqrt{2 - x}}.$

4. $\int \dfrac{dx}{\sqrt{2 - x^2}}.$ 5. $\int \dfrac{x\ dx}{(2 - x)^{\frac{3}{2}}}.$ 6. $\int x\sqrt[3]{2x - 5}\ dx.$

7. $\int \dfrac{2x\ dx}{(5 - 2x)^{\frac{3}{4}}}.$ 8. $\int \sqrt{x^3 - 3x^2}\ dx.$ 9. $\int x^3\sqrt{x^2 + 2}\ dx.$

10. $\int x\sqrt{x^2 + 2}\ dx.$ 11. $\int \dfrac{\sqrt{9 - x^2}}{x}\ dx.$ 12. $\int x^3\sqrt[3]{9 - x^2}\ dx.$

13. $\int \dfrac{(x^2 - 8)^{\frac{3}{2}}}{x}\ dx.$

14. $\int \dfrac{x^3\ dx}{(x^2 - 1)^2}$ (by putting $x^2 - 1 = u$. Verify using partial fractions).

Evaluate the following definite integrals:

15. $\int_1^4 \dfrac{t^2\, dt}{\sqrt{5-t}}.$

16. $\int_0^7 \dfrac{t\, dt}{(8-t)^{\frac{2}{3}}}.$

17. $\int_2^4 \dfrac{\sqrt{x^2-4}\, dx}{x}.$

18. $\int_3^5 x^3\sqrt{x^2-9}\, dx.$

Find the mean current i from $t = 0$ to $t = 2$ for the following:

19. $i = \dfrac{t}{\sqrt{4-t}}.$

20. $i = t(3-t)^{\frac{3}{2}}.$

21. Find the rms current from $t = 1$ to $t = 3$ for the current of Exercise 19.

22. Find the rms current from $t = 1$ to $t = 3$ for the current of Exercise 20.

13.5 INTEGRATION OF FURTHER TRIGONOMETRIC FUNCTIONS

Certain types of trigonometric functions require some manipulation before they are in a form that is easily integrated.

$\int \tan u\, du$ may be readily handled as $\int \sin u\, du/\cos u$, in which the numerator is the negative of the differential of the denominator. Therefore

$$\int \tan u\, du = -\ln \cos u + C = \ln (\cos u)^{-1} + C = \ln \sec u + C.$$

Similarly,

$$\int \cot u\, du = \ln \sin u + C \quad\text{or}\quad -\ln \operatorname{cosec} u + C.$$

$\int \sec u\, du$ presents a more difficult problem, and its solution is an illustration of the frequent necessity to make the expression more complicated in order to put it into a form that can be integrated.

$$\int \sec u\, du = \int \frac{1}{\cos u}\, du = \int \frac{\cos u}{\cos^2 u}\, du = \int \frac{\cos u}{1-\sin^2 u}\, du = \int \frac{d(\sin u)}{1-\sin^2 u}$$

$$= \frac{1}{2}\ln \frac{1+\sin u}{1-\sin u} + C.$$

This is usually converted to another form.

$$\frac{1}{2}\ln \frac{1+\sin u}{1-\sin u} = \frac{1}{2}\ln \frac{(1+\sin u)^2}{1-\sin^2 u} = \frac{1}{2}\ln \frac{(1+\sin u)^2}{\cos^2 u}$$

$$= \ln \frac{1+\sin u}{\cos u} = \ln (\sec u + \tan u).$$

Hence

$$\int \sec u \; du = \ln (\sec u + \tan u) + C.$$

Similarly, $\int \operatorname{cosec} u \; du$ may be proved equal to $-\ln (\operatorname{cosec} u + \cot u) + C$. $\int \sin^2 u \; du$ and $\int \cos^2 u \; du$ have already been discussed in Example 5 of Chapter 10. The formulas for $\sin^2 u$ and $\cos^2 u$ used for these integrations may also be used to integrate higher positive even powers of the sine and cosine functions.

EXAMPLE 18.   Find $\int \cos^4 x \; dx$.
Solution: By successive use of the formula $\cos^2 x = \frac{1}{2}(1 + \cos 2x)$ the fourth power is reduced to a first power.

$$\cos^4 x = (\cos^2 x)^2 = [\tfrac{1}{2}(1 + \cos 2x)]^2 = \tfrac{1}{4}(1 + 2 \cos 2x + \cos^2 2x)$$

$$= \tfrac{1}{4}[1 + 2 \cos 2x + \tfrac{1}{2}(1 + \cos 4x)]$$

$$= \tfrac{1}{8}(3 + 4 \cos 2x + \cos 4x).$$

$$\therefore \int \cos^4 x \; dx = \tfrac{1}{8}(3x + 2 \sin 2x + \tfrac{1}{4} \sin 4x) + C.$$

To integrate odd powers of the sine and cosine functions, the expression is adapted to fit the structure of $\int u^n \; du$.

EXAMPLE 19.   Find $\int \sin^5 x \; dx$.
Solution:

$$\int \sin^5 x \; dx = \int \sin^4 x \sin x \; dx = \int (1 - \cos^2 x)^2 \sin x \; dx$$

$$= \int (1 - 2 \cos^2 x + \cos^4 x) \sin x \; dx$$

$$= \int \sin x \; dx + 2 \int \cos^2 x \; d(\cos x) - \int \cos^4 x \; d(\cos x)$$

$$= -\cos x + \tfrac{2}{3} \cos^3 x - \tfrac{1}{5} \cos^5 x + C.$$

This same method may also be used to integrate the product of powers of $\sin x$ and $\cos x$ if the sum of the powers is an odd number. Thus for $\int \cos^5 x \sin^4 x \; dx$, the integral may be put in the form $\int \cos^4 x \sin^4 x \cos x \; dx$ and $\cos^4 x$ must be expressed in terms of $\sin x$. Then $\cos x \; dx$ is $d(\sin x)$.

The integration of the product of two sine or cosine functions to the first power or the product of a sine and a cosine is performed by using a standard trigonometric formula to change these products to sums or differences.

EXAMPLE 20.   Find $\int \sin 3x \cos 5x \, dx$.

Solution: The integral is

$$\frac{1}{2} \int [\sin 8x + \sin(-2x)] \, dx = \frac{1}{2} \int (\sin 8x - \sin 2x) \, dx$$

$$= \frac{1}{2} \left(-\frac{\cos 8x}{8} + \frac{\cos 2x}{2} \right) + C.$$

Integrals involving powers of the tangent function may in some cases be calculated by using the formula $\sec^2 x = 1 + \tan^2 x$ and using the fact that $\sec^2 x \, dx$ is the differential of $\tan x$.

EXAMPLE 21.   Find $\int \tan^3 x \, dx$.

Solution:

$$\int \tan^3 x \, dx = \int \tan^2 x \tan x \, dx = \int (\sec^2 x - 1) \tan x \, dx$$

$$= \int \tan x \, d(\tan x) - \int \tan x \, dx = \frac{\tan^2 x}{2} - \ln \sec x + C.$$

Some other integrals of combinations of trigonometric functions will be discussed in Section 13.7.

EXERCISES

Find the following integrals:

1. $\int \tan \theta \sec^2 \theta \, d\theta$.

2. $\int \cos^4 \theta \sin \theta \, d\theta$.

3. $\int \cosh^4 x \sinh x \, dx$.

4. $\int \operatorname{cosec} x \, dx$.

5. $\int \cos^2 2\theta \sin^2 2\theta \, d\theta$.

6. $\int \cosh^2 2x \sinh^2 2x \, dx$.

7. $\int \cos^3 x \, dx$.

8. $\int \sin^4 2x \, dx$.

9. $\int \sin^6 x \, dx$.

10. $\int \sinh^7 x \, dx$.

11. $\int \cos^2 x \sin^5 x \, dx$.

12. $\int \cos^2 (50\pi t + \pi/3) \cos^3 (50\pi t - \pi/6) \, dt$.

13. $\int \sin^4 x \cos^2 x \, dx$.

14. $\int \tan^2 x \, dx$.

15. $\int \tan^4 \theta \, d\theta$.

16. $\int \sec^4 \theta \, d\theta$.

17. $\int \sec^4 \theta \tan^2 \theta \, d\theta$.

18. $\int \dfrac{\sin^3 x \, dx}{\cos^5 x}$.

19. $\int \dfrac{\sin^3 x \, dx}{\cos^3 x}$.

20. $\int \tan^5 x \, dx$.

21. $\int \sec^5 x \sqrt{1 + \tan^2 x} \, dx.$

22. $\int \cos 2x \cos 5x \, dx.$

23. $\int \sin 2x \cos 2x \, dx.$

24. $\int \sin 3x \sin 5x \, dx.$

25. $\int \sin^2 3x \cos^2 5x \, dx.$

26. $\int \sin (\theta - \pi/3) \cos (\theta + \pi/3) \, d\theta.$

27. $\int \sin (\theta + \pi/3) \cos (\theta - \pi/6) \, d\theta.$

28. $\int (1 - \sin \theta) \cos^2 \theta \, d\theta.$

29. $\int \sqrt{\sin^4 \theta - \sin^6 \theta} \, d\theta.$

Evaluate the following definite integrals:

30. $\int_0^{\pi/2} \sin^5 \theta \, d\theta.$

31. $\int_0^{\pi/6} \tan^3 2\theta \, d\theta.$

32. $\int_0^{1/400} (\cos 100\pi t + \cos 200\pi t)^2 \, dt.$

33. $\int_0^{\pi/4} \frac{\sec^2 x \, dx}{\sqrt{1 - \tan^2 x}}.$

34. Find the rms current i for one cycle of the current $i = 6 \sin 200\pi t + 2 \sin 600\pi t$. One cycle corresponds to one cycle of $6 \sin 200\pi t$.

35. Find the energy consumed when a current $i = 6 \sin 100\pi t + 3 \sin (200\pi t - \pi/3)$ amperes flows in a resistance of 1.5 ohms for two hours.

13.6   TRIGONOMETRIC AND HYPERBOLIC SUBSTITUTIONS

If any one of three sides of a right-angled triangle is constant and one of the other sides is variable, the third side, calculated by the Pythagorian theorem, is expressed in such forms as $\sqrt{a^2 - x^2}$, $\sqrt{a^2 + 4x^2}$, $\sqrt{9x^2 - 4}$, etc. When these types of expressions occur in functions to be integrated, they may be represented as a side of a right-angled triangle, and various functions of the independent variable may be expressed as trigonometric functions of an angle of this triangle. The resulting expression is often in a form that may be readily integrated. This is the technique known as *trigonometric substitution*.

EXAMPLE 22.   Find $\int x^2 \sqrt{4 - x^2} \, dx.$

Solution: The clue for determining what method to use is the form of the expression $\sqrt{4 - x^2}$, which is the length of one side of a right-angled triangle in which the hypotenuse is 2 and the other side x, as illustrated in Fig. 13-1. In Fig. 13-1, the sides represented by x and $\sqrt{4 - x^2}$ may be interchanged without affecting the final result. The method now consists of substituting for all expressions involving x and for dx as functions of θ and $d\theta$.

$$\sqrt{4 - x^2} = 2 \cos \theta.$$
$$x = 2 \sin \theta.$$
$$dx = 2 \cos \theta \, d\theta.$$

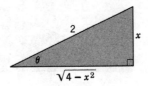

Fig. 13-1. Diagram for a trigonometric substitution.

Hence
$$\int x^2\sqrt{4 - x^2}\, dx = \int 4\sin^2\theta\, 2\cos\theta\, 2\cos\theta\, d\theta$$

$$= 16\int \sin^2\theta\cos^2\theta\, d\theta$$

$$= 16\int (\tfrac{1}{2}\sin 2\theta)^2\, d\theta$$

$$= 4\times\tfrac{1}{2}\int (1 - \cos 4\theta)\, d\theta$$

$$= 2\theta - \tfrac{1}{2}\sin 4\theta + C.$$

In order to express the answer in terms of the original variable x, we may again use Fig. 13-1 to advantage. We see that $\theta = \arcsin(x/2)$. However, $\sin 4\theta$ must first be expressed in terms of functions of θ.

$$\sin 4\theta = 2\sin 2\theta\cos 2\theta = 4\sin\theta\cos\theta(\cos^2\theta - \sin^2\theta).$$

Thus the integral is

$$2\theta - 2\sin\theta\cos\theta(\cos^2\theta - \sin^2\theta) + C$$

$$= 2\arcsin\frac{x}{2} - \frac{2x}{2}\frac{\sqrt{4 - x^2}}{2}\left(\frac{4 - x^2}{4} - \frac{x^2}{4}\right) + C$$

$$= 2\arcsin\frac{x}{2} - \frac{x\sqrt{4 - x^2}(2 - x^2)}{4} + C.$$

EXAMPLE 23.   Find
$$\int_0^1 \frac{dx}{(1 + x^2)^{3/2}}.$$

Solution: The expression $(1 + x^2)^{3/2}$ provides the clue for representing $(1 + x^2)^{1/2}$ as the hypotenuse of the right-angled triangle shown in Fig. 13-2.

Let $\sqrt{1 + x^2} = \sec\theta$. Then $(1 + x^2)^{3/2} = \sec^3\theta$, $x = \tan\theta$, and $dx = \sec^2\theta\, d\theta$.

When $x = 0$, $\theta = 0$ and when $x = 1$, $\theta = \pi/4$.

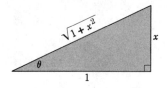

Fig. 13-2.

Hence the integral is

$$\int_0^{\pi/4} \frac{\sec^2 \theta}{\sec^3 \theta}\, d\theta \;=\; \int_0^{\pi/4} \frac{d\theta}{\sec \theta}$$

$$= \int_0^{\pi/4} \cos \theta\, d\theta$$

$$= \big[\sin \theta\big]_0^{\pi/4}$$

$$= 0.707.$$

Occasionally, a hyperbolic substitution offers the easiest solution to an integration. Such expressions as $\sqrt{a^2 + x^2}$, $(x^2 - a^2)^{3/2}$, etc., may be put into hyperbolic form with proper substitutions by using the formula $\cosh^2 u - \sinh^2 u = 1$ and its transposed forms $1 + \sinh^2 u = \cosh^2 u$ and $\cosh^2 u - 1 = \sinh^2 u$. The expression $a^2 + x^2$ must be made to fit the formula $1 + \sinh^2 u = \cosh^2 u$, which may be accomplished by substituting $x = a \sinh u$; then

$$a^2 + x^2 = a^2 + a^2 \sinh^2 u = a^2(1 + \sinh^2 u) = a^2 \cosh^2 u.$$

The expression $x^2 - a^2$ could be made to fit the formula $\cosh^2 u - 1 = \sinh^2 u$ by letting x equal $a \cosh u$. Hence

$$x^2 - a^2 = a^2 \cosh^2 u - a^2 = a^2(\cosh^2 u - 1) = a^2 \sinh^2 u.$$

These substitutions should not be attempted unless a trigonometric substitution produces a difficult integral. In other words, a trial and error procedure is often necessary, and no fixed rules can be given regarding when to use a trigonometric substitution and when to use a hyperbolic substitution, or, for that matter, when a substitution for the root of an expression such as that in Example 17 will be most advantageous.

EXAMPLE 24. Find $\int_0^1 \sqrt{x^2 + 4}\, dx$.

Solution: A trigonometric substitution yields the difficult integral $\int \sec^3 \theta\, d\theta$. For a hyperbolic substitution, let $x = 2 \sinh u$. Then

$$x^2 + 4 = 2\sqrt{\sinh^2 u + 1} = 2 \cosh u \quad \text{and} \quad dx = 2 \cosh u\, du.$$

When $x = 0$, $\sinh u = 0$ and $u = 0$, and when $x = 1$, $\sinh u = 0.5$ and $u = 0.48$. The integral is

$$\int_0^{0.48} 4 \cosh^2 u \, du = 2 \int_0^{0.48} (1 + \cosh 2u) \, du$$

$$= \left[2u + \sinh 2u\right]_0^{0.48}$$

$$= 0.96 + \sinh 0.96 - 0$$

$$= 0.96 + 1.11 = 2.07.$$

EXERCISES

Find the following integrals:

1. $\int \sqrt{9 - x^2} \, dx.$

2. $\int \sqrt{9 - x} \, dx.$

3. $\int x\sqrt{9 - x^2} \, dx.$

4. $\int \frac{\sqrt{4 - x^2}}{x^2} \, dx.$

5. $\int \frac{dx}{(4 + x^2)^2}.$

6. $\int \frac{dx}{(4 + x^2)^{5/2}}.$

7. $\int x^3 \sqrt{4 - x^2} \, dx.$

8. $\int \sqrt{1 - 4e^{2x}} \, e^x \, dx.$

9. $\int x^3 \sqrt{9 - 4x^2} \, dx.$

The following Exercises can be calculated by at least three different methods. Work each of them by as many separate methods as you can:

10. $\int x\sqrt{x^2 - 1} \, dx.$

11. $\int x^3 \sqrt{x^2 - 1} \, dx.$

12. $\int \frac{x \, dx}{\sqrt{4 + x^2}}.$

13. $\int \frac{x \, dx}{9 - 4x^2}.$

Find the following integrals:

14. $\int \sqrt{4 - (x - 1)^2} \, dx.$

15. $\int x\sqrt{4 - (x - 1)^2} \, dx.$

16. $\int \sqrt{6x - x^2} \, dx.$

17. $\int \sqrt{x^2 - 6x} \, dx.$

18. $\int \sqrt{\frac{x}{1 - x}} \, dx.$ (*Hint:* Put $\sqrt{x} = \cos \theta.$)

19. $\int \sqrt{\frac{x}{1 + x}} \, dx.$ (*Hint:* Use a hyperbolic substitution for $\sqrt{x}.$)

Evaluate the following integrals:

20. $\int_0^2 \frac{x^2 \, dx}{(4 - x^2)^{3/2}}.$

21. $\int_0^{2\sqrt{3}} \frac{x^3 \, dx}{(4 + x^2)^{3/2}}.$

22. $\int_0^2 \frac{x^2 \, dx}{\sqrt{x^2 + 4}}.$

23. $\int_3^6 \frac{x^2 \, dx}{\sqrt{x^2 - 9}}.$

24. Find by calculus the area of a circle of radius R, using the equation $x^2 + y^2 = R^2$ for the circle.

25. Find by calculus the area of an ellipse with equation $x^2/a^2 + y^2/b^2 = 1$.

26. Find the area enclosed between the hyperbola $x^2/a^2 - y^2/b^2 = 1$ and the line $x = 2a$, assuming a to be positive.

27. Find the field intensity of a solenoid, expressed as

$$\int_{-k/2}^{k/2} \frac{2\pi r^2 In}{k(r^2 + x^2)^{3/2}} \, dx.$$

13.7 INTEGRATION BY PARTS

We first establish a formula, and then show how it may be used to integrate certain functions.

The product formula in differential form is $d(uv) = u \, dv + v \, du$.

Taking the integral of both sides, $uv = \int u \, dv + \int v \, du$. This gives us the formula for *integration by parts*, which is $\int \boldsymbol{u} \, \boldsymbol{dv} = \boldsymbol{uv} - \int \boldsymbol{v} \, \boldsymbol{du}$.

If this formula is to be used to integrate any expression, one factor of the expression will be represented by u and the other factor (the balance of the expression) by dv. The formula does not accomplish a complete integration, since $\int v \, du$ still remains to be calculated. However, if the formula has been properly applied, the integral that still remains to be calculated will usually be a fairly standard integral. Thus the principal advantage of using the formula for integration by parts is in reducing a complicated integral to a simpler integral.

Looking more closely into the structure of this formula, we see that the factor represented by u does not have to be integrated, since the right-hand side of the formula involves u only in the form u and du, which are the original factor and its differential, respectively. In other words, u must be differentiated, but not integrated. This formula will thus be very useful if the integrand contains a factor for which there is no standard integral, as long as there is a standard differential for this factor. Examples of this type of function are the ln function and the inverse functions. No standard integration formulas have been developed for these functions, and yet they may be differentiated by standard formulas.

EXAMPLE 25.   Find $\int x^3 \ln x \, dx$.

Solution: Since $\ln x$ cannot be integrated by standard formulas, we use the formula for integration by parts and represent $\ln x$ by the "u" of this formula. Let $\ln x = u$. Hence $x^3 \, dx = dv$, $(1/x) \, dx = du$, and $v = \int dv = x^4/4$. (*Note:* The integration constant should be omitted here. If it is not omitted, it becomes involved later on with other factors and eventually disappears or becomes part of the final integration constant.) Thus (using the formula)

$$\int \ln x(x^3 \, dx) = \frac{x^4}{4} \ln x - \int \frac{x^4}{4} \frac{1}{x} \, dx$$

$$= \frac{x^4 \ln x}{4} - \int \frac{x^3}{4} \, dx$$

$$= \frac{x^4 \ln x}{4} - \frac{x^4}{16} + C.$$

The student should verify this answer.

The device of integration by parts provides us with a method of integrating $\ln x$ itself.

EXAMPLE 26. Find $\int \ln x \, dx$.

Solution: Let $\ln x = u$ and $dx = dv$ in the formula for integration by parts. Then $(1/x) \, dx = du$ and $x = v$. Hence

$$\int \ln x \, dx = x \ln x - \int x \frac{1}{x} \, dx = x \ln x - \int dx = x \ln x - x + C.$$

Integration by parts is the usual method of integrating the product of two functions, provided that at least one of these functions can be integrated by a direct formula and the remaining integral (after one or more steps of integration by parts) may be calculated by some other method. Thus integration by parts often replaces a product formula for integration, although, as in Example 26, the integrand is not always a product of two functions.

We must be careful in deciding which function is to be represented by the "u" of the formula and which the "dv." The "dv" must be integrated to get "v" and this was the deciding factor in representing $\ln x$ by "u" in the previous two examples. Often, however, both of the functions involved in a product can be integrated or differentiated, and the choice of the factor to represent by "u" is governed by the principle that the remaining integral should be simpler than the original. Thus a mental or written calculation is frequently necessary to predict the form of the remaining integral before the choices of "u" and "dv" are finally made.

EXAMPLE 27. At t seconds, a current $i = 4t^2 e^{-100t}$ amperes flows in an *RLC* circuit. Find an expression for the charge q on the capacitor as a function of t, assuming that, at $t = 0$, $q = 0$.

Solution: $q = \int i \, dt = \int 4t^2 e^{-100t} \, dt$. Since the integrand is the product of two functions, integration by parts is indicated. Both of the functions can be integrated or differentiated as many times as necessary, and the remaining integral will involve the product of a power of t and the exponential e^{-100t}. If the power function of t were to disappear in the final integral, then the integrand would be an exponential function only, and the integral could be calculated by a standard formula. Since differentiation of a power function reduces the power by one order, the power function may be made to disappear with two steps of integration by parts provided t^2 is the "u" of the formula. Let

$$t^2 = u \quad \text{and} \quad e^{-100t} \, dt = dv.$$

Then

$$2t \, dt = du \quad \text{and} \quad \frac{-e^{-100t}}{100} = v.$$

Thus

$$q = 4\left(\frac{-t^2 e^{-100t}}{100} + \int \frac{t e^{-100t}}{50} \, dt \right).$$

Now let

$$t = u_1 \quad \text{and} \quad e^{-100t} \, dt = dv.$$

Then

$$dt = du_1 \quad \text{and} \quad \frac{-e^{-100t}}{100} = v.$$

Hence

$$q = 4\left[\frac{-t^2 e^{-100t}}{100} + \frac{1}{50} \left(\frac{-t e^{-100t}}{100} + \int \frac{e^{-100t}}{100} \, dt \right) \right]$$

$$= \frac{-t^2 e^{-100t}}{25} - \frac{t e^{-100t}}{1250} - \frac{e^{-100t}}{125,000} + C.$$

At $t = 0$, $q = 0$. Therefore $C = 1/125,000$, and

$$q = \frac{1}{125,000} - \frac{e^{-100t}}{25} \left(t^2 + \frac{t}{50} + \frac{1}{5000} \right).$$

EXAMPLE 28.   Find $\int e^{-50t} \sin 100t \, dt$.

Solution: Here, both functions can be integrated or differentiated as many times as necessary, and, irrespective of which one is "u" and which "dv" the remaining integral will be some constant times $\int e^{-50t} \cos 100t \, dt$. This integral is not any simpler than the original. If one more step of integration by parts is taken, the remaining integral will be some constant times $\int e^{-50t} \sin 100t \, dt$, and the original integral repeats itself in two steps of integration by parts. It may at first appear that this calculation leads nowhere. However, on further thought we see that with these two steps an equation is developed involving the original integral, and algebraic methods may be used to solve for the integral. Let $\sin 100t = u$ and $e^{-50t} dt = dv$. Then,

$$100 \cos 100t \, dt = du \qquad \text{and} \qquad \frac{-e^{-50t}}{50} = v.$$

Hence the integral is

$$\frac{-e^{-50t} \sin 100t}{50} + 2 \int e^{-50t} \cos 100t \, dt.$$

Now let $\cos 100t = u_1$ and $e^{-50t} dt = dv$. Then

$$-100 \sin 100t = du_1 \qquad \text{and} \qquad \frac{-e^{-50t}}{50} = v.$$

(*Caution:* In the first step it does not matter which of the two functions is "u." However, having decided, in the second step we must stick to this same choice; that is, in this case the trigonometric function must be the new u and the exponential the dv. Otherwise, the calculations up to this point would simply be done backwards and the final result would show merely that the integral equals itself.) The integral is

$$\frac{-e^{-50t} \sin 100t}{50} + 2 \left[\frac{-e^{-50t} \cos 100t}{50} - 2 \int e^{-50t} \sin 100t \, dt \right].$$

Representing the original integral by I, we now have the equation

$$I = \frac{-e^{-50t} \sin 100t}{50} - \frac{e^{-50t} \cos 100t}{25} - 4I.$$

$$5I = \frac{-e^{-50t}}{50} (\sin 100t + 2 \cos 100t).$$

$$I = \frac{-e^{-50t}}{250} (\sin 100t + 2 \cos 100t) + C.$$

The method of solution used in Example 28 is known as a *cyclic case* of integration by parts. The student should try the same example using u for the exponential function and dv for the trigonometric function and verify that the same result is obtained. Any combination of two factors involving two of the functions $\sin u$, $\cos u$, $\sinh u$, $\cosh u$, and e^u may usually be integrated by this cyclic method, unless the integral remaining after two steps of integration by parts has unity for its constant multiplier. When this occurs, a cyclic method only proves that the integral equals itself. This could happen when one of the functions in product is a sinh or a cosh, as in $\int e^x \sinh x \, dx$.

Fortunately, these problems can be solved by replacing the hyperbolic function by its exponential equivalent.

EXAMPLE 29.   Find $\int \sec^3 x \, dx$.

Solution: $\int \sec^3 x \, dx = \int \sec x \sec^2 x \, dx$. Let

$$\sec x = u \qquad \text{and} \quad \sec^2 x \, dx = dv.$$

Then

$$\sec x \tan x \, dx = du \qquad \text{and} \quad \tan x = v.$$

Use the formula for integration by parts. Then

$$\int \sec^3 x \, dx = \sec x \tan x - \int \tan^2 x \sec x \, dx$$

$$= \sec x \tan x - \int (\sec^2 x - 1) \sec x \, dx$$

$$= \sec x \tan x - \int \sec^3 x \, dx + \int \sec x \, dx$$

$$= \sec x \tan x - \int \sec^3 x \, dx + \ln (\sec x + \tan x).$$

The original integral appears again here (it is cyclic in one step). Representing the original integral by I,

$$I = \sec x \tan x - I + \ln (\sec x + \tan x).$$

$$I = \frac{\sec x \tan x + \ln (\sec x + \tan x)}{2} + C.$$

EXERCISES

Show that the method of integration by parts may be used for the following integrals, and complete the calculations:

1. $\int x^2 \, dx$.
2. $\int e^{2x} \, dx$.
3. $\int \sin x \cos x \, dx$.
4. $\int x \sqrt{1 - x} \, dx$.

Find the following integrals:

5. $\int x^2 \ln x \, dx$.
6. $\int x e^x \, dx$.
7. $\int x \cos x \, dx$.
8. $\int (x + 3) \ln x \, dx$.
9. $\int x \ln (1 + x) \, dx$.
10. $\int \arcsin x \, dx$.
11. $\int_0^1 x \arctan x \, dx$.
12. $\int x^2 \sin x \, dx$.
13. $\int_0^2 \ln (1 + x) \, dx$.
14. $\int \ln (1 + x^2) \, dx$.
15. $\int x^2 \ln (1 + x^2) \, dx$.
16. $\int e^{-20t} \cos 50t \, dt$.

17. $\int e^{2x} \sinh 2x \, dx.$ **18.** $\int \sin x \cosh x \, dx.$ **19.** $\int \sinh x \cosh x \, dx.$

20. $\int \sin (\ln x) \, dx.$ **21.** $\int (\ln x)^2 \, dx.$ **22.** $\int_0^4 \dfrac{x \, dx}{\sqrt{4 - x}}.$

23. $\int (2x - 1) \cos x \, dx.$ **24.** $\int_0^1 \dfrac{\arcsin \sqrt{x} \, dx}{\sqrt{1 - x}}.$

25. $\int \sqrt{1 - x} \arcsin \sqrt{x} \, dx.$ **26.** $\int \sec x \tan^2 x \, dx.$

27. $\int \tan x \sec^2 x \, dx.$ **28.** $\int \sec^5 x \, dx.$

29. $\int e^{-50t} \cos^2 50t \, dt.$ **30.** $\int_0^\infty e^{-st} \sin mt \, dt.$

The following four reduction formulas are often used to find the integral of an expression involving a whole power of an independent variable or of a trigonometric function in terms of the integral of a lower power of the same function. Prove the validity of each formula:

31. $\int x^n e^{ax} \, dx = \dfrac{1}{a} x^n e^{ax} - \dfrac{n}{a} \int x^{n-1} e^{ax} \, dx.$

32. $\int x^n \sin ax \, dx = -\dfrac{x^n}{a} \cos ax + \dfrac{n}{a} \int x^{n-1} \cos ax \, dx.$

33. $\int \sin^n ax \, dx = -\dfrac{1}{na} \sin^{n-1} ax \cos ax + \dfrac{n-1}{n} \int \sin^{n-2} ax \, dx.$

(*Hint:* Let $\sin^{n-1} ax = u$.)

34. $\int \cos^n ax \, dx = \dfrac{1}{na} \cos^{n-1} ax \sin ax + \dfrac{n-1}{n} \int \cos^{n-2} ax \, dx.$

35. Use the formula of Exercise 33 to find $\int \sin^2 3x \, dx$ and verify your answer by integrating independently.

36. Use the formula of Exercise 34 to find $\int \cos^3 2x \, dx$ and verify by an independent integration.

37. Use the formula of Exercise 34 as many times as necessary to find $\int \cos^4 2x \, dx.$

Find the charge q on a capacitor in an *RLC* series circuit as a function of t in seconds, assuming that $q = 0$ at $t = 0$, if the current at t seconds is:

38. $2te^{-20t}$ amperes.   **39.** $2e^{-50t} \sin 20t$ amperes.

40. $(2t + 3)e^{-50t}$ amperes.   **41.** $4 \sin 200t + 2e^{-50t} \cos 500t$ amperes.

42. Find the total energy dissipated from $t = 0$ to $t = \infty$ when, at t seconds, a current of $i = e^{-100t} \sin 100t$ amperes flows through a 1.5-ohm resistor.

Summary. In this chapter, most of the frequently-used techniques of integration have been discussed. With these techniques, the scope of expressions that may be integrated has been considerably increased. There will still be some expressions that are either very difficult or impossible to integrate. In such instances, if the problem is to evaluate a definite integral, the method of power series (Chapter 15) may often be used to get an answer to any desired degree of accuracy.

For each of these integration techniques, certain suggestions and clues have been given to assist the student in deciding which technique to use. In many problems, several methods may be effective, each yielding the same answer. In the following Exercises, the method of integration will usually vary from question to question, and one of the main problems confronting the student will be to use the suggestions and clues provided in this chapter to decide what integration technique to use. Rules and clues cannot possibly cover all problems; they should be used as a guide, but they cannot replace a certain amount of ingenuity and analysis.

For convenience, a list of standard integration formulas, most of which have been previously developed in this text, follows.

STANDARD INTEGRATION FORMULAS

$$\int u^n \, du = \frac{u^{n+1}}{n+1} + C. \qquad (n \neq -1.)$$

$$\int \sin u \, du = -\cos u + C.$$

$$\int \cos u \, du = \sin u + C.$$

$$\int \sec^2 u \, du = \tan u + C.$$

$$\int \operatorname{cosec}^2 u \, du = -\cot u + C.$$

$$\int \tan u \, du = \ln \sec u + C.$$

$$\int \cot u \, du = \ln \sin u = -\ln \operatorname{cosec} u + C.$$

$$\int \sec u \, du = \ln (\sec u + \tan u) + C.$$

$$\int \operatorname{cosec} u \, du = -\ln (\operatorname{cosec} u + \cot u) + C.$$

$$\int \sec u \tan u \, du = \sec u + C.$$

$$\int \operatorname{cosec} u \cot u \, du = -\operatorname{cosec} u + C.$$

$$\int e^u \, du = e^u + C.$$

$$\int \frac{du}{u} = \ln u + C.$$

$$\int \sinh u \, du = \cosh u + C.$$

$$\int \cosh u \, du = \sinh u + C.$$

$$\int \operatorname{sech}^2 u \, du = \tanh u + C.$$

$$\int \frac{du}{\sqrt{a^2 - u^2}} = \arcsin \frac{u}{a} + C. \qquad (u^2 \le a^2.)$$

$$\int \frac{du}{a^2 + u^2} = \frac{1}{a} \arctan \frac{u}{a} + C.$$

$$\int \frac{du}{\sqrt{a^2 + u^2}} = \sinh^{-1} \frac{u}{a} + C = \ln \frac{u + \sqrt{a^2 + u^2}}{a} + C.$$

$$\int \frac{du}{\sqrt{u^2 - a^2}} = \cosh^{-1} \frac{u}{a} + C = \ln \frac{u + \sqrt{u^2 - a^2}}{a} + C. \qquad \left(\frac{u}{a} \ge 1.\right)$$

$$\int \frac{du}{a^2 - u^2} = \frac{1}{a} \tanh^{-1} \frac{u}{a} + C = \frac{1}{2a} \ln \frac{a + u}{a - u} + C. \qquad (u^2 \le a^2.)$$

$$\int u \, dv = uv - \int v \, du.$$

MISCELLANEOUS EXERCISES

Find the following integrals:

1. $\int \frac{\arctan x}{1 + x^2} \, dx.$ 2. $\int x\sqrt{1 - 4x} \, dx.$ 3. $\int \frac{dx}{x^3 - 4x}.$

4. $\int \cos^7 10t \, dt.$ 5. $\int x^3 \sqrt{1 - 4x^2} \, dx.$ 6. $\int_1^2 \frac{dt}{\sqrt{10t - t^2}}.$

7. $\int_0^1 \frac{dx}{4 - x^2}.$ 8. $\int_3^4 \frac{dx}{4 - x^2}.$ 9. $\int_0^\infty 5te^{-10t} \, dt.$

10. $\int \frac{x^3 \, dx}{x^4 - 8}.$ 11. $\int (x^2 + 1) \ln x \, dx.$ 12. $\int \frac{x + 3}{x^2 + 9} \, dx.$

13. $\int \frac{x^2 + 3}{x + 9} \, dx.$ 14. $\int \frac{x + 2}{x^2 + 2x + 2} \, dx.$ 15. $\int (3 + 2x)^{3/2} \, dx.$

16. $\int_0^\pi \sin x \sec^2 x \, dx.$ 17. $\int_0^{\pi/4} \frac{\sin^2 \theta}{\cos^4 \theta} \, d\theta.$ 18. $\int \frac{x \, dx}{(x - 2)^2}.$

19. $\int \dfrac{dx}{(4 + x^2)^{3/2}}.$

20. $\int \dfrac{\text{arc tan } 3x}{1 + 9x^2}\, dx.$

21. $\int_0^{0.5} x \tanh^{-1} x\, dx.$

22. $\int \dfrac{e^{2x} + 1}{e^{2x} - 1}\, dx.$

23. $\int_1^e \dfrac{1}{x} \ln \dfrac{2}{x}\, dx.$

24. $\int_0^1 \dfrac{x^2\, dx}{\sqrt{1 + 4x^6}}.$

25. $\int_0^2 \dfrac{x^2\, dx}{\sqrt{1 + 4x}}.$

26. $\int \dfrac{x^4\, dx}{x^3 - 8}.$

CHAPTER 14

Partial Derivatives

14.1 FUNCTIONS WITH MORE THAN ONE INDEPENDENT VARIABLE

A variable z may depend for its value on two or more independent variables. In other words, there may be two or more quantities which may vary in a completely independent way, with another variable depending for its value

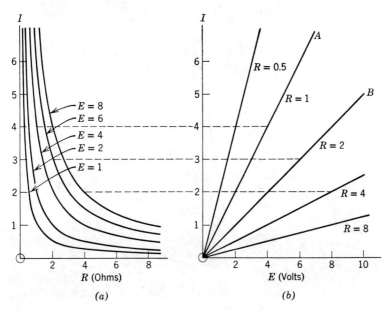

Fig. 14-1. Graphs of current in a resistive circuit, R and E being independently adjusted.

on the independent values of each of them. This relationship may be expressed as $z = f(x,y)$.

Suppose, for example, in the interpretation of Ohm's law for a conductor of resistance R ohms connected to a d-c source of voltage E volts, both the resistance and the source voltage may be varied independently either continuously or in steps. Then the current I varies with each independent setting of R and E, and the equation $I = E/R$ has two independent variables and one dependent variable I.

Graphically, this equation might often be represented by a family of curves representing I as a function of R, each corresponding to a particular value of E. Another graph of I against E might also be drawn showing the variation for several particular values of R. Figure 14-1 shows the two types of graphs. Of course, the graphs shown in Fig. 14-1 do not present a complete picture, since in each instance one of the independent variables does not vary

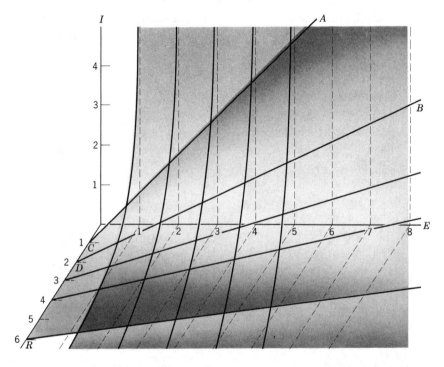

Fig. 14-2. Continuous graph of $I = \dfrac{E}{R}$, E and R being two independent variables.

continuously but in steps. A great deal of approximate interpolation would be necessary to evaluate I for specified values of E and R not appearing simultaneously on either graph.

If a complete graph is to be constructed, there must be a separate axis for each independent variable and an axis for the dependent variable; a third dimension is then required. If each axis is perpendicular to the other two axes, as shown in the foreshortened illustration in Fig. 14-2, the continuous graph is the curved surface represented in this diagram.

14.2   MEANING AND CALCULATION OF PARTIAL DERIVATIVES

When we now consider the rate of change of I with respect to E, there must be some stipulation regarding the other independent variable R. The least complicated stipulation is that R be considered as remaining constant. Thus the rate of change of I with respect to E when $R = 1$ is the slope of the straight line CA in Fig. 14-2, or the slope of the line OA in Fig. 14-1b; moreover, the rate of change of I with respect to E when $R = 2$ is the slope of the straight line DB in Fig. 14-2, or the slope of the line OB in Fig. 14-1b. Graphically, the rate of change of I with respect to E with a constant R will be the slope of the two-dimensional graph resulting from the intersection of the continuous curved surface with a plane surface parallel to the EI plane (on which the value of R is constant).

The symbol used for this rate of change is $\partial I/\partial E$ (often briefly read as die-I by die-E, or simply die-I die-E) and this is called the *partial derivative* of I with respect to E, since it does not present the full picture of variation. In its calculation, R must be considered as remaining constant. We can see that the value of $\partial I/\partial E$ depends on the constant value of R assumed, and therefore R will appear in the answer to $\partial I/\partial E$. Thus, given $I = E/R$, where E and R are independent variables, $\partial I/\partial E = (1/R)(dE/dE) = 1/R$; at $R = 1$, $\partial I/\partial E = 1$, and at $R = 2$, $\partial I/\partial E = \frac{1}{2}$, etc.

Similarly, the graphical interpretation of $\partial I/\partial R$ is the instantaneous slope of the two-dimensional graph resulting from the intersection of the continuous curved surface with a plane surface parallel to the RI plane on which the value of E is constant.

$$\frac{\partial I}{\partial R} = \frac{\partial}{\partial R}\left(\frac{E}{R}\right) = E\frac{d}{dR}R^{-1} = \frac{-E}{R^2}.$$

At $E = 1$, $\partial I/\partial R = -1/R^2$, which is the general expression for the slope of the curve corresponding to $E = 1$ in Fig. 14-1a.

For any specified simultaneous values of E and R, $\partial I/\partial E$ represents the slope of the curved surface at this point in a direction parallel to the EI plane, and $\partial I/\partial R$ represents the slope of the curved surface at the same point in a direction parallel to the RI plane.

There may be, of course, more than two independent variables and a single dependent variable, depending for its value on the values of all of the independent variables. To represent a continuous graph of such a condition would be very difficult, since there would be more than three axes and it would be impossible to arrange that each pair of axes be perpendicular to the other. However, the meaning of the partial derivatives and the principles involved in their calculation may be inferred from those used in the simpler instance involving only two independent variables. Thus, if $z = f(x,y,w)$, $\partial z/\partial x$ is calculated assuming that y and w remain constant; $\partial z/\partial y$ is calculated assuming that x and w remain constant, etc.

EXAMPLE 1. Find all partial derivatives for the function given by $z = x^3 - e^{-2xy} + \cos \pi(3x - y)$, assuming x and y are two independent variables. Find their values when $x = 0$ and $y = 1$.

Solution: Each derivative is calculated as an ordinary derivative with respect to one of the variables, the other variable being treated as a constant. The variable being treated as a constant is often indicated as a type of subscript. Thus $\dfrac{\partial z}{\partial x}\bigg]_y$ indicates that y is treated as a constant. Therefore

$$\frac{\partial z}{\partial x}\bigg]_y = 3x^2 + 2ye^{-2xy} - 3\pi \sin \pi(3x - y).$$

At $x = 0$ and $y = 1$, $\partial z/\partial x = 2e^0 - 3\pi \sin(-\pi) = 2$

$$\frac{\partial z}{\partial y}\bigg]_x = 2xe^{-2xy} + \pi \sin \pi (3x - y).$$

At $x = 0$ and $y = 1$, $\partial z/\partial y = 0 + \pi \sin(-\pi) = 0$.

We must take care to avoid using certain relationships between partial derivatives that are more or less taken for granted when dealing with ordinary derivatives, which involve only two variables. The reciprocal rule $\left(\dfrac{dy}{dx} = \dfrac{1}{dx/dy}\right)$ will ordinarily apply to partial derivatives. $\dfrac{\partial y}{\partial x} = \dfrac{1}{\partial x/\partial y}$

provided the same equation is used to find $\partial y/\partial x$ and $\partial x/\partial y$. However, the *chain rule* relationship involving partial derivatives in product does not apply; that is, $\partial z/\partial x$ is not equal to $(\partial z/\partial y)(\partial y/\partial x)$, since each of the three derivatives is calculated by assuming that a different variable remains constant. Since the assumptions for each derivative are in conflict, there is no general relationship between the three derivatives. The chain rule, however, must often be used in the calculation of a single partial derivative, where the consideration of the variable assumed to remain constant in the calculation occurs only in the last step of the calculation.

Thus $\dfrac{\partial}{\partial x} \sin^3 xy = 3 \sin^2 xy (\cos xy)(y)$.

In a more detailed analysis, if we let $\sin xy = u$ and $xy = v$, the partial derivative is calculated as $3u^2(du/dv)(\partial v/\partial x)$. A chain rule has been used in which all steps but the last have been calculated as ordinary derivatives. To generalize on this calculation

$$\frac{\partial}{\partial x} f(u) = \frac{d}{du} f(u) \frac{\partial u}{\partial x}.$$

EXAMPLE 2.   For   the   function $z = \arcsin x^2 y$:   (a)   show   that $\dfrac{\partial z}{\partial x} = \dfrac{1}{(\partial x/\partial z)}$ by calculating each of the two derivatives directly, (b) find $\partial z/\partial y$, (c) find $\partial x/\partial y$, and (d) show that $\partial z/\partial x$ is not equal to $(\partial z/\partial y)(\partial y/\partial x)$.

Solution: (a) $\qquad\qquad \left.\dfrac{\partial z}{\partial x}\right]_y = \dfrac{1}{\sqrt{1 - x^4 y^2}} \, 2xy.$

$\partial x/\partial z$ may be calculated either implicitly or explicitly.
By using implicit differentiation of the whole equation with respect to z, and noting that y is to be assumed as constant, we find

$$1 = \frac{1}{\sqrt{1 - x^4 y^2}} \, 2xy \frac{\partial x}{\partial z}, \qquad \text{and} \qquad \frac{\partial x}{\partial z} = \frac{\sqrt{1 - x^4 y^2}}{2xy};$$

$$\therefore \quad \frac{\partial z}{\partial x} = \frac{1}{\dfrac{\partial x}{\partial z}}.$$

Alternatively, by expressing the function explicitly,

$$x^2 y = \sin z. \qquad x = \frac{(\sin z)^{\frac{1}{2}}}{y^{\frac{1}{2}}}. \qquad \left.\frac{\partial x}{\partial z}\right]_y = \frac{1}{2} y^{-\frac{1}{2}}(\sin z)^{-\frac{1}{2}} \cos z.$$

Since $\sin z = x^2 y$ and $\cos z = \sqrt{1 - \sin^2 z}$,

$$\frac{\partial x}{\partial z} = \frac{1}{2} y^{-\frac{1}{2}} x^{-1} y^{-\frac{1}{2}} \sqrt{1 - x^4 y^2} = \frac{\sqrt{1 - x^4 y^2}}{2xy}, \text{ the reciprocal of } \frac{\partial z}{\partial x}.$$

The reciprocal rule is true because y is assumed constant in the calculation of both derivatives.

(b)
$$\left.\frac{\partial z}{\partial y}\right]_x = \frac{1}{\sqrt{1 - x^4 y^2}} x^2.$$

(c) Differentiating the whole equation with respect to y and assuming z as constant for the calculation of $\partial x / \partial y$,

$$0 = \frac{1}{\sqrt{1 - x^4 y^2}} \left(2xy \frac{\partial x}{\partial y} + x^2\right); \qquad 2xy \frac{\partial x}{\partial y} = -x^2; \qquad \frac{\partial x}{\partial y} = \frac{-x}{2y}.$$

(d) $\partial z / \partial y$ was calculated in (b).

$$\frac{\partial y}{\partial x} = \frac{1}{\partial x / \partial y} = -\frac{2y}{x}.$$

Therefore

$$\frac{\partial z}{\partial y} \frac{\partial y}{\partial x} = \frac{x^2}{\sqrt{1 - x^4 y^2}} \left(\frac{-2y}{x}\right) = \frac{-2xy}{\sqrt{1 - x^4 y^2}}.$$

This does not equal $\partial z / \partial x$, calculated in (a).

EXERCISES

1. If $z = 3x^2 - 2y^2$, find (a) $\partial z / \partial x$, (b) $\partial z / \partial y$, (c) the value of $\partial z / \partial x$ at $x = 2$, $y = 1$, and (d) the value of $\partial z / \partial y$ at $x = 2$, $y = 1$. Distinguish clearly between the meanings of (c) and (d).

2. If $Z^2 = R^2 + X^2$, find $\partial Z / \partial X$ and $\partial Z / \partial R$ where $R = 4$ and $X = 3$.

3. If $z = x^2 y + e^{x-y} - \sin xy$, find (a) $\partial z / \partial x$ and (b) $\partial z / \partial y$.

4. If $z = \ln (x/y)$, evaluate for the point where $x = 2$ and $y = 1$, (a) $\partial z / \partial x$, (b) $\partial z / \partial y$, and (c) $\partial y / \partial x$.

5. If $z = x^n y$, show that $x(\partial z / \partial x) + y(\partial z / \partial y) = (n + 1)z$.

6. A d-c voltage E is applied across a resistance R. The power in this circuit is given by $P = EI$. Since $E = IR$, the power is also given by $P = I^2 R$. Since $I = E/R$, the power is also given by $P = E^2/R$. Assuming that E and R may be varied independently:

   (a) Find $\partial P / \partial E$ from the equation $P = EI$ and also from the equation $P = E^2/R$. Why are the two expressions not equal?

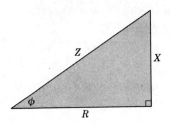

Fig. 14-3.

(b) Find $\partial P/\partial R$ from the two equations involving P and R. Explain the basic difference in meaning of the two answers.

7. Figure 14-3 is a representation of the impedance triangle for a-c circuits. It is seen from this triangle that (a) $\phi = \arctan{(X/R)}$, and (b) $R = Z\cos\phi$. By using (a), show that $\partial\phi/\partial R = -(\sin\phi)/Z$. Using (b), find $\partial R/\partial\phi$. Explain clearly why $\partial R/\partial\phi$ from (b) is not the reciprocal of $\partial\phi/\partial R$ from (a).

8. Given $P = e^{-3y}x + \ln y - \sin y + x$, and $Q = -3x^2 e^{-3y}/2 + x/y - x\cos y$, show that $\partial P/\partial y = \partial Q/\partial x$.

Find $\partial z/\partial x$ and $\partial z/\partial y$ for the following functions:

9. $z = xy\ln\dfrac{x}{y}$. **10.** $z = \dfrac{x}{y}\ln xy$. **11.** $z = \dfrac{\ln{(x/y)}}{\ln xy}$.

12. For the equation of Exercise 3, (a) find $\partial x/\partial y$, and (b) prove that $(\partial x/\partial y)(\partial y/\partial z)(\partial z/\partial x) = -1$.

13. Find the two partial derivatives of z for the function $z = (\cos xy)/(\tan xy)$.

14. Find the three partial derivatives of u for the function

$$u = \frac{x^2 - y^2 + z^2}{y^2}.$$

15. Find $\partial Z/\partial R$ and $\partial Z/\partial C$ for the equation

$$Z = \sqrt{R^2 + \frac{1}{4\pi^2 f^2 C^2}}.$$

16. Find $\partial I/\partial R$ and $\partial I/\partial\omega$ for the equation

$$I = \frac{E}{\sqrt{R^2 + \left(\omega L - \dfrac{1}{\omega C}\right)^2}}.$$

17. Evaluate $\partial P/\partial R$ at the moment when $R = 0.5$ and $E = 4$, if

$$P = \frac{E^2 R}{(0.1 + R)^2}.$$

18. If x and y are two independent variables, and $z^2 = a^2 - x^2$ for all values of y, (a) find $\partial z/\partial y$, and (b) evaluate

$$\int_0^a \sqrt{1 + \left(\frac{\partial z}{\partial x}\right)^2 + \left(\frac{\partial z}{\partial y}\right)^2} \; dx.$$

19. If $z = (xy)^y$, use a logarithmic method to find (a) $\partial z/\partial x$ and (b) $\partial z/\partial y$.

14.3 HIGHER-ORDER PARTIAL DERIVATIVES

A second derivative expresses the rate of change of the first derivative. If y is a function of x only, d^2y/dx^2 represents the rate at which the slope is increasing with respect to x, in the direction of increasing x.

Higher-order partial derivatives are also used where more than one independent variable is present in the function. The significance of a second partial derivative may be inferred from the meaning of an ordinary second derivative. For a function $z = f(x,y)$, there are two first partial derivatives: $\partial z/\partial x$ representing the rate of change of z with respect to x in the direction of increasing x, and $\partial z/\partial y$ the rate of change of z with respect to y in the direction of increasing y. In the same way, each of these two first derivatives may be differentiated again either with respect to x or with respect to y. The partial derivative of $\partial z/\partial x$ with respect to x is designated as $\partial^2 z/\partial x^2$, and it represents the rate of change of the slope of the surface graph in any plane with a constant y with respect to x, in the direction of increasing x. This direction of increasing x is in the same two-dimensional plane as the two-dimensional graph itself.

The partial derivative of $\partial z/\partial x$ with respect to y is designated as $\partial^2 z/(\partial y\,\partial x)$ [its meaning being $\partial/\partial y(\partial z/\partial x)$]. $\partial^2 z/(\partial y\,\partial x)$ represents the rate of change of the slope of the surface graph in any plane with a constant y, with respect to y, in the direction of increasing y. Here, the direction in which the increase of slope is considered is at right angles to the plane of the two-dimensional graph itself. In more general terms, $\partial^2 z/(\partial y\,\partial x)$ considers the change taking place in the slope of the graph as we move on the surface in a direction at right angles to the plane of the slope itself.

There will be two more second derivatives of $z = f(x,y)$, which are the derivatives of $\partial z/\partial y$, designated as $\partial^2 z/\partial y^2$ and $\partial^2 z/(\partial x\,\partial y)$. $\partial^2 z/(\partial x\,\partial y)$ designates that $\partial z/\partial y$ is first calculated, and then its derivative taken with respect to x. We should note that in this two-step calculation, x is assumed constant in the first step and y is considered constant in the second step. For the calculation of $\partial^2 z/\partial y^2$, x is assumed constant in both steps.

EXAMPLE 3. Find all second-order partial derivatives of the function $z = 2y^2e^{3x} - e^{-2y} \cos 3x$.

Solution:

$$\left.\frac{\partial z}{\partial x}\right]_y = 6y^2e^{3x} + 3e^{-2y} \sin 3x.$$

$$\nearrow \left.\frac{\partial^2 z}{\partial x^2}\right]_y = 18y^2e^{3x} + 9e^{-2y} \cos 3x.$$

$$\searrow \left.\frac{\partial^2 z}{\partial y \, \partial x}\right]_x = 12ye^{3x} - 6e^{-2y} \sin 3x.$$

$$\left.\frac{\partial z}{\partial y}\right]_x = 4ye^{3x} + 2e^{-2y} \cos 3x.$$

$$\nearrow \left.\frac{\partial^2 z}{\partial y^2}\right]_x = 4e^{3x} - 4e^{-2y} \cos 3x.$$

$$\searrow \left.\frac{\partial^2 z}{\partial x \, \partial y}\right]_y = 12ye^{3x} - 6e^{-2y} \sin 3x.$$

It is seen in the preceding example that $\partial^2 z/(\partial x \, \partial y) = \partial^2 z/(\partial y \, \partial x)$. The function representing these two "mixed" derivatives is always the same for both. This function may be used to evaluate these second derivatives at all points on the graph except where there is a point of discontinuity. At a point of discontinuity, usually detected by either an infinite value or an indeterminate value (such as 0/0) for the function, a more fundamental method must be used to calculate these second derivatives. This method has little practical value and is therefore not included in this analysis.

Partial derivatives of higher order than the second may be calculated in a manner similar to that used in second-order derivatives. The function in Example 3 would have eight third derivatives, although several of them would be identical functions. For example, $\partial^3 z/(\partial x^2 \, \partial y)$ is equivalent to $\partial^3 z/(\partial x \, \partial y \, \partial x)$ and to $\partial^3 z/(\partial y \, \partial x^2)$.

EXERCISES

1. Find all second derivatives of the function $z = \ln (x + y) - e^{-2x} \sin 3y$.

2. For the function $z = (x - y)/(x + y)$: (a) calculate $\partial^2 z/(\partial y \, \partial x)$, (b) calculate $\partial^2 z/(\partial x \, \partial y)$, (c) evaluate $\partial^2 z/(\partial x \, \partial y)$ at the point where $x = 2$ and $y = 1$, and (d) explain why the function representing $\partial^2 z/(\partial x \, \partial y)$ cannot be used to evaluate this derivative at $x = 0$ and $y = 0$.

3. For the function $z = e^{-2xy} + (x^2 + y^2)^{1/2}$:
   (a) Find the equation of the tangent at the point where $x = 0$ and $y = 2$ in the plane of constant y.
   (b) Find the rate of change of the slope of this tangent with respect to x.
   (c) Find the rate of change of the slope of this tangent with respect to y.

4. If $z = A \cos (x + \omega t) + B \sin (x + \omega t)$, prove that $\partial^2 z/\partial t^2 = \omega^2(\partial^2 z/\partial x^2)$.

5. If $z = \ln(x^2 + 2y^2)$, prove that $2(\partial^2 z/\partial x^2) + \partial^2 z/\partial y^2 = 0$.

6. (a) Find all second derivatives of the function $z = e^{xy} + \sin xy$.

(b) Evaluate these second derivatives where $x = 2$ and $y = 0$.

7. Find all second derivatives of the function $z = uvw + \ln(u + v + w)$, where u, v, and w are three independent variables.

8. (a) Find $\partial^3 z/\partial x^3$ and $\partial^3 z/(\partial x\ \partial y\ \partial x)$ for the function $z = x^2 y - \cos(x + 2y)$.

(b) Show that $\partial^3 z/(\partial x\ \partial y\ \partial x) = \partial^3 z/(\partial x^2\ \partial y)$ for the function of (a).

9. (a) Find all second derivatives of the function $z = xy\dfrac{x + y}{x - y}$ by using logarithms.

(b) At what point would the expression for $\partial^2 z/(\partial x\ \partial y)$ calculated in (a) be useless?

14.4 THE TOTAL DIFFERENTIAL

The idea concerning differentials established in Section 5.1 may be extended to analyze the meaning of a differential applied to a function of more than one independent variable. We recall that, in the simpler case where y is a function of x only and the graph is two dimensional, a differential of x (that is, dx) may be chosen independently as any horizontal length starting at the point of contact $P(x,y)$ and therefore ending at the point $Q(x + dx,y)$ (see Fig. 14-4). Then the corresponding differential of y is the rise measured from the point $(x + dx, y)$ up to R, on the tangent. The coordinates of this point on the tangent are therefore $(x + dx, y + dy)$. $dy = f'(x)\,dx$, where dx is an independent variable.

The continuous graph of $z = f(x,y)$ is a three-dimensional surface such as that shown in Fig. 14-5, the surface $PABC$. At any point (x,y,z) on this surface, the tangent is a plane surface $PDEF$, and any straight line on this

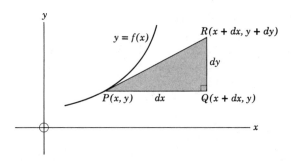

Fig. 14-4.

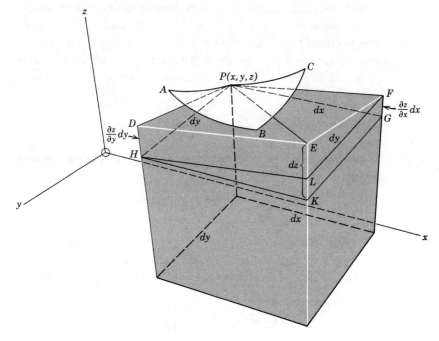

Fig. 14-5.

plane surface through the point of contact in any direction, such as the straight line *PE*, is a tangent to the surface. The point *E* on the tangential plane may be fixed by increasing the *x*-coordinate of the point *P* by an arbitrary amount *dx*, increasing the *y*-coordinate of the point *P* by an arbitrary amount *dy*, and locating *E* on the tangential plane directly above the point $K(x + dx, y + dy, z)$. The differential of *z*, *dz*, corresponding to these arbitrarily chosen differentials *dx* and *dy* will then be the increase in the *z* coordinate between the point *P* and the point *E*. If the surface *PHKG* represents a horizontal surface of constant *z*, *dz* is the length of the line *KE*. The point *L* is chosen such that $KL = GF$. We may see that triangle *FEL* is congruent to triangle *PDH* (since $FL = PH = dy$; angle FLE = angle $PHD = 90°$, foreshortened in the diagram; and angle EFL = angle *DPH*, angles between two intersecting planes on parallel surfaces). This means that $LE = HD$. Now $\partial z/\partial x$ is the slope of the line $PF = FG/dx$; therefore $FG = (\partial z/\partial x)\,dx$. Similarly, $HD = (\partial z/\partial y)\,dy$. Therefore $dz = KL + LE = GF + HD = (\partial z/\partial x)\,dx + (\partial z/\partial y)\,dy$, where *dx* and *dy* may be arbitrarily chosen.

Another way of explaining this is that the point E is reached by moving from P to F in a plane of constant y in which x is increased by an amount dx, and then moving from F to E in a plane of constant x in which y is increased by an amount dy. In the first motion, z is increased by an amount $(\partial z/\partial x)\, dx$ and in the second motion, z is further increased by an amount $(\partial z/\partial y)\, dy$. Therefore *the total differential* is

$$dz = \frac{\partial z}{\partial x}\, dx + \frac{\partial z}{\partial y}\, dy.$$

In general, the total differential dz will have as many components as there are independent variables. If $z = f(x,y,w)$,

$$dz = \frac{\partial z}{\partial x}\, dx + \frac{\partial z}{\partial y}\, dy + \frac{\partial z}{\partial w}\, dw.$$

One application of the total differential is to problems on related rates. If z is a function of x and y and both x and y are changing at an established rate with respect to time t, then we may assume from the form of the total differential that

$$\frac{dz}{dt} = \frac{\partial z}{\partial x}\frac{dx}{dt} + \frac{\partial z}{\partial y}\frac{dy}{dt}.$$

Another application of the total differential is where two or more independent variables take on small changes. We may find an approximation of the resulting small change in the dependent variable by assuming the small changes are approximately equal to the differentials, in much the same way as we did in problems on small changes with only one independent variable. Thus

$$\Delta z \simeq \frac{\partial z}{\partial x}\, \Delta x + \frac{\partial z}{\partial y}\, \Delta y,$$

provided the function is continuous over the intervals Δx and Δy, and that Δx and Δy are small. Graphically, in Fig. 14-5, this means that $\Delta x = dx$, $\Delta y = dy$, and $\Delta z \simeq dz$. The exact Δz is the length of the vertical line from K to the point where KE intersects the surface graph, and the approximation consists in assuming this length equals KE. dz may be considered as the exact change in z corresponding to small changes in the independent variables if the partial rates of change were maintained constant over the intervals of these independent changes.

EXAMPLE 4.   If $I = \dfrac{E}{\sqrt{R^2 + 1/\omega^2 C^2}}$, use differentials to determine the approximate percent change in I when E is decreased by 1%, R is increased by 1%, C is decreased by 1%, and ω remains constant.

Solution: Representing $\sqrt{R^2 + 1/\omega^2 C^2}$ as Z for simplicity,

$$\frac{\partial I}{\partial E} = +1/Z.$$

$$\frac{\partial I}{\partial R} = -\frac{1}{2} E(Z^2)^{-3/2} 2R = -ERZ^{-3}.$$

$$\frac{\partial I}{\partial C} = -\frac{1}{2} E(Z^2)^{-3/2}\left(\frac{-2C^{-3}}{\omega^2}\right) = +EZ^{-3}C^{-3}\omega^{-2}.$$

It is given that $\Delta E = -0.01E$, $\Delta R = 0.01R$, and $\Delta C = -0.01C$.

$$\therefore \ \Delta I \simeq \frac{\partial I}{\partial E}\Delta E + \frac{\partial I}{\partial R}\Delta R + \frac{\partial I}{\partial C}\Delta C$$

$$\simeq \frac{1}{Z}(-0.01E) + (-ERZ^{-3})(0.01R) + (EZ^{-3}C^{-3}\omega^{-2})(-0.01C)$$

$$= -\frac{0.01E}{Z} - \frac{0.01ER^2}{Z^3} - \frac{0.01E}{Z^3C^2\omega^2}$$

$$= -\frac{0.01E}{Z}\left(1 + \frac{R^2 + \dfrac{1}{\omega^2 C^2}}{Z^2}\right)$$

$$= -\frac{0.01E}{Z}\left(1 + \frac{Z^2}{Z^2}\right) = -\frac{0.02E}{Z} = -0.02I.$$

Therefore the approximate change in I is a decrease of 2%.

The substitution of a single letter for a recurring expression often helps with the problem of seeing structure and eliminating detail.

EXERCISES

1. For the function $z = xy$, find dz, (a) by using partial derivatives, and (b) by the product formula.

2. For the function $z = 3 \cos (2x - y) - 2e^{2y - 4x} + xy$, find the total differential dz corresponding to the following values of independent variables: $x = 1$, $y = 2$, $dx = 0.2$, and $dy = 0.5$.

3. For the function $z = 3xy - \ln (x - y)$:
(a) Find z at the point where $x = 10$ and $y = 9$.
(b) Find Δz by an exact calculation if x changes from 10 to 10.2 and y changes from 9 to 9.2.
(c) Find Δz approximately for the changes in (b), using the approximate relationship $\Delta z \doteq (\partial z/\partial x) \Delta x + (\partial z/\partial y) \Delta y$.

4. The impedance in an RL series circuit is given by $Z = \sqrt{R^2 + \omega^2 L^2}$. Find the approximate change in Z if R changes from 5.0 to 5.2, ω changes from 300 to 295, and L remains constant at 4.0×10^{-2}.

5. If $Z = \sqrt{R^2 + \omega^2 L^2}$, and R is increasing at a rate of 2 ohms per minute and ω is decreasing at a rate of 20 rad per second per minute, find the rate at which Z is changing at the moment when $R = 5.0$ ohms and $\omega = 300$ rad per second, assuming L remains constant at 4.0×10^{-2} henry.

6. The resonant frequency in an RLC circuit is given by $f_0 = 1/(2\pi \sqrt{LC})$. Find the approximate percent change in f_0 if L is increased by 3% and C is decreased by 1%.

7. The reactance of a circuit is given as $X = \omega L - 1/\omega C$. Assuming that X is positive, find the approximate percent change in X if ω is decreased by 2%, L is increased by 3%, and C is increased by 1%.

8. Exercises 6 and 7 are designed to give a simple numerical answer. This will not always be the case. For the equation of 7, express the percent change in X in terms of ω, L, and C if ω, L, and C are each increased by 2%.

9. If $I = E/\sqrt{R^2 + X^2}$, find (a) the approximate change in I if E changes from 6.0 to 6.1 volts, R changes from 6.0 to 5.9 ohms, and X changes from 8.0 to 7.8 ohms; (b) the approximate percent change in I if E changes from 8.0 to 7.8 volts, R changes from 6.0 to 5.9 ohms, and X changes from 8.0 to 8.2 ohms, and (c) the approximate percent change in I if E is decreased by 2% and R and X are both increased by 2%.

10. The average power in an a-c circuit may be expressed as $P = \dfrac{E^2 R}{R^2 + X^2}$.
(a) Find expressions for $\partial P/\partial E$, $\partial P/\partial X$, and $\partial P/\partial R$.
(b) If E changes from 4.0 to 4.1 volts, X changes from 8.0 to 7.9 ohms, and R remains constant at 4.0 ohms, find the approximate percent change in P.
(c) If E increases by 2% and R and X each decrease by 3%, find the approximate percent change in P.

11. The formula $A = \frac{1}{2}ab \sin C$ is used to calculate the area of a triangle. If a is measured 2% too high, b is measured 3% too high, and C is measured as $62°$ instead of its correct value of $60°$, find the percent error in the calculated area.

12. The resonant period of an LC circuit is given by $T = 2\pi\sqrt{LC}$. If there is a maximum possible error of 2% in the measurement of L and 3% in the measurement of C, find the maximum possible error in T as calculated from the given function.

13. The balancing condition for a bridge circuit is $R_x/R_2 = R_3/R_4$. If $R_2 = 10 \pm 0.2$, $R_3 = 50 \pm 1.2$, and $R_4 = 20 \pm 0.5$, find the approximate maximum error in the calculation of R_x.

14. If $R = \dfrac{R_1 R_2}{R_1 + R_2}$ and R_1 and R_2 are both decreased by 2%, find the percent change in R.

14.5   APPLICATIONS IN PROBLEMS OF MAXIMUM AND MINIMUM

The principles established for maximum and minimum where there is only one independent variable may be extended to cover problems with more than one independent variable.

If there are two independent variables, a three-dimensional graph of the function can be constructed, and we readily see that at a maximum or minimum point on this graph, assuming continuity through such points, the tangential plane is horizontal; if $z = f(x,y)$, it is a plane of constant z. Thus $dz = 0$ for any independently chosen values of dx and dy. Since $dz = (\partial z/\partial x)\, dx + (\partial z/\partial y)\, dy$ and dx and dy may have any values, the conclusion is that

both $\partial z/\partial x$ and $\partial z/\partial y$ must be zero at a maximum or minimum point.

In most cases, for a two-dimensional graph, the second derivative is negative at a maximum and positive at a minimum point. For the function $z = f(x,y)$, the second derivative test is not quite so simple. In most cases, at a maximum point, $\partial^2 z/\partial x^2$ and $\partial^2 z/\partial y^2$ are both negative, whereas, at a minimum point, they are both positive. However, even though we may establish that both first derivatives are zero and the two unmixed second derivatives have the same sign at a particular point, this point may not be either a maximum or a minimum. The values of the mixed second derivatives must also be obtained, and these are not necessarily negative at a maximum point or positive at a minimum point. It can be shown that an additional condition necessary at a maximum or minimum point is that

$$\frac{\partial^2 z}{\partial x^2}\frac{\partial^2 z}{\partial y^2} \geq \left(\frac{\partial^2 z}{\partial x\, \partial y}\right)^2.$$

If these two quantities are equal, the point will not always be either a maximum or a minimum. The following tabulation shows the necessary and sufficient conditions for maximum and minimum points in the function $z = f(x,y)$:

Quantity Evaluated	For Maximum	For Minimum
$\dfrac{\partial z}{\partial x}$ and $\dfrac{\partial z}{dy}$	0	0
$\dfrac{\partial^2 z}{\partial x^2}$ and $\dfrac{\partial^2 z}{\partial y^2}$	Negative	Positive
$\dfrac{\partial^2 z}{\partial x^2}\dfrac{\partial^2 z}{\partial y^2} - \left(\dfrac{\partial^2 z}{\partial x\,\partial y}\right)^2$	Positive	Positive

In many questions where it is awkward to apply all these tests, if the first derivatives have both been equated to zero, other more direct tests may often be applied to determine whether the point is an actual maximum or minimum and which of these it is.

EXAMPLE 5.   Find the maximum and minimum values of the function $4x - 6y - 2x^2 - y^2 + 2xy + 2$, assuming x and y are both independent variables.

Solution: Representing the function by z,

$$\frac{\partial z}{\partial x} = 4 - 4x + 2y = 0 \qquad \text{for turning points.}$$

$$\frac{\partial z}{\partial y} = -6 - 2y + 2x = 0 \qquad \text{for turning points.}$$

The solution of these two equations is $x = -1$, $y = -4$.
We know that, *if a maximum or minimum exists*, it occurs at this point.

$$\frac{\partial^2 z}{\partial x^2} = -4; \qquad \frac{\partial^2 z}{\partial y^2} = -2; \qquad \frac{\partial^2 z}{\partial x\,\partial y} = 2.$$

Hence, referring to the above tabulation, all conditions for a maximum are satisfied. The values $x = -1$ and $y = -4$ must produce a maximum value of the function. By substitution, this maximum value is

$$4(-1) - 6(-4) - 2(-1)^2 - (-4)^2 + 2(-1)(-4) + 2 = 12.$$

EXAMPLE 6.   The product of four positive numbers is 400 and one of the numbers is four times another one. For all sets of four numbers satisfying these conditions, find the set whose sum is a minimum.

Solution: Let the numbers be x, $4x$, y, and $400/4x^2y$ (these satisfy the first two conditions). The sum S is

$$x + 4x + y + \frac{400}{4x^2y} = 5x + y + \frac{100}{x^2y}.$$

For minimum S,

$$\frac{\partial S}{\partial x} = 5 - \frac{200}{x^3y} = 0 \quad \text{and} \quad \frac{\partial S}{\partial y} = 1 - \frac{100}{x^2y^2} = 0.$$

These equations are equivalent to $x^3y = 40$ and $x^2y^2 = 100$.
By division $y/x = 2.5$ and $y = 2.5x$.
By substitution, $2.5x^4 = 40$ and $x = \pm 2$.
Since the restriction is to positive numbers, $x = 2$, and $y = 2.5 \times 2 = 5$.

$$\frac{\partial^2 S}{\partial x^2} = +\frac{600}{x^4y}. \qquad \frac{\partial^2 S}{\partial y^2} = +\frac{200}{x^2y^3}. \qquad \frac{\partial^2 S}{\partial x\,\partial y} = +\frac{200}{x^3y^2}.$$

Substituting $x = 2$, $y = 5$,

$$\frac{\partial^2 S}{\partial x^2} = \frac{15}{2}. \qquad \frac{\partial^2 S}{\partial y^2} = \frac{2}{5}. \qquad \frac{\partial^2 S}{\partial x\,\partial y} = 1.$$

All conditions for a minimum are satisfied. Therefore the numbers are 2, 8, 5, and 5.

14.6 ELECTRONIC APPLICATIONS

Certain electron tube characteristics are defined as partial derivatives. The plate current is a function of two independent variables, plate voltage and grid voltage. Figures 14-6 and 14-7 are typical graphs of this variation.

The following definitions apply for electron tubes:

Let v_p represent plate voltage, v_g represent grid voltage, and i_p represent plate current.

Then mutual conductance is $g_m = \partial i_p/\partial v_g$,

plate resistance is $r_p = \partial v_p/\partial i_p$, and

amplification factor is $\mu = -\partial v_p/\partial v_g$.

For a transistor, the following definition holds:
Let i_c represent collector current, and i_e represent emitter current.

Then current gain is $\alpha = \dfrac{\partial i_c}{\partial i_e}$.

RCA 6SK7
AVERAGE PLATE CHARACTERISTICS

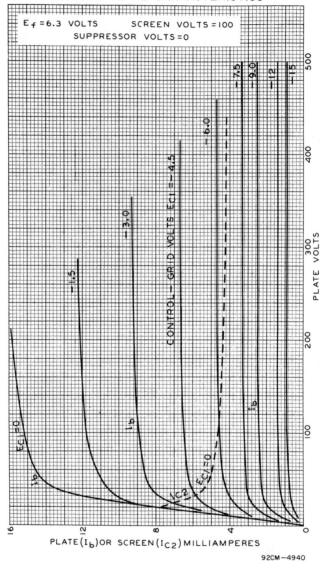

E_f = 6.3 VOLTS   SCREEN VOLTS = 100
SUPPRESSOR VOLTS = 0

CONTROL - GRID VOLTS E_{C1} = -4.5

PLATE VOLTS

PLATE(I_b)OR SCREEN(I_{C2})MILLIAMPERES

92CM—4940

Fig. 14-6.

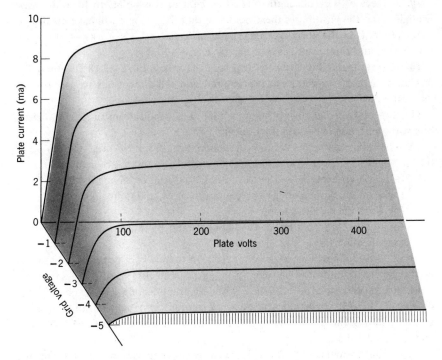

Fig. 14-7. Typical pentode characteristics.

EXERCISES

Examine the following four functions for maximum and minimum:

1. $z = 2x^2 + y^2 - 4xy - 3x + 4y - 3$.

2. $z = 2x^2 + y^2 - xy - 3x + 4y - 3$.

3. $z = 2x^2 - y^2 - 4xy - 3x + 4y - 3$.

4. $z = 1 + 4x^2 + 4x^2y^2 - x^4 - 8y^4$.

5. Find three positive numbers whose sum is 12 such that the sum of their squares is a minimum.

6. Find three positive numbers whose product is 216 such that their sum is a minimum.

7. Find the maximum value of $(2x - x^2) \sin \theta + (2x - x^2) \cos \theta$, where x and θ may both vary.

8. Investigate the function $(x + y)e^{-x^2 - 2y^2}$ for maximum.

9. A piece of wire of length k is to be bent at two points to form a closed triangle. Find the position of these points so that the triangle will have maximum area. [*Hint:* Use the formula: area of a triangle $= \sqrt{s(s - a)(s - b)(s - c)}$, where a, b, and c are the sides and $s = (a + b + c)/2$.]

10. A rectangular box with no top is to have a volume of 18 ft³. Find the minimum area of surface material required and the dimensions to use for this minimum area.

11. Repeat Exercise 10, but assume that the specifications are for a closed box with double thickness on the bottom.

12. The tube characteristics of a triode are given by the relation $i_p = 0.28 (v_g + v_p/12)^{3/2}$, where i_p is in milliamperes and v_g and v_p are in volts.

  (a) Find an expression for the mutual conductance, giving its units.

  (b) Find an expression for the plate resistance, giving its units.

  (c) Find the amplification factor.

  (d) Show that mutual conductance times plate resistance equals the amplification factor.

13. If the amplification factor of an electron tube is kept constant, what must be the general relationship between (a) v_g and v_p and (b) r_p and g_m? Use 12(d).

14. For a certain type of triode with an amplification factor of 30, $i_p = 4.44[v_g + (v_p/30)]^{3/2}$ milliamperes. Find the mutual conductance and plate resistance, stating the units of each, for the operating point where $v_p = 300$ volts and $v_g = -2$ volts.

15. A 20-ft trough is to be formed from a rectangular piece of tin 20 ft long and 15 in. wide by bending the 20-ft length along two longitudinal lines equidistant from the two edges to form equal angles. Find how this should be done to produce the greatest possible trough volume.

16. Examine the following function for maximum:

$$z = 3x + 2y + \frac{4}{(x + 2y)^2} + \frac{2}{x - 1}.$$

17. When two experimental variables x and y are known to be related linearly, a standard method of finding the best straight-line graph to represent the variation of y with x is the method of "least squares"; this consists of plotting the line $y = mx + b$, where m and b are determined such that the expression

$$S = [y_1 - (mx_1 + b)]^2 + [y_2 - (mx_2 + b)]^2 + [y_3 - (mx_3 + b)^2] \cdots$$

is a minimum, where (x_1,y_1) (x_2,y_2), $(x_3,y_3) \cdots$ are the simultaneous experimental values of the variables. Find by this method the values of m and b and the best linear function to represent the following pairs of experimental values of x and y:

x	2	4	6	8
y	41	48	51	60

Illustrate the points and the straight line graphically.

18. In long-distance, three-phase power transmission, a standard level of system stability must be maintained. This may easily be accomplished provided there are enough transmission lines, which are a large item of expense. If the number of lines is reduced, special equipment (also expensive) must be provided to maintain the stability level; specially designed generators, extra switching stations along the transmission right-of-way, more synchronous condensers, etc., are examples of the extra equipment that will probably be required. The decision on transmission design depends on what is most economical. There are basically two variables: the number of lines and the number of intermediate switching stations.

Suppose that a workable formula to cover cost of 230-kv three-phase transmission for a required 170-mile distance has been found to be

$$C = 6.6L + 3.4 \times 10^{-3}P(s + 1) + \frac{2.2 \times 10^{-7}P^3}{(L - 0.7)^2} + \frac{2.8 \times 10^{-5}P^2}{(0.3L + s)^2},$$

where C = cost in millions of dollars
L = number of three-phase lines
s = number of intermediate switching stations
P = power transmitted in megawatts

Find the number of lines and the number of intermediate switching stations for the most economical transmission of (a) 1000 megawatts and (b) 500 megawatts.

CHAPTER 15

Power Series

15.1 THE BINOMIAL AND EXPONENTIAL SERIES

A series is the sum of a succession of terms in which each term is always formed from the previous term by a definite law.

Thus $x + 3x^2 + 5x^3 + 7x^4$ is a series, whereas $a - 5a^3 + 3a^2 + 9a^5$ is probably only a succession of terms.

A series may always be represented in $\sum$ notation. For example, the preceding series may be represented as $\sum_1^4 (2n - 1)x^n$.

Certain types of series should already be familiar to the student, in particular *arithmetic, geometric, binomial,* and *exponential series.* If a series progresses with constantly increasing powers of a variable, it is called a power series. An arithmetic series is not a power series, but a geometric series may be a power series. In this chapter, we are not particularly concerned with these two series. The binomial and exponential series, however, are good preliminary examples of *power series.*

The Binomial Series. A binomial series may always be put in the form

$$(1 + x)^n \equiv 1 + \frac{n}{1!} x + \frac{n(n - 1)}{2!} x^2 + \frac{n(n - 1)(n - 2)}{3!} x^3 \cdots$$

to an infinite number of terms.

The series is identically equal to the binomial expression on the left-hand side, provided the absolute value of x is less than 1. We will investigate the

292

reason for this later. In general, all of the power series considered in this chapter have an infinite number of terms. Even the instance

$$(1 + x)^3 = 1 + 3x + 3x^2 + x^3,$$

which terminates in four terms, can still be considered as an infinite series. The fifth term would be

$$\left(\frac{3 \times 2 \times 1 \times 0}{4!}\right)x^4;$$

the sixth term is

$$\left(\frac{3 \times 2 \times 1 \times 0 \times -1}{5!}\right)x^5, \text{ etc.,}$$

and therefore all of the terms of the infinite series beyond the fourth become zero. We see that the series terminates only if n is a positive integer. It still may be considered an infinite series with all its later terms equal to zero.

In a nonterminating series, we may readily see that if the absolute value of x is greater than 1, the absolute value of x^n will increase as n increases, and successive terms in the series may become progressively greater and finally approach infinity. In such series the sum of the terms cannot be determined, and the series is rendered useless. This type of series is called a *diverging series*. If the absolute value of later terms of a series becomes progressively smaller and finally approaches zero, the series is a *converging series*, and its sum can be determined, if not exactly, at least to any degree of accuracy required.

One important application of any power series was implied in the last paragraph: that is, to express quantities that cannot be determined exactly (usually irrational quantities) to as high a degree of accuracy as required. Thus, by either a binomial series or other power series that we will investigate later, such quantities as $\sqrt{26.1}$, $1/\sqrt[3]{26.1}$, $\sin 31°$, $\ln 1.5$, e, π, etc., can be evaluated to any degree of accuracy.

EXAMPLE 1.   Find the value of $1/\sqrt[3]{26.1}$ to 5 decimal places.

Solution: $1/\sqrt[3]{26.1} = (26.1)^{-\frac{1}{3}}$, and 26.1 may be split into two terms to give a binomial expression. For convergence, the second term should be less than the first, and the first term should preferably be 1, in order to use the standard binomial form given previously. However, the form $(1 + 25.1)^{-\frac{1}{3}}$ cannot be used. (Why?)

The clue on the proper split-up is that $(27)^{-\frac{1}{3}}$ equals exactly $\frac{1}{3}$, a rational number which does not limit the accuracy of the answer in any way.

Thus

$$(26.1)^{-\frac{1}{3}} = (27 - 0.9)^{-\frac{1}{3}} = 27^{-\frac{1}{3}}\left(1 - \frac{0.9}{27}\right)^{-\frac{1}{3}}$$

$$= \frac{1}{3}\left(1 - \frac{1}{30}\right)^{-\frac{1}{3}}$$

$$= \frac{1}{3}\left[1 + \left(-\frac{1}{3}\right)\left(-\frac{1}{30}\right) + \frac{(-\frac{1}{3})(-\frac{4}{3})}{2!}\left(-\frac{1}{30}\right)^2 \right.$$

$$\left. + \frac{(-\frac{1}{3})(-\frac{4}{3})(-\frac{7}{3})}{3!}\left(-\frac{1}{30}\right)^3 \cdots \right]$$

$$= \frac{1}{3}\left(1 + \frac{1}{90} + \frac{2}{8100} + \frac{14}{81 \times 30^3} \cdots\right)$$

$$= 0.333333 + 0.003704 + 0.000082 + 0.000002\cdots$$

$$\simeq 0.33712.$$

Note that each term is rounded off at the sixth decimal, not the fifth; all terms that do not affect the sixth decimal are neglected in order to get 5-decimal accuracy. The third and fourth terms could be determined by slide rule, since accuracy to the sixth decimal amounts to only two-figure accuracy for the third term, and one-figure accuracy for the fourth term.

The Exponential Series (See Section 9.3). The identity

$$e^x \equiv 1 + x + \frac{x^2}{2!} + \frac{x^3}{3!} + \cdots \qquad \text{to an infinite number of terms}$$

is another familiar example of an infinite power series. Since this is an identity, it is true for every value of x. Actually this series converges for any finite value of x, although, if it is to be used to get an answer of stated accuracy, any value of x greater than 1 results in slow convergence and a tedious calculation.

EXERCISES

1. Without using the binomial theorem, find an infinite power series for $1/(1 + x)$.

2. Show that, if x is much smaller than 1, $\sqrt{1 \pm x}$ is approximately equal to $1 \pm x/2$.

3. Using the general approximation of Exercise 2, find a quick approximate answer to the following: (a) $\sqrt{1.02}$, (b) $\sqrt{0.95}$, (c) $\sqrt{0.84}$, and (d) $\sqrt{27}$.

4. Using the binomial theorem, calculate Exercises 3a and 3c to three decimal places.

5. Show that, if x is much smaller than 1, $1/\sqrt{1 \pm x}$ is approximately equal to $1 \mp x/2$.

6. Using the general approximation of Exercise 5, find a quick approximation to the following: (a) $1/\sqrt{1.05}$ and (b) $1/\sqrt{0.93}$.

7. Show that, if X is much smaller than R, $\sqrt{R^2 + X^2}$ is approximately equal to $R(1 + X^2/2R^2)$.

8. By using the first two terms only of the binomial series, find an approximation for $\sqrt{R^2 + X^2}$ if R is much smaller than X.

9. By using the first two terms of a binomial series, find an approximation for $E/\sqrt{R^2 + X^2}$ if R is much smaller than X.

Evaluate the following expressions to four-figure accuracy:

10. $\sqrt{26.10}$.

11. $\sqrt[3]{9.000}$.

12. $\dfrac{1}{\sqrt[5]{29.00}}$.

13. $\dfrac{1}{\sqrt[4]{24.00}}$.

14. $\dfrac{1}{(8.000)^{1.5}}$.

Evaluate the following expressions to four-figure accuracy:

15. $e^{0.1}$.

16. $e^{-0.4}$.

17. $2.000(1 - e^{-0.5})$.

18. If Rt is small compared to L, find an approximate equivalent for $(E/R)(1 - e^{-Rt/L})$ by taking only the first two terms of the exponential series.

19. For small values of t, which do not approach the magnitude of RC, find an approximate equivalent for $EC(1 - e^{-t/RC})$.

20. If L is measured 2% too high and C is measured 3% too high, find to two-figure accuracy the percent error in the calculation of f from the formula $f = 1/2\pi\sqrt{LC}$: (a) by the binomial series and (b) by a total differential using partial derivatives.

15.2   CONVERGENCE AND DIVERGENCE

This topic has been mentioned in Section 15.1 and many texts will go into considerable detail on tests for convergence. For most practical applications, these tests are not necessary. We find that a power series will converge in all practical cases if the absolute value of x is less than one. Actually, rapidity of convergence is just as important as theoretical convergence, and if a series does not converge fairly rapidly, it is not of much practical use without the aid of a computer.

15.3   THE MACLAURIN SERIES

This is a technique that was developed about 1740 by Colin Maclaurin, a Scottish mathematician. It would be of tremendous advantage if all the so-called transcendental functions ($\sin x$, $\ln x$, e^x, $\cosh x$, $\arctan x$, etc.) could be expressed as a power series. Not only could this series be used in estimating the values of these types of functions to any desired degree of accuracy, but the derivatives and integrals of these functions at any particular point could be evaluated to any desired degree of accuracy, using the power formula only. This is of particular value when an exact integration either is impossible or involves a long calculation.

Let us try to find a power series for $\sin x$, as an example. We may begin by using unknown constants for coefficients of the "x" terms, and try to solve for these constants.

Thus let $$\sin x \equiv a_0 + a_1 x + a_2 x^2 + a_3 x^3 \cdots.$$

Since this is to be identically true, any value of x may be substituted on both sides. The appropriate value of x for substitution is one that results in an equation from which one of the constant coefficients may be determined. We see that a substitution of $x = 0$ will make all terms on the right-hand side except a_0 equal to zero, and a_0 may immediately be determined. Thus $\sin 0 = a_0$; that is,

$$a_0 = 0 \quad (1).$$

Now, in the evaluation of a_1, some preliminary calculation will be necessary to make the term in a_1 the only term on the right-hand side that is independent of x. This is accomplished by taking the derivative of both sides with respect to x. Thus $\cos x \equiv a_1 + 2a_2 x + 3a_3 x^2 + 4a_4 x^3 \cdots$ to ∞.
Let $x = 0$. Then $\cos 0 = a_1$; that is,

$$a_1 = 1 \quad (2).$$

Now, taking a derivative of both sides of this latest equation and then replacing x by zero results in the a_2 term remaining alone on the right-hand side. Therefore $-\sin x \equiv 2a_2 + 3 \times 2a_3 x + 4 \times 3a_4 x^2 \cdots$ to ∞
and $-\sin 0 = 2a_2$. Hence

$$a_2 = 0 \quad (3).$$

By repeating the process,

$$-\cos x \equiv 3 \times 2a_3 + 4 \times 3 \times 2a_4 x \cdots \qquad \text{to } \infty$$

$$-\cos 0 = 3 \times 2a_3.$$

$$a_3 = \frac{-1}{3!} \qquad (4).$$

The pattern is established by this time, and by putting in the values of the coefficients, we see that

$$\sin x = x - \frac{x^3}{3!} + \frac{x^5}{5!} - \frac{x^7}{7!} \cdots \qquad \text{to } \infty.$$

This series, of course, is true only if x is the numerical value of the angle in radians, since the differentiation formulas used in the calculations are valid only for radians.

EXAMPLE 3.   Find $\sin 15°$ to four decimal places.
Solution: Using the sine series as an identity, we obtain

$$\sin 15\,° \equiv \sin \frac{\pi}{12}$$

$$= \frac{\pi}{12} - \left(\frac{\pi}{12}\right)^3 \frac{1}{3!} + \left(\frac{\pi}{12}\right)^5 \frac{1}{5!} \cdots .$$

This is a rapidly converging series, which gives

$0.26180 - 0.00299 + 0.00001$ (and later negligible terms) $\simeq 0.2588$.

Note that the accuracy of the answer depends on the accuracy of π. Later we will develop a series for π itself, enabling us to get any required accuracy for its value.
We now begin to see how the familiar sine tables were constructed. For angles beyond $30°$ ($\pi/6$ radians) the series would converge too slowly, and modifications of this series must be used.

An analysis of the techniques used in establishing the series for $\sin x$ leads to the general conclusion that a power series may be established for any function provided that the function is continuously differentiable and that the successive derivatives may be evaluated as finite quantities at $x = 0$. Since both these conditions are true about most of the common transcendental functions, it is possible to represent these functions as a standard power series.

We see further that the coefficients in the power series are determined as follows:

a_0 is the value of the function at $x = 0$,
a_1 is the value of the first derivative at $x = 0$,
a_2 is the value of the second derivative at $x = 0$ divided by 2!,
a_3 is the value of the third derivative at $x = 0$ divided by 3!, etc.

The Maclaurin series may now be stated in symbolic form as follows:

$$f(x) = f(0) + f'(0)x + \frac{f''(0)}{2!} x^2 + \frac{f'''(0)}{3!} x^3 \cdots \qquad \textbf{to an infinite number of terms.}$$

The Maclaurin series may even be used to establish the binomial and exponential series. Similarly, a series for all the standard transcendental functions may be established easily.

The logarithmic series is the only one of these that may present some difficulty. ln x and all of its derivatives are infinite at $x = 0$. This difficulty is overcome by using 1 rather than 0 as a reference value of x, and finding a series for ln $(1 + x)$.

If $f(x) = \ln (1 + x)$, $f(0) = \ln 1 = 0$.

$$f'(x) = \frac{1}{1 + x}, \qquad f'(0) = 1.$$

$$f''(x) = \frac{-1}{(1 + x)^2}, \qquad f''(0) = -1.$$

$$f'''(x) = \frac{2}{(1 + x)^3}, \qquad f'''(0) = 2.$$

$$f''''(x) = \frac{-3 \times 2}{(1 + x)^4}, \qquad f''''(0) = -3 \times 2.$$

$$\ln (1 + x) = x - \frac{x^2}{2!} + \frac{2x^3}{3!} - \frac{3 \times 2x^4}{4!} \cdots$$

$$= x - \frac{x^2}{2} + \frac{x^3}{3} - \frac{x^4}{4} \cdots.$$

This logarithmic series in its simplest form contains no factorials.

Following is a list of the standard power series, developed from the Maclaurin series:

$$\sin x \quad = x - \frac{x^3}{3!} + \frac{x^5}{5!} - \frac{x^7}{7!}\cdots.$$

$$\cos x \quad = 1 - \frac{x^2}{2!} + \frac{x^4}{4!} - \frac{x^6}{6!}\cdots.$$

$$e^x \quad\quad = 1 + x + \frac{x^2}{2!} + \frac{x^3}{3!}\cdots.$$

$$\ln (1 + x) = x - \frac{x^2}{2} + \frac{x^3}{3} - \frac{x^4}{4}\cdots.$$

$$\sinh x \quad = x + \frac{x^3}{3!} + \frac{x^5}{5!} + \frac{x^7}{7!}\cdots.$$

$$\cosh x \quad = 1 + \frac{x^2}{2!} + \frac{x^4}{4!} + \frac{x^6}{6!}\cdots.$$

The student should develop each of these from the Maclaurin series.

Many standard identities may be established, or simply verified, by using a standard power series. Following is a list of evaluations and identities easily verified from a standard power series:

$$\sin 0 = 0. \quad\quad\quad \cos 0 = 1. \quad\quad\quad \sin (-x) = -\sin x.$$

$$\cos (-x) = \cos x. \quad \frac{d}{dx} \sin x = \cos x. \quad \frac{d}{dx} \cos x = -\sin x.$$

$$\frac{d}{dx} \cosh x = \sinh x. \quad\quad \int \sin x \, dx = -\cos x + k.$$

$$\int \cos x \, dx = \sin x + k. \quad \frac{d}{dx} \ln (1 + x) = \frac{1}{1 + x}.$$

EXAMPLE 3.   Establish, by a series method, the fact that

$$\frac{d}{dx} \ln (1 + x) = \frac{1}{1 + x}.$$

Solution:

$$\ln (1 + x) = x - \frac{x^2}{2} + \frac{x^3}{3} - \frac{x^4}{4}\cdots.$$

Hence

$$\frac{d}{dx} \ln (1 + x) = 1 - x + x^2 - x^3 \cdots.$$

Also $1/(1 + x)$, by simple long division, equals $1 - x + x^2 - x^3 \cdots$. The given identity has been proved.

EXAMPLE 4.   Find the first four terms of a single power series for the expression

$$\ln \left(\frac{1 - x}{1 + 2x} \right)^2.$$

Solution: The expression is

$$2[\ln (1 - x) - \ln (1 + 2x)]$$

$$= 2\left[-x - \frac{x^2}{2} - \frac{x^3}{3} - \frac{x^4}{4} - \cdots - \left(2x - \frac{4x^2}{2} + \frac{8x^3}{3} - \frac{16x^4}{4} \cdots \right) \right]$$

$$= -6x + 3x^2 - 6x^3 + \frac{15x^4}{2} \cdots.$$

This is a series, even though the law by which each term is formed from the previous term is fairly complicated. The series could be expressed as

$$2 \sum_{n=1}^{\infty} \frac{-1 + (-2)^n}{n} x^n.$$

EXAMPLE 5.   Given $\ln (E - iR) - \ln E = -Rt/L$, in which $E = 50$, $R = 2.0$, and $L = 4.0 \times 10^{-2}$, find the time t at which $i = 10.0$.
Solution:

$$\ln \frac{E - iR}{E} = -\frac{Rt}{L} \quad \text{or} \quad \ln \left(1 - \frac{iR}{E} \right) = -\frac{Rt}{L}.$$

Therefore $\ln (1 - 0.04i) = -50t$, thus

$$-0.04i - \frac{0.0016i^2}{2} - \frac{0.000064i^3}{3} \cdots = -50t.$$

If $i = 10.0$, $-0.4 - 0.08 - 0.021 \cdots = -50t$. $t = 0.010$ to two significant figures.

EXAMPLE 6.   The current in a circuit is given by $i = 2.50e^{-50t} \sin 100t$ amperes. Find by a series method the value of i at $t = 1/200$.

Solution:

$i = 2.50e^{-0.25} \sin 0.50 \text{ rad}$

$$= 2.50\left(1 - 0.25 + \frac{0.0625}{2!} - \frac{0.0156}{3!}\cdots\right)\left(0.50 - \frac{0.50^3}{3!} + \frac{0.50^5}{5!}\cdots\right)$$

$= 2.50(0.779)(0.479) = 0.933$ ampere to three-figure accuracy.

EXERCISES

1. Find the power series for cos x by using the power series for sin x developed in Section 15.3 and the fact that cos $x = (d/dx) \sin x$.
  2. Establish the power series for sinh x:
    (a) By evaluating $f'(0)$, $f''(0)$, etc., and using the Maclaurin series.
    (b) By using the exponential form of sinh x and two exponential series.

By using a standard power series in each case, evaluate the following expressions to four-figure accuracy:

3. sin 12°.   4. cos 6°.   5. ln 1.25.
  6. ln 0.80.   7. sinh 0.25.   8. cosh 0.40.
  9. sinh (-0.20)   10. cosh (-0.25).   11. ln 3.21 $-$ ln 3.00.

12. ln $\sqrt{0.75}$.

Find a single power series for each of the following functions:

13. sin ($-2x$).   14. ln ($e - x^2$).

15. ln $\dfrac{1 + x}{1 - x}$.   16. ln $(1 + x)^2(1 - x)$.

17. ln $\dfrac{1 - 2x}{e(1 + x)^3}$.   18. ln $\dfrac{(1 + ax)^b}{(1 - cx)^a}$.

19. Find the first three terms of the power series for $\sin^2 x$:
    (a) By evaluating $f(0)$, $f'(0)$, $f''(0)$, etc., and using the Maclaurin series.
    (b) By expressing $\sin^2 x$ in terms of cos $2x$ and using the standard cosine series.
  20. Given ln $(E - iR) - \ln E = -Rt/L$, find the time t in terms of L and R at which $iR = 30\%$ of E, assuming four-figure accuracy.
  21. Given ln $(EC - q) - \ln EC = -t/RC$ and assuming q is small enough compared to EC so that second and higher powers of q/EC may be neglected, find an approximate value of q as a function of t: (a) by using a logarithmic power series, and (b) by first expressing q as a function of t and using an appropriate series.

Solve the following equations approximately by a series method, neglecting any powers of x higher than the second, and reject any solutions inconsistent with this approximation:

22. $e^x = \cos x.$ **23.** $3e^x = 7 - 3\cos x.$ **24.** $e^{-x} = \sin x.$

25. $\cosh x = e^x - 5x^2.$ **26.** $\ln(2 - 4x) = \sinh x.$

27. A curve passes through the point $P(0,2)$ at which point the values of the first three derivatives are as follows: $y' = -2$, $y'' = 2$, and $y''' = -4$:

(a) What is the equation of the curve if it is known to be a cubic?

(b) If the curve is not known to be a cubic, in what vicinity will the equation of (a) be a good approximation for the equation of the curve?

(c) Find the approximate height of the curve at $x = 0.25$.

15.4 SPECIAL TECHNIQUES FOR OBTAINING POWER SERIES

Often the calculation of successive derivatives of a function becomes increasingly difficult, even though it may be theoretically possible. In such cases one of the following techniques may be used to simplify the calculations in establishing a power series for a function.

Multiplication

EXAMPLE 7.   Find a series for $e^x \sinh x$.

Solution: If successive derivatives are taken, since the original function is a product, the first derivative will contain two terms and each successive derivative will contain twice as many terms as the previous derivative. The calculations become unwieldy. Here the series may be obtained as the product of two series as follows:

$$e^x \sinh x = \left(1 + x + \frac{x^2}{2!} + \frac{x^3}{3!}\cdots\right)\left(x + \frac{x^3}{3!} + \frac{x^5}{5!}\cdots\right)$$

$$= x + \frac{x^3}{6} + \frac{x^5}{120} + \cdots + x^2 + \frac{x^4}{6} + \frac{x^6}{120} + \cdots$$

$$+ \frac{x^3}{2} + \frac{x^5}{12} + \cdots + \frac{x^4}{6} + \frac{x^6}{36} + \cdots + \cdots$$

$$= x + x^2 + \frac{2x^3}{3} + \frac{x^4}{3}\cdots.$$

It is usual in these cases where the series is complicated to give only the first few terms, with no indication of the general term.

Division

EXAMPLE 8.   Find a series for tan x.

Solution: If a Maclaurin series is used, the second derivative is a product and successive derivatives become increasingly bulky. However,

$$\tan x = \frac{\sin x}{\cos x} = \frac{x - \dfrac{x^3}{6} + \dfrac{x^5}{120}\cdots}{1 - \dfrac{x^2}{2} + \dfrac{x^4}{24}\cdots}.$$

By dividing out, the first three or four terms are fairly readily obtained. Hence

$$\tan x = x + \tfrac{1}{3}x^3 + \tfrac{2}{15}x^5\cdots.$$

Integration or Differentiation

EXAMPLE 9.   Find a series for $\sinh^{-1} x$.

Solution: Since

$$\frac{d}{dx}\sinh^{-1} x = \frac{1}{\sqrt{1 + x^2}}$$

$$\sinh^{-1} x = \int \frac{1}{\sqrt{1 + x^2}}\, dx$$

$$= \int (1 + x^2)^{-\frac{1}{2}}\, dx.$$

By binomial series, this is

$$\int \left(1 - \frac{x^2}{2} + \frac{3}{8}x^4 - \frac{5}{16}x^6\cdots\right) dx = x - \frac{x^3}{6} + \frac{3x^5}{40} - \frac{5x^7}{112} + \cdots + K.$$

But, since $\sinh^{-1} 0 = 0$, $K = 0$, and

$$\sinh^{-1} x = x - \frac{x^3}{6} + \frac{3x^5}{40} - \frac{5x^7}{112}\cdots.$$

EXAMPLE 10.   Find a series for $\ln (1 + x)$ without using the Maclaurin expansion.

Solution:

$$\frac{d}{dx}\ln (1 + x) = \frac{1}{1 + x}.$$

$$\therefore\ \ln (1 + x) = \int \frac{1}{1 + x}\, dx.$$

This, by direct division, equals

$$\int (1 - x + x^2 - x^3 \cdots) \, dx = x - \frac{x^2}{2} + \frac{x^3}{3} - \frac{x^4}{4} + \cdots + K.$$

Since $\ln 1 = 0$, $K = 0$, and we have the previously established series for $\ln (1 + x)$.

EXAMPLE 11.   Find a power series for π.
Solution:

$$\frac{\pi}{6} = \text{arc sin } \tfrac{1}{2} \text{ exactly} \qquad \text{or} \qquad \frac{\pi}{4} = \text{arc tan } 1 \text{ exactly.}$$

These are exact, since they are derived geometrically.

$$\text{arc sin } x = \int \frac{1}{\sqrt{1 - x^2}} \, dx = \int \left(1 + \frac{x^2}{2} + \frac{3x^4}{8} + \frac{5x^6}{16} \cdots \right) dx$$

$$= x + \frac{x^3}{6} + \frac{3x^5}{40} + \frac{5x^7}{112} + \cdots + K$$

and $K = 0$. (Why?) Thus

$$\frac{\pi}{6} = \text{arc sin } \frac{1}{2} = \frac{1}{2} + \frac{1}{48} + \frac{3}{1280} + \frac{5}{14,336} \cdots.$$

Hence π is 6 times this series, that is,

$$\pi = 3 + 0.125 + 0.0141 + 0.0021 \cdots$$
$$\doteq 3.1412.$$

The value of π (the ratio of circumference to diameter in any circle) was originally estimated by actual measurement of the circumference and diameter of large circles. Recently, by programming a series for π for a computer, π was determined to 3,000,000 decimal places.

Question. Why is the arc sin series more suitable than the arc tan series for determining an approximate value for π?

EXERCISES

Find the first three terms of a single power series to represent each of the following functions:

1. $e^x \sin x$.
2. $\ln (1 - x)^{x^3}$.
3. $\cos x \cosh x$.
4. $\tan x \cos^2 x$.
5. $e^{-5t} \sin 10t$.
6. $\sqrt{1 - x^2} \cos x$.

7. $\dfrac{\sin x}{1 - x^2}$.

8. cot x.

9. tanh x.

10. sec x.

11. $\dfrac{\sin x}{\sinh x}$.

12. $\dfrac{\ln (1 + 2x)}{\ln (1 + x^2)}$.

Evaluate each of the following functions at the specified value of the independent variable, using a series method:

13. $e^{-4t} \cos 5t$ at $t = 0.05$, to three-figure accuracy.
14. sech $2x$ at $x = 0.2$, to three-figure accuracy.
15. Find a single power series for $\sec^2 x$ by expressing it as a derivative of a more standard power series.

By using the method of Example 9, find a power series for the following functions:

16. arc tan x.

17. $\tanh^{-1} x$.

18. $\cosh^{-1} x$.

19. $[\ln (1 + x)]^2$.

20. Find a series for π by establishing and using a power series for arc tan x. Approximately how many terms of this series are necessary to determine π to one decimal place?

Evaluate each of the following infinite series to three-figure accuracy by expressing each as a transcendental function and using tables or slide rule:

21. $0.8 - \dfrac{(0.8)^2}{2} + \dfrac{(0.8)^3}{3} - \dfrac{(0.8)^4}{4} \cdots$ to an infinite number of terms.

22. $2 - \dfrac{2}{2!} + \dfrac{2}{4!} - \dfrac{2}{6!} \cdots$ to an infinite number of terms.

23. $1 - \dfrac{1}{3!} + \dfrac{1}{5!} - \dfrac{1}{7!} + \dfrac{1}{9!} \cdots$ to an infinite number of terms.

24. $\dfrac{\pi}{4} + \dfrac{1}{2!} \left(\dfrac{\pi}{4}\right)^2 - \dfrac{1}{3!} \left(\dfrac{\pi}{4}\right)^3 - \dfrac{1}{4!} \left(\dfrac{\pi}{4}\right)^4 + \dfrac{1}{5!} \left(\dfrac{\pi}{4}\right)^5 + \dfrac{1}{6!} \left(\dfrac{\pi}{4}\right)^6 - - + + \cdots$ to an infinite number of terms.

25. Prove that $-1 + \frac{1}{2} - \frac{1}{3} + \frac{1}{4} \cdots$ to an infinite number of terms is exactly equal to

$$-0.5 - \dfrac{(0.5)^2}{2} - \dfrac{(0.5)^3}{3} - \dfrac{(0.5)^4}{4} \cdots \qquad \text{to an infinite number of terms.}$$

15.5 APPLICATIONS TO INTEGRATION

If a definite integral is difficult or impossible to evaluate exactly, very often a value for the integral can be obtained to as close an approximation as

required by expressing some or all of the functions as a power series and integrating the power series term by term.

EXAMPLE 12.   Evaluate $\displaystyle\int_0^1 x^{\frac{2}{3}} \sin x \, dx.$

Solution: The exact integration here is very involved. Using a series for
$\sin x$, the integral is

$$\int_0^1 x^{\frac{2}{3}}\left(x - \frac{x^3}{3!} + \frac{x^5}{5!} - \frac{x^7}{7!}\cdots\right) dx$$

$$= \int_0^1 \left(x^{5/3} - \frac{x^{11/3}}{6} + \frac{x^{17/3}}{120} - \frac{x^{23/3}}{5040}\cdots\right) dx$$

$$= \left[\frac{3}{8}x^{8/3} - \frac{1}{28}x^{14/3} + \frac{1}{800}x^{20/3} - \frac{1}{43,680}x^{26/3}\cdots\right]_0^1$$

$$= 0.375 - 0.03571 + 0.00125 - 0.00002\cdots$$

$$= 0.3405 \qquad \text{to four figures.}$$

EXERCISES

Evaluate the following definité integrals to four-figure accuracy, and verify
the answers to Exercises 1, 2, and 3 by an integration by parts:

1. $\displaystyle\int_0^1 x \cos x \, dx.$ **2.** $\displaystyle\int_0^{0.2} x^2 e^{-x} \, dx.$

3. $\displaystyle\int_0^{0.3} x \ln(1 + x) \, dx.$ **4.** $\displaystyle\int_0^1 \frac{x - \sin x}{x} \, dx.$

5. $\displaystyle\int_0^{0.5} \frac{1 - \cos x}{x} \, dx.$ **6.** $\displaystyle\int_0^{0.01} e^{-20t} \sin 20t \, dt.$

7. $\displaystyle\int_0^{0.2} \sqrt{e^x(\sinh x + \cosh x)} \, dx.$ **8.** $\displaystyle\int_0^{0.5} x \arctan x \, dx.$

9. $\displaystyle\int_0^1 x^2 \sin x \sinh x \, dx.$

10. Find to three significant figures the area bounded by the curve $y = \dfrac{\cos x}{x}$
the x-axis, and the lines $x = 0.5$ and $x = 1$.

11. Find to three-figure accuracy the mean value of i from $t = 0$ to $t = 0.1$ if
$i = 5te^{-10t}$.

12. Find the rms current i to three significant figures for Exercise 11.

13. Find to three significant figures the rms current i between $t = 0$ and $t = 0.01$ if $i = 4e^{-5t} \cos 10t$.

15.6 TAYLOR SERIES AS AN AID TO CONVERGENCE

If such expressions as $\sin 63°$, $e^{2.5}$ or $\ln 3.7$ are to be evaluated by using a Maclaurin series, the value of x in the series is greater than 1 and the series either diverges or converges too slowly. The Maclaurin series uses $f(0)$, $f'(0)$, $f''(0)$, etc., in the coefficients, and thus $x = 0$ is the point of reference.

The *Taylor series* may be adapted to use any value of x as a point of reference. If a particular fixed value of x, x_1, is used as a point of reference, the function for which a power series is required may be designated as $f(x_1 + h)$, where h is the variable portion of x as measured from x_1. Since $x_1 + h$ represents the total variable x, $dh/dx = 1$ and derivatives with respect to x are identical to derivatives with respect to h.

Let

$$f(x_1 + h) = a_0 + a_1 h + a_2 h^2 + a_3 h^3 \cdots \qquad \text{to an infinite number of terms.}$$

$$f'(x_1 + h) = a_1 + 2a_2 h + 3a_3 h^2 \cdots$$

$$f''(x_1 + h) = 2a_2 + 3!a_3 h \cdots$$

$$f'''(x_1 + h) = 3!a_3 \cdots.$$

If h is replaced by 0 in these identities the values of a_0, a_1, a_2, ... are successively determined:

$$a_0 = f(x_1). \qquad a_1 = f'(x_1). \qquad a_2 = \frac{f''(x_1)}{2!}. \qquad a_3 = \frac{f'''(x_1)}{3!} \qquad \text{etc.}$$

The Taylor series results from these substitutions.

$$f(x_1 + h) = f(x_1) + f'(x_1)h + \frac{f''(x_1)}{2!} h^2 + \frac{f'''(x_1)}{3!} h^3 \cdots.$$

The Maclaurin series is now seen to be a particular case of the Taylor series where $x_1 = 0$ and h is the total x. Essentially, the Taylor series takes a first approximation of $f(x_1)$ for the function $f(x_1 + h)$, and the successive terms after the first give a closer approximation. If x_1 is chosen to be close to the total x, the first approximation will be close, h will be small, and the series will converge rapidly.

EXAMPLE 13.   Find sin 63° to four decimals.

Solution: 60° is a logical reference point and sin 60° a first approximation.

$$\sin 63° = \sin (60° + 3°) = \sin \left(60° + \frac{\pi}{60}\right).$$

$$f(60°) = \sin 60° = \frac{\sqrt{3}}{2} = 0.86603.$$

$$f'(60°) = \cos 60° = 0.50000.$$

$$f''(60°) = -\sin 60° = -0.86603.$$

$$f'''(60°) = -\cos 60° = -0.50000 \qquad \text{etc.}$$

Adapting the Taylor series to this case,

$$\sin 63° = 0.86603 + 0.50000 \left(\frac{\pi}{60}\right) - \frac{0.86603}{2}\left(\frac{\pi}{60}\right)^2 - \frac{0.50000}{6}\left(\frac{\pi}{60}\right)^3 \cdots.$$

$$= 0.86603 + 0.02618 - 0.00119 - 0.00001$$

$$= 0.8910 \qquad \text{to four decimals.}$$

The method by which mathematical tables have been calculated now begins to become apparent. For the sine tables, as an example, a Maclaurin standard sine series could be used with rapid convergence from 0 to 5°. Beyond this, to give more rapid convergence, 5° could be used as a reference point for a Taylor series, using the previously calculated values of sine 5° and cos 5° as a basis. By constantly changing the reference point as the table progresses, a quick convergence is achieved.

Two approximations frequently used in mathematics may now be seen to actually be the first two terms of a Taylor series. The theory of small changes (Section 5.2) states that $\Delta y \simeq y' \Delta x$; in functional notation this becomes

$$f(x + \Delta x) - f(x) \simeq f'(x)\,\Delta x.$$

The Taylor series to fit this expression could be stated as

$$f(x + \Delta x) = f(x) + f'(x)\,\Delta x + \frac{f''(x)}{2!}(\Delta x)^2 \cdots,$$

and if the third and succeeding terms of this series are neglected, the result is the same as the approximate law of small changes. To make the law of small changes more accurate, a Taylor series must be used.

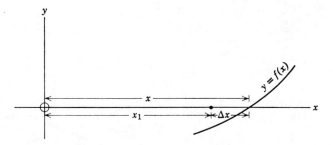

Fig. 15-1. Newton's approximation to the root of $f(x) = 0$.

The other approximation resulting from a partial Taylor series is Newton's approximation to the root of an equation. Suppose one of the roots of the equation $f(x) = 0$ is to be found. Figure 15-1 represents a magnified portion of the graph of $y = f(x)$ close to the x-axis. One root of the equation $f(x) = 0$ is the value of x where this graph crosses the x-axis. Suppose that the first approximation of this root is x_1, estimated from the graph, and that x_1 is actually Δx less than the true root x. A Taylor series indicates that

$$f(x) = f(x_1 + \Delta x) = f(x_1) + f'(x_1)(\Delta x) + \frac{f''(x_1)}{2!}(\Delta x)^2 \cdots.$$

But $f(x) = 0$, since x is the exact value of the independent variable at which the graph crosses the x-axis. Therefore, neglecting the third and successive terms of the series,

$$0 \simeq f(x_1) + f'(x_1)(\Delta x).$$

$$\Delta x \simeq -\frac{f(x_1)}{f'(x_1)}.$$

Since $x = x_1 + \Delta x$, a second approximation to x is $x_1 - f(x_1)/f'(x_1)$, provided x_1 is a close first approximation. This is known as Newton's Approximation. This approximation applies for positive or negative slopes and for positive and negative Δx. In Fig. 15-1, $f(x_1)$ is negative and $f'(x_1)$ is positive, so that a small amount is added to x_1, not subtracted, to get the closer approximation to x. This fact is substantiated by reference to the graph.

EXAMPLE 14.   By a graphical method, approximate the value of x for which $2e^{-x} = x$. Use Newton's approximation to find a more accurate value of x.

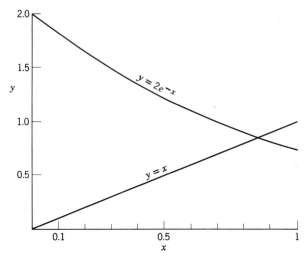

Fig. 15-2.

Solution: The graph of $y = 2e^{-x} - x$ may be drawn, and the estimated value of x for the equation would be the value of x where the graph crosses the x-axis.

A better method here is to draw the graphs of $y = 2e^{-x}$ and $y = x$ separately as demonstrated in Fig. 15-2, and the approximate value of x is the value of x at their point of intersection. The graphical approximation might be $x_1 = 0.8$.

Since

$$2e^{-x} - x = 0, \qquad f(x) = 2e^{-x} - x; \qquad \text{and} \quad f'(x) = -2e^{-x} - 1.$$

$$x \doteq x_1 - \frac{f(x_1)}{f'(x_1)} = 0.8 - \frac{2e^{-0.8} - 0.8}{-2e^{-0.8} - 1}$$

$$= 0.8 - \frac{2(0.4493) - 0.8}{-2(0.4493) - 1} \simeq 0.852.$$

$f(x_1)$ will be close to zero if x_1 is a close approximation. In this example $f(x_1)$ is the difference of two numbers close to one another, and each of these two numbers should be calculated with considerable accuracy (more than slide rule accuracy) so that their difference is quite accurate. A repetition of Newton's method using 0.852 or even 0.85 as a starting value of x will produce an even more accurate answer. The accuracy of the answer is therefore limited only by the number of repetitions of Newton's Approximation that are performed.

EXERCISES

Choosing an appropriate reference value for the independent variable, evaluate the following expressions to four-figure accuracy by using a Taylor series:

1. $\cos 48°$.

2. $\sin 25.5°$.

3. $\ln 9.0000$ (by using $\ln 10$).

4. $e^{-1.25}$ (by assuming e^{-1}).

5. $\sinh 2.2$ (by using $\sinh 2$ and $\cosh 2$).

6. $\tan 42°$.

7. arc $\tan 1.02$.

8. $\sin 50°$ (by using $\sin 60°$).

9. arc $\sin 0.52$.

10. $\ln 3.0000$ (by using $\ln e$).

11. The graph of a function passes through the point $(2, -1)$ and at this point $y' = 4$, $y'' = -2$, and $y''' = -3$. Find the approximate value of the function at $x = 2.2$.

12. (a) Find a Taylor series using $X = 3$ as reference for the function $\sqrt{16 + X^2}$, as far as the first three terms.

(b) Use the series of (a) to find the approximate value of $\sqrt{16 + 4^2}$.

For each of the following equations, estimate all real values of the independent variable by a graphical method; then find a closer approximation by using Newton's approximation:

13. $x^3 - x^2 - x - 1 = 0$.

14. $x^4 - 3x^2 - 5 = 0$.

15. $x^3 - 3x = 17$.

16. $2x^3 - 4x^2 - 10x + 11 = 0$.

17. $x^4 + x^3 + x^2 - 2x = 1$.

18. $e^x = x^3$.

19. $3 \ln x = x$.

20. $\sin x + 0.5 = x$.

21. $\tan x = \ln (2 + x)$.

22. $e^{x-2} = \ln (x + 1)$.

15.7   RELATIONSHIPS INVOLVING $e^{\pm jx}$

Certain previously unsuspected relationships between the transcendental functions themselves may be established by making use of the standard power series for these functions. The quantity e^{jx}, where j is the complex operator $\sqrt{-1}$, apparently involves an imaginary power. It can be shown to have a very real interpretation, which is particularly important in certain types of differential equations.

By using the e^x power series,

$$e^{\pm jx} = 1 \pm jx + \frac{j^2 x^2}{2!} \pm \frac{j^3 x^3}{3!} + \frac{j^4 x^4}{4!} \cdots$$

$$= 1 \pm jx - \frac{x^2}{2!} \mp \frac{jx^3}{3!} + \frac{x^4}{4!}$$

$$= 1 - \frac{x^2}{2!} + \frac{x^4}{4!} \cdots \pm j\left(x - \frac{x^3}{3!} + \frac{x^5}{5!} \cdots\right).$$

Hence
$$e^{\pm jx} = \cos x \pm j \sin x.$$

The right-hand side is the familiar representation of a vector of 1 at an angle $\pm x$ by its horizontal and vertical components. Thus $e^{\pm jx}$ represents a vector of 1 at an angle of $\pm x$ radians.

There are therefore three ways of representing a vector of 1 at an angle x:

$\cos x + j \sin x$ (component form).

$1\underline{/x}$ (polar form).

e^{jx} (exponential form).

This is why, in multiplying two vectors in polar form, the angles are added. This simply obeys the laws of indices, since the two vectors can be represented exponentially.

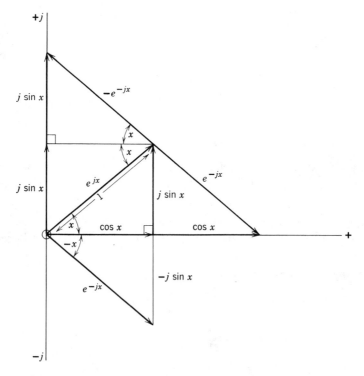

Fig. 15-3. Relationships between $e^{\pm jx}$, $\cos x$, and $\sin x$.

Further relationships may be established as follows:

$$e^{jx} = \cos x + j \sin x.$$

$$e^{-jx} = \cos x - j \sin x.$$

Add, and divide by 2, then

$$\cos x = \frac{e^{jx} + e^{-jx}}{2}.$$

Subtract, and divide by $2j$, then

$$\sin x = \frac{e^{jx} - e^{-jx}}{2j}.$$

Note here the similarity between these formulas and the definitions of sinh x and cosh x.

The symbols e^{jx} and e^{-jx} represent conjugate vectors, and the truth of the aforementioned relationships may be clearly seen from Fig. 15-3.

Any imaginary quantity can be expressed by using the j notation.

EXAMPLE 15.   ln (-1), which has no real value, equals

$$\ln (1\underline{/\pi}) = \ln e^{j\pi} = j\pi.$$

Moreover, certain expressions containing imaginary quantities may often actually be real quantities.

EXAMPLE 16. $j^j = \left(1\underline{/\frac{\pi}{2}}\right)^j = (e^{j\pi/2})^j = e^{-\pi/2}$, a real quantity.

EXAMPLE 17. $(1 + j\sqrt{3})e^{-j\pi/3} = 2\underline{/\frac{\pi}{3}}\, e^{-j\pi/3} = 2e^{j\pi/3}e^{-j\pi/3} = 2e^0 = 2.$

EXAMPLE 18.   Simplify $-4je^{(-250 + j200)t} + 4je^{(-250 - j200)t}$.
Solution: The given expression is

$$4j(-e^{-250t}e^{j200t} + e^{-250t}e^{-j200t})$$
$$= 4je^{-250t}(-e^{j200t} + e^{-j200t})$$
$$= 4je^{-250t}(-2j \sin 200t)$$
$$= 8e^{-250t} \sin 200t \qquad \text{a damped oscillation.}$$

EXERCISES

1. Express the following exponentials in polar and component form, illustrating the vector:

(a) $e^{j\pi/3}$.   (b) $2e^{-j4\pi}$.   (c) $e^{-j\pi/4}$.   (d) $3e^{j}$.

2. Express each of the following in exponential form:

   (a) j.   (b) $-j$.   (c) -1.

   (d) $2\underline{/30°}$.   (e) $1 + j\sqrt{3}$.   (f) $2 - j2$.

   (g) $j \sin x - \cos x$.   (h) $\sinh x - \cosh x$.

3. By using the exponential forms of $\cos x$ and $\sin x$, prove $\cos^2 x + \sin^2 x = 1$.

Prove the following identities:

   4. $\cosh jx = \cos x$.   **5.** $\sinh jx = j \sin x$.

   6. $\cos jx = \cosh x$.   **7.** $\sin jx = j \sinh x$.

   8. $\cosh jx + \sinh jx = e^{jx}$.   **9.** $\cos jx + j \sin jx = e^{-x}$.

10. Express the following imaginary quantities in simplest form with the j operator: (a) $\ln(-10)$, and (b) $\log_{10}(-10)$.

Simplify each of the following expressions to real expressions or values:

11. j^{2j}.

12. $\dfrac{1}{2}\left(\dfrac{1}{s - jw} + \dfrac{1}{s + jw}\right)$.

13. $j \ln(-1)$.

14. $e^{1 + j\pi}$.

15. $\dfrac{e^{-j5\pi/2}}{2j}$.

16. $\dfrac{e^{j\pi/3}(1 + j\sqrt{3})}{1\underline{/300°}}$.

17. $\cosh j\pi + \sinh j\pi$.

18. $(e^{j\pi/2} - 1)(e^{j\pi/2} + 1)$.

19. $e^{(-10 - j20)t}(\cos 20t + j \sin 20t)$.

20. $-j(e^{-2t + j\pi/4})^2$.

21. $e^{3 + j\pi/2}(2e^{-3 + j\pi/4} - 2e^{-3 - j\pi/4})$.

22. $2je^{(-3 - j4)t} - 2je^{(-3 + j4)t}$.

23. $(2 - 3j)e^{(-4 + j2)t} + (2 + 3j)e^{(-4 - j2)t}$.

24. Find $\int e^{-2x} \cos 3x\, dx$ by first expressing $\cos 3x$ in exponential form and then integrating the exponential function and simplifying.

25. Find $\int e^{-5t} \sin 3t\, dt$ by the method suggested in Exercise 24.

CHAPTER 16

First-Order Differential Equations

16.1 INTRODUCTION

A *differential equation* (often abbreviated to D.E.), is an equation expressing a relationship between two or more variables, involving one or more derivatives.

EXAMPLES:

(a) $\dfrac{dy}{dx} + 3y = 8x$ (or $dy + 3y\,dx = 8x\,dx$).

(b) $Ri + L\dfrac{di}{dt} = 10 \sin \omega t.$

(c) $L\dfrac{d^2q}{dt^2} + R\dfrac{dq}{dt} + \dfrac{q}{C} = 4.$

(d) $\left(\dfrac{d^3y}{dx^3}\right)^2 + \dfrac{d^2y}{dx^2} - 2\left(\dfrac{dy}{dx}\right)^3 = 4x.$

To solve a differential equation means to find an equation expressing the relationship between the variables with no derivatives occurring.

The *order* of a differential equation is the highest order among the derivatives appearing in the equation. Thus, Examples (a) and (b) are first-order differential equations, (c) is second-order, and (d) is third-order.

The *degree* of a differential equation is the power of the highest-order derivative occurring. Thus Examples (a), (b), and (c) are of the first degree, (d) is of the second.

The subject of differential equations is one of wide scope, which has been pursued exhaustively in many texts. We will confine ourselves to the most basic types of differential equations, with stress on applications.

16.2 TYPES SOLVED BY DIRECT INTEGRATION

Consider the first-order differential equation, $dy/dx = 3x^2 - 2$. It is apparent that $y = \int (3x^2 - 2)\, dx = x^3 - 2x + K$, which is the *general solution* of the differential equation. If any particular simultaneous values of x and y are given, we can solve for K and get an *exact solution*, just as we have already done in integration problems. Thus, if $y = 2$ when $x = 1$, $K = 3$; and the exact solution is $y = x^3 - 2x + 3$.

This direct integration may be used to solve some types of differential equations of higher order.

EXAMPLE 1.   Solve the differential equation

$$\frac{d^3y}{dx^3} = \sin 2x.$$

Solution:

$$\frac{d^2y}{dx^2} = \int \sin 2x\, dx = -\frac{1}{2}\cos 2x + K_1.$$

Thus

$$\frac{dy}{dx} = \int \left(-\frac{1}{2}\cos 2x + K_1 \right) dx$$

$$= -\frac{1}{4}\sin 2x + K_1 x + K_2$$

and

$$y = \int \left(-\frac{1}{4}\sin 2x + K_1 x + K_2 \right) dx$$

$$= +\frac{1}{8}\cos 2x + \frac{K_1 x^2}{2} + K_2 x + K_3.$$

Three further specific pieces of information would be required to solve for the *arbitrary constants* K_1, K_2, and K_3. If, when $x = 0$, $y = 2$, $dy/dx = 1$, and $d^2y/dx^2 = -3$, then, by direct substitutions, $2 = \frac{1}{8} + K_3$. Hence $K_3 = \frac{15}{8}$.

$$1 = K_2, \qquad -3 = -\frac{1}{2} + K_1, \qquad K_1 = -\frac{5}{2}.$$

Therefore the exact solution is

$$y = \frac{\cos 2x}{8} - \frac{5}{4}x^2 + x + \frac{15}{8}.$$

The student may, at this stage, think that the solution of all differential equations consists only of successive integrations. Two examples here will point out the fallacy of this reasoning.

Given $dy/dx = 5y$, then $y = \int 5y\, dx$.

This cannot be integrated unless y is known as a function of x; this is the very solution we are seeking.

Given $d^2y/dx^2 - 5(dy/dx) = 2x$.

One integration gives $dy/dx - 5y = x^2 + K_1$. A second integration is impossible since $\int -5y\, dx$ cannot be calculated without knowing y as a function of x.

We may conclude that direct integration will solve a differential equation if and only if only one derivative is involved and the dependent variable does not occur directly in the equation.

One characteristic of all differential equations has been demonstrated by the foregoing examples; that is, the general solution of any differential equation will contain the same number of arbitrary constants as the order of the differential equation. Whether the solution is achieved by direct integration or by other more involved processes, one integration is directly or indirectly involved each time the order of the derivatives is reduced by one. Thus, for example, a third-order differential equation will have three arbitrary constants in its solution. These are not in general called integration constants, since the solution is not always achieved by direct integration.

16.3 SOME INTUITIVE METHODS

Many integration problems become apparent by looking backwards at the process of differentiation. In the same way, many differential equations may be solved intuitively by a thorough knowledge of differentiation. Note in particular the following derivatives:

If $y = Ae^{\pm mx}$,

$$\frac{dy}{dx} = \pm mAe^{\pm mx}$$

$$= \pm my.$$

If $y = \sin(mx + \phi)$,

$$\frac{dy}{dx} = m \cos(mx + \phi)$$
$$= m\sqrt{1 - \sin^2(mx + \phi)}$$
$$= m\sqrt{1 - y^2}.$$

If $y = \cos(mx + \phi)$,

$$\frac{dy}{dx} = -m \sin(mx + \phi)$$
$$= -m\sqrt{1 - \cos^2(mx + \phi)}$$
$$= -m\sqrt{1 - y^2}.$$

The first of this group of derivatives is particularly noteworthy. It states symbolically that the first derivative of an exponential function is proportional to the function itself, the proportionality factor being the multiple of x in the exponential. This property is true only of the exponential function. Note that A is any arbitrary constant, the value of which does not affect the conclusion. Note also that this property does not hold true for $y = Ae^{\pm mx} + K$. Thus, if a differential equation states that the first derivative of a function is proportional to the function, an intuitive solution can quickly be determined.

EXAMPLE 2. If $dy/dx = -3y$, then $y = Ae^{-3x}$, and A is the arbitrary constant necessary in the general solution of a differential equation of the first order.

EXAMPLE 3. If $Ri + L(di/dt) = 0$, $di/dt = (-R/L)i$. Hence

$$i = Ae^{(-R/L)t}.$$

VERIFICATION: Using this value of i,

$$\frac{di}{dt} = -\frac{AR}{L}e^{-Rt/L} = -\frac{Ri}{L}.$$

Thus $Ri + L(di/dt) = 0$, which is the original differential equation.

This type of differential equation should be comprehended to such an extent that it becomes completely familiar, since it occurs frequently in other more extensive and involved differential equations.

EXAMPLE 4. Solve the differential equation

$$\frac{dq}{dt} = 4\sqrt{1 - q^2}.$$

Solution: Following reverse reasoning on the second of the three derivatives calculated above, $q = \sin(4t + \phi)$, ϕ being the arbitrary constant. This solution should be verified.

EXERCISES

1. How many arbitrary constants should appear in the general solution of each of the following differential equations:

(a) $\dfrac{d^2y}{dx^2} = -5y.$ (b) $\left(\dfrac{dy}{dx}\right)^2 - 3x = 2.$ (c) $y' - 3y''' = e^{-x}.$

2. State whether each of the following differential equations can or cannot be solved by a method of direct integration:

(a) $\dfrac{d^2y}{dx^2} = 2y.$ (b) $\dfrac{d^2y}{dx^2} - 3\dfrac{dy}{dx} = x^2.$ (c) $\dfrac{d^2y}{dx^2} = 0.$

(d) $y' = 3y - 2x.$ (e) $Dy - e^{5x} = \sin 2x.$

Find the general solution to each of the following differential equations:

3. $\dfrac{dy}{dt} = 5 \sin 10t.$ **4.** $\dfrac{di}{dt} = 3e^{-6t}.$

5. $y' - 2 \cos 4x = \dfrac{3}{x}.$ **6.** $Dy = 4e^{-8x}e^{-2}.$

7. $\dfrac{dy}{dx} = xe^{-x^2-5}\, dx.$ **8.** $\dfrac{dy}{dt} - te^{-50t} = 0.$

9. $(y')^2 = \dfrac{4}{6 - 4x^2}.$ **10.** $y' = 2\sqrt{9 - 4x^2}.$

11. $Dy + \dfrac{x}{x^2 + 6x + 8} = 0.$

Find the exact solution to the following differential equations:

12. $\dfrac{dy}{dx} = 6 \sin 3x + 6 \cos 3x$ (at $x = \pi/6$, $y = 1$).

13. $\dfrac{dy}{dt} = \dfrac{1}{6 + 2t^2}$ (at $t = 0$, $y = -2$).

14. $\dfrac{dy}{dt} = 2 \ln 3t$ (at $t = \tfrac{1}{3}$, $y = 1$).

15. $\cos^4 x \dfrac{dy}{dx} = \sin^3 x$ (at $x = \pi/3$, $y = 8/3$).

$\left[\text{*Hint:* } \displaystyle\int \frac{\sin^3 x}{\cos^4 x}\, dx = \int \frac{(1 - \cos^2 x)}{\cos^4 x} \sin x \, dx.\right]$

Find the general solution to each of the following:

16. $\dfrac{dy}{dt} + 7y = 0.$ **17.** $\dfrac{dy}{dt} + 20jy = 0.$

18. $\dfrac{v}{C} + R\dfrac{dv}{dt} = 0.$ **19.** $8\,dy - 3y\,dx = 0.$

20. $\dfrac{d}{dx}(y - 3) = -5y + 15.$ **21.** $\dfrac{dy}{dx} = -5y + 15$ (by using Exercise 20).

22. $\dfrac{di}{dt} = 50\sqrt{1 - i^2}.$

Find the exact solution to each of the following:

23. $\dfrac{dy}{dx} - 5y = 0$ (at $x = 0$, $y = 2$).

24. $q + RC\dfrac{dq}{dt} = 0$ (at $t = 0$, $q = EC$).

25. $L\dfrac{d}{dt}\left(i - \dfrac{E}{R}\right) + Ri = E$ (at $t = 0$, $i = 0$).

26. $y' = 20\sqrt{1 - y^2}$ (at $x = 0$, $y = -0.5$).

16.4   METHOD OF SUPERPOSITION

Newton's law of cooling states that the rate of cooling of a body surrounded by cooler air is proportional to the temperature difference between the body and the air. This will serve as an example of a frequently occurring principle in differential equations. Figure 16-1 is the graph of temperature of a cooling body.

It is apparent that the rate of cooling is the same whether the actual temperature of the body is used or the difference between its temperature and the room temperature. That is,

$$\frac{dT}{dt} = \frac{d(T - T_a)}{dt}.$$

Newton's law states that $dT/dt = -k(T - T_a)$, the negative sign indicating a decreasing T. Therefore

$$\frac{d(T - T_a)}{dt} = -k(T - T_a),$$

and $T - T_a$ has a first derivative proportional to itself. Thus

$$T - T_a = Ae^{-kt}, \quad \text{and} \quad T = T_a + Ae^{-kt}.$$

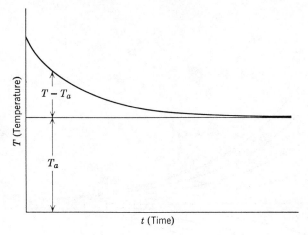

Fig. 16-1. Temperature T of a cooling body plotted against time t.

T is therefore an exponential decay superimposed on T_a. The constant T_a in the solution is the value of T after there is no longer any change in T, that is, the value of T when $dT/dt = 0$ (at $t = \infty$ here). This is known as the *sustained* or *steady-state* value. The variable value in the solution, Ae^{-kt} is the general solution of $dT/dt = -kT$.

More generally, consider the D.E. $dy/dx + ky = Q$, where Q may be either a constant or a function of x. It is often possible to find a *particular solution*, say y_1, with no arbitrary constants. Then the general solution is $y = y_1 + y_2$, where y_2 involves an arbitrary constant. By substituting $y = y_1 + y_2$,

$$\frac{d(y_1 + y_2)}{dx} + k(y_1 + y_2) = Q, \quad \text{or} \quad \frac{dy_1}{dx} + ky_1 + \frac{dy_2}{dx} + ky_2 = Q.$$

However, $dy_1/dx + ky_1 = Q$ (since y_1 satisfies the D.E.). Therefore $dy_2/dx + ky_2 = 0$.

Hence, y_2 is found by finding the general solution of $dy/dx + ky = 0$ (that is, the original D.E. with $Q = 0$).

Therefore in such equations:

(a) Find a particular solution y_1.

(b) Find the general solution of $dy/dx + ky = 0$, y_2.

(c) The general solution is $y = y_1 + y_2$.

EXAMPLE 5.   Solve the D.E. $dy/dx + 2y = 6$.

Solution: Here, as in Newton's law of cooling, when Q is constant, the particular solution is a steady-state or sustained value of y, when

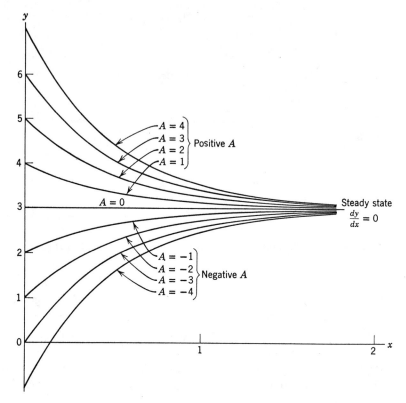

Fig. 16-2. Graphs of $y = 3 + Ae^{-2x}$ for positive, negative, and zero values of A.

$dy/dx = 0$. By substituting $dy/dx = 0$ in the D.E., $y_1 = 3$. The fact that the particular solution is constant with a derivative of 0 justifies the assumption that dy/dx becomes zero at some value of x.

y_2 is the solution of the D.E. $dy/dx + 2y = 0$. $dy/dx = -2y$, and $y_2 = Ae^{-2x}$.

Therefore $y = 3 + Ae^{-2x}$. $y = 3$ is the particular solution where $A = 0$. Figure 16-2 shows the graphs of the solution to this D.E. for positive, negative, and zero values of A. The student should verify that the solution satisfies the original D.E.

EXAMPLE 6. Find the exact solution of the D.E. $dy/dx - 2y = x$ if, at $x = 0, y = 1$.

Solution: Here y can never attain a steady-state value, since, if $dy/dx = 0$, $y = -\frac{1}{2}x$ for which dy/dx is not zero. Keeping in mind that the particular solution must satisfy the D.E., we soon see that the particular solution will usually contain a term in x and a term in each of its derivatives. This means only a term in x and a constant term, so that the constant term in dy/dx minus twice the constant term in y total up to zero.

The method of undetermined coefficients may be used to find the particular solution.

Let $y_1 = kx + m$. Then $dy_1/dx = k$. By substituting in the D.E.,

$$k - 2kx - 2m = x$$

for all values of x.

Thus $-2k = 1$, $k - 2m = 0$; $k = -\frac{1}{2}$, and $m = -\frac{1}{4}$.

Hence $y_1 = -x/2 - \frac{1}{4}$. The fact that y_1 satisfies the D.E. should be checked. y_2 is the solution of the D.E. $dy/dx = +2y$. Thus $y_2 = Ae^{2x}$, and

$$y = -\frac{x}{2} - \frac{1}{4} + Ae^{2x}.$$

The particular solution is this general solution where $A = 0$. For the exact solution, by substituting $(0,1)$ for (x,y),

$$1 = -\tfrac{1}{4} + A \qquad \text{and} \quad A = \tfrac{5}{4}.$$

$$\therefore \; y = -\frac{x}{2} - \frac{1}{4} + \frac{5}{4}e^{2x} \qquad \text{(verify this).}$$

FURTHER EXAMPLES: (Particular solution may involve original Q and all its derivatives.)

$$\frac{dy}{dx} + 2y = x^3 - 3x^2. \qquad \text{(Let } y_1 = kx^3 + mx^2 + nx + p.)$$

$$2\frac{dy}{dx} + 6y = \cos \omega x. \qquad \text{(Let } y_1 = k\cos \omega x + m\sin \omega x.)$$

$$\frac{dy}{dx} + 3y = 2e^{-2x}. \qquad \text{(Let } y_1 = ke^{-2x}.)$$

$$\frac{dy}{dx} + 2y = e^{-3x}\sin x. \qquad \text{(Let } y_1 = ke^{-3x}\sin x + me^{-3x}\cos x.)$$

$$\frac{dy}{dx} - 4y = xe^{-5x}. \qquad \text{(Let } y_1 = kxe^{-5x} + me^{-5x}.)$$

EXAMPLE 7.   Solve the equation $dy/dt + 2y = 8te^{-4t}$, if, at $t = 0$, $y = 1$.

Solution: The right-hand side is a product in which all successive derivatives can include only terms in te^{-4t} and e^{-4t}.
Let $y_1 = kte^{-4t} + me^{-4t}$. Then $dy_1/dt = ke^{-4t} - 4kte^{-4t} - 4me^{-4t}$.
By substituting in the D.E., $ke^{-4t} - 4kte^{-4t} - 4me^{-4t} + 2kte^{-4t} + 2me^{-4t} = 8te^{-4t}$.
By equating like terms, $-4k + 2k = 8$, and $k - 4m + 2m = 0$.
By algebra, $k = -4$ and $m = -2$. Therefore

$$y_1 = -4te^{-4t} - 2e^{-4t}.$$

y_2 is the solution of the equation $dy/dt + 2y = 0$, which is $y_2 = Ae^{-2t}$.

Therefore $\qquad y = Ae^{-2t} - 4te^{-4t} - 2e^{-4t}.$

At $t = 0$, all exponentials $= 1$. By substituting $(0,1)$ for (t,y), $1 = A - 2$, and $A = 3$. Thus $\quad y = 3e^{-2t} - 4te^{-4t} - 2e^{-4t}.$
(This is worth verifying.)

EXAMPLE 8.   When a switch connects an *RL* series circuit to an a-c voltage source, application of Kirchhoff's law of voltages produces the differential equation $2.5 \times 10^{-3}(di/dt) + 2i = 2 \sin 400t$. At $t = 0$, $i = 0$. Solve for i as a function of t.
Solution: Let $i_1 = k \sin 400t + m \cos 400t$.

Then $di/dt = 400k \cos 400t - 400m \sin 400t$.

By substituting in the D.E.,

$$k \cos 400t - m \sin 400t + 2k \sin 400t + 2m \cos 400t = 2 \sin 400t.$$

By equating coefficients, $k + 2m = 0$, and $-m + 2k = 2$.
By algebra, $k = \frac{4}{5}$ and $m = -\frac{2}{5}$. Hence, $i_1 = \frac{4}{5} \sin 400t - \frac{2}{5} \cos 400t$.
i_2 is found by solving the equation $2.5 \times 10^{-3} \, di_2/dt + 2i_2 = 0$; $i_2 = Ae^{-800t}$. Therefore

$$i = Ae^{-800t} + \tfrac{4}{5} \sin 400t - \tfrac{2}{5} \cos 400t.$$

At $t = 0$, the exponential is 1, the sin is 0, and the cos is 1. $A = \frac{2}{5}$.
Hence

$$i = \tfrac{2}{5}e^{-800t} + \tfrac{4}{5} \sin 400t - \tfrac{2}{5} \cos 400t,$$

a rapid decay superimposed on a sinusoidal oscillation, as graphed in Fig. 16-3.

If the form of the particular solution becomes too difficult to devise or work out, there are often other methods of solving the D.E. No rule can be

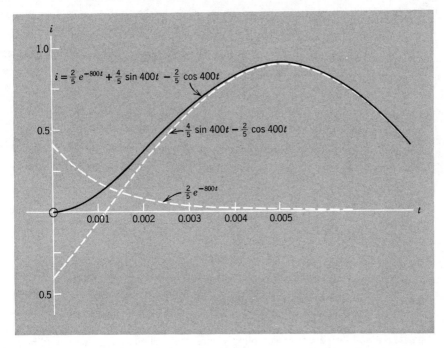

Fig. 16-3. Graph of $i = \frac{2}{5}e^{-800t} + \frac{4}{5}\sin 400t - \frac{2}{5}\cos 400t$.

given for the form of the particular solution to cover all types, but with a little analysis a solution can usually be devised.

The method of superposition also applies to differential equations of higher order, and it will be discussed again in Chapter 17.

EXERCISES

1. What is the steady-state value of the dependent variable (if any occurs) for each of the following differential equations:

(a) $\dfrac{dy}{dx} + 5y = 30$.

(b) $2\dfrac{di}{dt} + 20i = 0$.

(c) $\dfrac{dy}{dt} - 7y - 3 = 0$.

(d) $\dfrac{dT}{dt} = -4.0(T - 60)$.

(e) $\dfrac{dy}{dt} = -3y - 2\cos 5t$.

(f) $Ri + L\dfrac{di}{dt} = E$.

Find the general solution to the following differential equations:

2. $2\dfrac{di}{dt} + 20i = 50$.

3. $\dfrac{dy}{dx} = 3(y + 5)$.

4. $\dfrac{dy}{dx} + 3y = 0$.

5. $R\dfrac{dq}{dt} + \dfrac{q}{C} = E$.

6. $\dfrac{dT}{dt} = -\dfrac{1}{4}(T - T_a)$.

7. $3\dfrac{dy}{dx} - 6y = -12x$.

8. $\dfrac{dy}{dx} + 2y = 4x - 4$.

9. $2\dfrac{dy}{dx} + 3x + 4y = 0$.

10. $\dfrac{dy}{dt} + 3y = 3t^2 - 6$.

11. $\dfrac{dy}{dx} + 5y = 2e^{-10x}$.

12. $\dfrac{dy}{dx} + 5x = 2e^{-10x}$.

13. $4\dfrac{dy}{dt} - 7y = 3e^{0.5t}$.

14. $\dfrac{dy}{dx} - 5y = 29 \sin 2x$.

15. $10^{-2}\dfrac{di}{dt} + 2i = 2 \cos 200t$.

16. $\dfrac{dy}{dx} + 6y = 2xe^{-5x}$.

17. $\dfrac{dv}{dt} + 4000v = 2 \cos 1000t + \sin 1000t$.

18. $\dfrac{dy}{dt} + 3y = e^{-3t}$. (*Note:* The correct solution must contain two or more individual terms, as for other similar D.E.'s. If the guiding principles for finding a particular solution are followed blindly here, the particular solution would be a multiple of e^{-3t}. But the second (transient) component is also a multiple of e^{-3t} and would simply combine with the particular solution to form one term. To point out the error, it is worth trying a particular solution of $y_1 = ke^{-3t}$ to see what happens. In such instances, an extra factor of the independent variable occurs in the correct particular solution. y_1 should therefore be of the form kte^{-3t}.)

19. A machine initially at a temperature of 300°F is cooling in a room at 70°F. After 10 min of cooling, its temperature is 240°F. Using the law that rate of cooling is proportional to the difference between body temperature and room temperature:

   (a) Set up the differential equation of temperature and time, using a proportionality constant, and measuring time in minutes.
   (b) Solve the differential equation mathematically, as a general solution.
   (c) Find the exact solution, using the two pieces of information in the problem.
   (d) Find the temperature after one hour of cooling.

Find the exact solution to the following:

20. $3\dfrac{dy}{dx} + 8y = 12$ (at $x = 0$, $y = 4$).

21. $Ri + L\dfrac{di}{dt} = E$ (at $t = 0$, $i = 0$).

22. $\dfrac{dy}{dt} - 8t = 32t^2$ (at $t = 0$, $y = 1$).

23. $5\dfrac{dv}{dt} + 10v = 9e^{-20t}$ (at $t = 0$, $v = 1$).

24. $\dfrac{dy}{dx} + 0.5y = 14xe^{-4x}$ (at $x = 0$, $y = 0$).

25. $\dfrac{dq}{dt} + 500q = 50 \sin 500t$ (at $t = 0$, $q = 0$).

26. $\dfrac{dy}{dx} + 5y = 4e^x \sin x$ (at $x = 0$, $y = 2$).

16.5 TRANSIENTS IN *RL* AND *RC* CIRCUITS

The transient conditions resulting from switching changes in *RL* and *RC* circuits were discussed in Section 11.3. Appendix I indicates the differential equations set up from electrical laws for these conditions. Each of the differential equations contains a term in the dependent variable, a term in its first derivative, and a third term which may be zero or any function of the independent variable, depending on the type of voltage source. All of these differential equations may now be solved either directly, as in Section 11.3, or by a method of superposition. Some of the previous exercises are examples of these transient conditions. One more general example follows.

EXAMPLE 9. Set up and solve the differential equation for current in the *RC* series circuit when an a-c voltage of $A \sin \omega t$ is switched in, assuming zero current at $t = 0$.

Solution: By using Kirchhoff's law of voltages, $Ri + q/C = A \sin \omega t$ at all values of t. The differential equation for current results from the derivative of this equation, in which $dq/dt = i$. Thus

$$R\frac{di}{dt} + \frac{i}{C} = A\omega \cos \omega t.$$

Let the particular $i_1 = k \cos \omega t + m \sin \omega t$; then $di_1/dt = -k\omega \sin \omega t + m\omega \cos \omega t$.
By substituting in the D.E.,

$$-kR\omega \sin \omega t + mR\omega \cos \omega t + \frac{k}{C}\cos \omega t + \frac{m}{C}\sin \omega t = A\omega \cos \omega t.$$

By equating coefficients,

$$-kR\omega + \frac{m}{C} = 0 \quad \text{and} \quad mR\omega + \frac{k}{C} = A\omega.$$

By solving for k and m in terms of the known constants,

$$k = \frac{AC\omega}{1 + R^2C^2\omega^2} \quad \text{and} \quad m = \frac{ARC^2\omega^2}{1 + R^2C^2\omega^2}.$$

Hence

$$i_1 = \frac{AC\omega}{1 + R^2C^2\omega^2}(\cos \omega t + RC\omega \sin \omega t).$$

This may be expressed as a single sine function by the method discussed in Section 9.1.

$$i_1 = \frac{AC\omega\sqrt{1 + R^2C^2\omega^2}}{1 + R^2C^2\omega^2} \sin (\omega t + \phi) = \frac{AC\omega}{\sqrt{1 + R^2C^2\omega^2}} \sin (\omega t + \phi),$$

where $\tan \phi = 1/RC\omega$.

The transient component i_2 is the solution of the D.E. $R(di_2/dt) + i_2/C = 0$. In the form $di_2/dt = (-1/RC)i_2$, the solution is seen to be $i_2 = Be^{-t/RC}$. Thus

$$i = Be^{-t/RC} + \frac{AC\omega}{\sqrt{1 + R^2C^2\omega^2}} \sin (\omega t + \phi),$$

where B must still be determined.

By substituting (0,0) for (t,i), $B = -(AC\omega/\sqrt{1 + R^2C^2\omega^2}) \sin \phi$. Hence

$$i = \frac{AC\omega}{\sqrt{1 + R^2C^2\omega^2}} [\sin (\omega t + \phi) - \sin \phi e^{-t/RC}],$$

where $\tan \phi = 1/RC\omega$.

i_1 may, of course, be kept in the form of the sum of a cosine and a sine, in which form the final solution is

$$i = \frac{AC\omega}{1 + R^2C^2\omega^2}(\cos \omega t + RC\omega \sin \omega t - e^{-t/RC}).$$

EXERCISES

Find the exact solution for each of the following differential equations:

1. $\dfrac{L}{R}\dfrac{di}{dt} + i = \dfrac{E}{R}$, where R, L, and E are constant (at $t = 0$, $i = 0$).

2. $L\dfrac{di}{dt} + Ri = Ae^{-50t}$, where R, L, and A are known constants (at $t = 0$, $i = 0$).

3. $\dfrac{dv}{dt} + \dfrac{R}{L}v = A\omega \sin \omega t$, where R, L, and A are known constants (at $t = 0$, $v = -A$).

4. $R\dfrac{dq}{dt} + \dfrac{q}{C} = A \sin \omega t$, where R, C, and A are known constants (at $t = 0$, $q = 0$).

5. $RC\dfrac{dv}{dt} + v = A \cos \omega t$, where R, C, and A are known constants (at $t = 0$, $v = 0$).

16.6   SEPARATION OF VARIABLES

Consider again the differential equation $dy/dx = -3y$ (already solved intuitively). This can also be readily solved by a direct calculation. It has been indicated that this cannot be integrated directly as $\int -3y \, dx$ since both variables are involved in this expression. A simple expedient for eliminating this difficulty would be to consider y as the independent variable and take the reciprocal of both sides. Therefore

$$\frac{dx}{dy} = -\frac{1}{3y} \quad \text{and} \quad x = \int -\frac{dy}{3y} = -\frac{1}{3}\ln y + C.$$

Now the integration constant does not have to be a single constant letter as long as the expression used for the integration constant is itself constant. Many times, it leads to a neater result if the integration constant is represented by a more involved expression originally. Here it is convenient to represent the integration constant by the natural log of a constant, since this leads to an application of log theory to give a simpler expression for y. Thus, to find y as an explicit function of x, $x = -\frac{1}{3}\ln y + \ln K$. Therefore

$$-3x = \ln y - 3 \ln K = \ln \frac{y}{K^3}.$$

Hence

$$\frac{y}{K^3} = e^{-3x} \quad \text{and} \quad y = K^3 e^{-3x} = Ae^{-3x},$$

where A is an arbitrary constant.

A further analysis of the method used at the beginning of this solution will show that taking the reciprocal is not an essential part of the technique.

We have merely changed the form of the original equation from a derivative to the differential form $dx = -dy/3y$, ensuring that each side of the equation involves only one variable; that is, we have separated the variables (x's and dx on one side, y's and dy on the other). In this form, one differential is equated to another, and both sides may be integrated, since there is only one variable involved on each side.

A quick way of separating variables is demonstrated here. If $\dfrac{a}{b} = \dfrac{c}{d}$, any one or more of the four factors can be taken diagonally across to the other side, and the equation still holds true. Thus

$$\frac{ad}{b} = c, \qquad \frac{ad}{bc} = 1, \qquad \frac{a}{bc} = d, \qquad \text{etc.}$$

The first of these transformed equations results from multiplying both sides by d; the second from multiplying both sides by d/c, etc. But, in practice, it is merely transferring factors diagonally to the other side.

EXAMPLE 10.   Solve the D.E. $dy/dx + 3y = 5$.
Solution: $dy = (5 - 3y)\, dx$.

$$\frac{dy}{5 - 3y} = dx \qquad \begin{array}{l}\text{(diagonal transfer of the factor } 5 - 3y \text{ to}\\ \text{place it on the same side as } dy\text{).}\end{array}$$

Therefore $\displaystyle\int \frac{dy}{5 - 3y} = \int dx$.

$-\tfrac{1}{3} \ln (5 - 3y) + C = x.$ (The integration constants on the two sides have been combined into one constant C on the left-hand side.)

By multiplying by -3, $\ln (5 - 3y) - \ln K = -3x$. Thus

$$\ln \left(\frac{5 - 3y}{K}\right) = -3x. \qquad \frac{5 - 3y}{K} = e^{-3x}. \qquad y = \frac{1}{3}(5 - Ke^{-3x}).$$

EXAMPLE 11.   Solve the D.E. $x(dy/dx) - 2x^2y = 3y$.
Solution: $x\, dy = (3y + 2x^2y)\, dx = y(3 + 2x^2)\, dx$. Therefore

$$\int \frac{dy}{y} = \int \frac{3 + 2x^2}{x}\, dx = \int \frac{3}{x}\, dx + \int 2x\, dx.$$

$\ln y + \ln K = 3 \ln x + x^2.$ Hence

$$\ln \frac{Ky}{x^3} = x^2. \qquad \frac{Ky}{x^3} = e^{x^2}. \qquad y = Ax^3 e^{x^2}.$$

Of course, it is not always possible to separate variables; for example,

$$x \frac{dy}{dx} - 2x^2 = 3y.$$

$$x \, dy = (3y + 2x^2) \, dx.$$

Here, both x and y are contained in the one factor, so that the variables cannot be separated.

EXAMPLE 12. Find an exact solution to the differential equation $3 \, dq/dt + 10q = 5$, if at $t = 0$, $q = 0$.
Solution: Two methods may be used.
Method 1: Superposition. For steady-state q_1, $dq/dt = 0$. Therefore $10q_1 = 5$ and $q_1 = 0.5$.

$$3 \frac{dq_2}{dt} + 10q_2 = 0. \qquad q_2 = Ae^{-10t/3}.$$

$$q = Ae^{-10t/3} + 0.5 \qquad \text{(the general solution)}.$$

To find the exact solution, substitute $t = 0$ and $q = 0$ in the D.E. and solve for A. $A = -0.5$. Therefore

$$q = 0.5(1 - e^{-10t/3}).$$

Method 2: Separation of Variables.

(a) $$\int \frac{3 \, dq}{1 - 2q} = \int 5 \, dt.$$

$$-\tfrac{3}{2} \ln (1 - 2q) + K = 5t.$$

To solve for K in this particular case, substitute $(0,0)$ for (t,q). Thus

$$-\tfrac{3}{2} \ln 1 + K = 0, \qquad K = 0,$$

and

$$\ln (1 - 2q) = -\frac{10t}{3}.$$

$$1 - 2q = e^{-10t/3}.$$

$$q = 0.5(1 - e^{-10t/3}).$$

(b) A variation of Method 2a, avoiding the integration constant when an exact solution is required, follows.

$$\int_0^q \frac{3 \, dq}{1 - 2q} = \int_0^t 5 \, dt.$$

Both integrals are transformed into definite integrals by inserting the variable letter as the upper limit on each side and the stated simultaneous values of the variables as the lower limits. Therefore

$$-\tfrac{3}{2} \ln (1 - 2q) + \tfrac{3}{2} \ln 1 = 5t - 0.$$

The calculation is then completed as in Method 2a.

EXAMPLE 13.   Solve the following for y as an exact function of x, if at $x = 1$, $y = 2$:

$$3 \frac{dy}{dx} - \sqrt{y^2 - 4} = 0.$$

Solution: Separating variables and using definite integrals,

$$\int_2^y \frac{3 \, dy}{\sqrt{y^2 - 4}} = \int_1^x dx.$$

$3 \cosh^{-1}(y/2) - 3 \cosh^{-1} 1 = x - 1$; but $\cosh^{-1} 1 = 0$, since $\cosh 0 = 1$. Therefore

$$\cosh^{-1} \frac{y}{2} = \frac{x - 1}{3}. \qquad \frac{y}{2} = \cosh \frac{x - 1}{3}. \qquad y = 2 \cosh \frac{x - 1}{3}.$$

This solution should be verified by substitution in the original differential equation.

EXERCISES

1. Where possible, separate the variables in preparation for integration in each of the following:

(a) $\dfrac{dy}{dx} = \dfrac{e^x}{2y}.$

(b) $2 \dfrac{dy}{dt} - 5y = 0.$

(c) $2 \dfrac{dy}{dt} - 5y = 8.$

(d) $2 \dfrac{dy}{dt} - 5y = 8t.$

(e) $2 \dfrac{dy}{dt} - 5 = 8t.$

(f) $\left(\dfrac{dy}{dx}\right)^2 + 3y = xy.$

(g) $\dfrac{dy}{dx} = \sqrt{x - 4xy^2}.$

(h) $\dfrac{dy}{dx} = \sqrt{1 - 4xy^2}.$

(i) $3 \dfrac{dy}{dx} = 2ye^{-x-y^2}.$ (*Hint:* $e^{-x-y^2} = e^{-x}e^{-y^2}.$)

Find the general solution to each of the following D.E.'s, expressing the dependent variable as a function of the independent variable:

2. $\dfrac{dy}{dx} = \dfrac{x}{y}$.

3. $x\,dy = y\,dx$.

4. $\dfrac{dy}{d\theta} = \dfrac{\sin\theta}{2y}$.

5. $\dfrac{dy}{dx} = -3y$.

6. $\dfrac{dy}{dx} - y = 3$.

7. $\dfrac{dy}{dx} + y = 3$.

8. $\dfrac{dy}{dx} - 3y + 5 = 0$.

9. $\dfrac{dy}{dx} + 3y - 5 = 0$.

10. $RC\dfrac{dv}{dt} + v = E$.

11. $\dfrac{di}{d\theta} = \sqrt{1 - i^2}$.

12. $\dfrac{di}{d\theta} = 4 + i^2$.

13. $2y\dfrac{dy}{dx} = 4 + y^2$.

14. $2y\dfrac{dy}{dx} = \sqrt{4 + y^2}$.

15. $\dfrac{dy}{dx} - \dfrac{4 - y^2}{y} = 0$.

16. $dy = (y^2 + 5y + 6)\,dx$.

17. $\cos 2y\dfrac{dy}{dx} = \sec^2\dfrac{x}{2}$.

18. $\dfrac{dy}{dx} = e^{-3y + x}$.

19. $\dfrac{dy}{dx} = xe^{-3y - x}$.

20. $dy = y\ln x\,dx$.
21. $dy(4 - x^2) = dx(4y - 8)$, where $x^2 < 4$.
22. $dy(4 - x^2) = dx(4y - 8)$, where $x^2 > 4$.

23. $\dfrac{dy}{dx} = \dfrac{\sqrt{4 - x^2}}{y}$.

24. $\dfrac{dy}{dx} = e^{0.5x}\sqrt{4y - y^2}$.

25. $\dfrac{dy}{dx} = \dfrac{y^2 + xy^2}{x^2y - x^2}$.

Find the exact solution to the following D.E.'s, expressing the dependent variable as a function of the independent variable:

26. $\dfrac{dy}{dt} = 4 - 0.001y$ (at $t = 0$, $y = 500$).

27. $2i + 0.002\dfrac{di}{dt} = 4$ (at $t = 0$, $i = 0$).

28. $Ri + L\dfrac{di}{dt} = E$ (at $t = 0$, $i = 0$).

29. $\dfrac{dv}{d\theta} = \sqrt{4 - v^2}$ (at $\theta = \pi/6$, $v = 2$).

30. $xy\,dy = (1 + y^2)\,dx$ (at $x = 1$, $y = 0$).

31. $dy = \sqrt{9 + y^2}\, dx$ (at $x = 1$, $y = 0$).

32. $\sin^2 x\, dy + \cos x \cos^2 y\, dx = 0$ (at $x = \pi/2$, $y = \pi/4$).

33. $(4y + 4) \sin 2\theta\, d\theta + \cos 2\theta\, dy = 0$ (at $\theta = 0$, $y = 0$).

34. $e^{-3x + 6y}\, dx = 2\, dy$ (at $x = 0$, $y = 0$).

35. $\dfrac{dy}{dx} = -x^2 \sqrt{\dfrac{1 + y}{1 - x^3}}$ (at $x = 1$, $y = -1$).

36. $\dfrac{dy}{dx} = \ln x$ (at $x = 1$, $y = 1$).

37. $dy = 2t\sqrt{6y - y^2}\, dt$ (at $t = 1$, $y = 3$).

38. $y^2 \left(\dfrac{dy}{dx}\right)^2 = 4 - y^2$ (at $x = 1$, $y = \sqrt{3}$).

16.7 EXACT DIFFERENTIALS

This type of differential equation depends on the product formula for differentiation.

$$\frac{d}{dx}(uv) = u\frac{dv}{dx} + v\frac{du}{dx} \qquad \text{(derivative form)}.$$

or

$$d(uv) = u\, dv + v\, du \qquad \text{(differential form)}.$$

By examining the differential form in reverse, we see that the sum of two differentials may be the exact differential of a product, and a solution to a differential equation involving this type of expression may usually be determined. The principal clue is that, in the expression $u\, dv + v\, du$, the second term is formed from the first by multiplying the differential of one factor by the integral of the other.

EXAMPLE 14.   Find the general solution to the differential equation

$$3xy^2 \frac{dy}{dx} + y^3 = \frac{1}{x + 1}.$$

Solution: In this type of problem, the differential form of the equation lends itself to easier solution. Therefore

$$3xy^2\, dy + y^3\, dx = \frac{dx}{x + 1}.$$

The clue to the solution is that $3y^2\, dy$ is the differential of y^3, and x is the integral of dx; or, alternatively, dx is the differential of x and y^3

is the integral of $3y^2\,dy$. Thus the two terms on the left-hand side constitute an exact differential, and the equation may be written

$$d(xy^3) = \frac{dx}{x+1}. \quad \text{(Verify that } d(xy^3) = 3xy^2\,dy + y^3\,dx.\text{)}$$

By integrating both sides

$$\int d(xy^3) = \int \frac{dx}{x+1}.$$

Therefore

$$xy^3 = \ln(x+1) + \ln A$$
$$= \ln A(x+1),$$

and

$$y = \sqrt[3]{\frac{\ln A(x+1)}{x}}.$$

Verifying,

$$x^{\frac13}y = [\ln A(x+1)]^{\frac13}.$$
$$xy^3 = \ln A(x+1).$$

By taking derivatives,

$$3xy^2\frac{dy}{dx} + y^3 = \frac{A}{A(x+1)}$$
$$= \frac{1}{x+1},$$

which is the original D.E.

EXAMPLE 15. Solve the D.E. $\cos x\,dy + dx = y\sin x\,dx$.
Solution: $\cos x\,dy - y\sin x\,dx + dx = 0$
$$d(y\cos x) + dx = 0.$$

Therefore

$$y\cos x + x = C \quad \text{and} \quad y = \frac{C-x}{\cos x}.$$

EXAMPLE 16. Solve the differential equation

$$\frac{dy}{dx} = -\frac{2xy - y^2 + 3x^2y}{x^2 - 2xy + x^3}.$$

given that, at $x = 1$, $y = 3$.

Solution: Note that each term in the denominator is formed from the corresponding term in the numerator by taking the derivative with respect to y of the factor involving y and multiplying by the integral with respect to x of the factor involving x. This suggests that the equation may be rearranged, grouping together pairs of terms, and it is likely that each such pair will constitute an exact differential. By cross-multiplying,

$$(x^2 - 2xy + x^3)\, dy = -(2xy - y^2 + 3x^2y)\, dx.$$

$$(x^2\, dy + 2xy\, dx) - (2xy\, dy + y^2\, dx) + (x^3\, dy + 3x^2y\, dx) = 0.$$

$$d(x^2y) - d(xy^2) + d(x^3y) = 0.$$

Now, by integrating both sides term by term,

$$x^2y - xy^2 + x^3y = K.$$

In solving for K, substitute $(1,3)$ for (x,y). $K = -3$. Therefore

$$x^2y - xy^2 + x^3y + 3 = 0.$$

This solution could be made explicit by solving for y in terms of x by means of the quadratic formula. However, the implicit solution is satisfactory.

The latter part of this solution may be calculated by using a definite integral, using simultaneous values of the two variables for lower limits as follows:

$$\int_{1,3}^{x,y} [d(x^2y) - d(xy^2) + d(x^3y)] = 0.$$

Hence

$$x^2y - xy^2 + x^3y - 3 + 9 - 3 = 0.$$

$$x^2y - xy^2 + x^3y + 3 = 0 \qquad \text{as before.}$$

Any exact differential equation may be put in the form $P\, dx + Q\, dy = 0$, where P and Q are functions of x and y. To test for exactness, it has already been indicated that the integral of factors involving x in $P\, dx$ equals the factors involving x in Q, and the differential of factors involving y in P equals the factors involving y in $Q\, dy$. This test for exactness is usually sufficient, provided the resulting exact differential is verified. Another way of stating this test for exactness is that $\partial P/\partial y = \partial Q/\partial x$. This partial-derivative test applies even when P and Q have more than one term. In Example 16, for instance,

$$(2xy - y^2 + 3x^2y)\, dx + (x^2 - 2xy + x^3)\, dy$$

fits the form $P\, dx + Q\, dy$, and $\partial P/\partial y = \partial Q/\partial x = 2x - 2y + 3x^2$. This fact indicates that the complete expression is an exact differential.

16.8   USE OF INTEGRATING FACTOR

A differential equation of the form $dy/dx + Py = Q$, or $dy + Py\,dx = Q\,dx$, where P and Q are constants or functions of x, is called a *linear* first-order differential equation. The left-hand side of this type of equation may always be transformed into an exact derivative or differential by multiplying by an appropriate factor called the *integrating factor* (I.F.). To determine this integrating factor, it is noted that $e^{\int P\,dx}\,dy + Pe^{\int P\,dx}\,y\,dx$ is an exact differential, since

$$d(e^{\int P\,dx}) = e^{\int P\,dx}\left(d\int P\,dx\right) = e^{\int P\,dx}\,P\,dx,$$

and the integral of dy is y. Thus, if the equation $dy + P\,dx = Q\,dx$ is multiplied by $e^{\int P\,dx}$, the resulting equation is

$$e^{\int P\,dx}\,dy + Pe^{\int P\,dx}\,y\,dx = Qe^{\int P\,dx}\,dx.$$

The left-hand side is now an exact differential and may be expressed as $d(e^{\int P\,dx}\,y)$. The integrating factor is therefore $e^{\int P\,dx}$.

The constant of integration should be omitted in the calculation of $\int P\,dx$.

EXAMPLE 17.   Solve: $dy/dx - 2y = 5x$.
Solution: $dy - 2y\,dx = 5x\,dx$. The I.F. is $e^{\int -2\,dx} = e^{-2x}$. Therefore
$e^{-2x}\,dy - 2ye^{-2x}\,dx = 5xe^{-2x}\,dx$;   that is $d(e^{-2x}y) = 5xe^{-2x}\,dx$.
Thus

$$e^{-2x}y = \int 5xe^{-2x}\,dx = -5\left(\frac{xe^{-2x}}{2} + \frac{e^{-2x}}{4}\right) + K \qquad \text{(by parts)}$$

and

$$y = \frac{-5x}{2} - \frac{5}{4} + Ke^{2x}.$$

(As a check, solve this equation by superposition.)

EXAMPLE 18.   Solve: $x^2(dy/dx) + 3xy = 3x - 1$, if at $x = 1$, $y = -2$.
Solution: To state the equation in the standard linear form, divide by x^2 and change to differential form. This is necessary to determine P.

$$dy + \frac{3}{x}y\,dx = \left(\frac{3}{x} - \frac{1}{x^2}\right)dx.$$

I.F. is

$$e^{\int (3/x)\,dx} = e^{3\ln x} = e^{\ln x^3} = x^3.$$

Thus

$$x^3 \, dy + 3x^2 y \, dx = (3x^2 - x) \, dx,$$

and

$$d(x^3 y) = (3x^2 - x) \, dx.$$

$$\int_{1,-2}^{x,y} d(x^3 y) = \int_{1}^{x} (3x^2 - x) \, dx.$$

Hence

$$x^3 y - (1^3)(-2) = x^3 - \frac{x^2}{2} - \left(1 - \frac{1}{2}\right).$$

$$x^3 y + 2 = x^3 - \frac{x^2}{2} - \frac{1}{2}.$$

$$y = 1 - \frac{1}{2x} - \frac{5}{2x^3}.$$

(Check this by substitution in the original differential equation.)

EXAMPLE 19.   The transient current in a charging *RL* circuit is given by $Ri + L(di/dt) = E$, and at $t = 0$, $i = 0$. Solve this equation for i using an integrating factor. Enumerate other methods of solving this equation.

Solution: By using the standard linear differential form,

$$di + \frac{R}{L} i \, dt = \frac{E}{L} \, dt.$$

$$\text{I.F.} = e^{\int (R/L) \, dt} = e^{Rt/L}.$$

Hence

$$e^{Rt/L} \, di + \frac{R}{L} i e^{Rt/L} \, dt = \frac{E}{L} e^{Rt/L} \, dt.$$

$$d(e^{Rt/L} i) = \frac{E}{L} e^{Rt/L} \, dt.$$

Therefore

$$\int_{0,0}^{t,i} d(e^{Rt/L} i) = \frac{E}{L} \int_{0}^{t} e^{Rt/L} \, dt.$$

Thus

$$e^{Rt/L} i - 0 = \frac{E}{L} \left(\frac{L}{R} e^{Rt/L} - \frac{L}{R} \right) \quad \text{and} \quad i = \frac{E}{R}(1 - e^{-Rt/L}).$$

This equation may also be solved by superposition where the steady-state value of $i = E/R$, and the transient component is found by solving the equation $L(di/dt) + Ri = 0$. The value of the arbitrary constant must be found by substituting $(0,0)$ for (t,i) in the general solution.

A third method of solution is by separating the variables and using a definite integral to find the exact solution.

EXERCISES

1. Determine which of the following expressions are exact differentials:

(a) $x \, dy + y \, dx$.   (b) $x \, dy - y \, dx$.   (c) $2x^3y \, dy + 3x^2y^2 \, dx$.

(d) $\dfrac{3y^2}{x} \, dy - \dfrac{y^3}{x^2} \, dx$. (e) $\dfrac{3y^2}{x} \, dx + 6y \ln x \, dy$. (f) $ye^{-5x} \, dx - 5e^{-5x} \, dy$.

(g) $2 \sin x \cos 2y \, dx - \cos x \sin 2y \, dy$.

(h) $\ln y \arc \sin x \, dx + \dfrac{dy}{y\sqrt{1 - x^2}}$.

2. Express each of the following as a single differential:

(a) $x^2y \, dy + y^2x \, dx$.   (b) $2e^{-3x} \sec^2 2y \, dy - 3e^{-3x} \tan 2y \, dx$.

(c) $3 \ln y \sin 3x \, dx - \dfrac{\cos 3x \, dy}{y}$. (d) $2y \arc \tan 2x \, dy + \dfrac{2y^2}{1 + 4x^2} \, dx$.

(e) $e^{-3x} \sin y \, dy + 3y^2 \ln x \, dy + 3e^{-3x} \cos y \, dx + \dfrac{y^3}{x} \, dx$.

3. (a) Using functional notation and the prime notation for derivatives, find $d[f(x) F(y) + \phi(x)]$ in the form $P \, dx + Q \, dy$.

   (b) For the answer to (a), show that $\partial P/\partial y = \partial Q/\partial x$.

Find the general solution to each of the following:

4. $2x \, dy + 2y \, dx = e^{3x} \, dx$.

5. $\dfrac{1}{x^2} \dfrac{dy}{dx} - \dfrac{2y}{x^3} = \cos 5x$.

6. $\ln t \, dv + \dfrac{v}{t} \, dt = \ln t \, dt$.

7. $\dfrac{dy}{dx} + \dfrac{y}{x} = x^2 - 1$.

8. $\dfrac{dy}{dx} + 3y = e^{-3x}$.

9. $e^{3x} \dfrac{dy}{dx} + 3ye^{3x} = x \sin x$.

10. $t \dfrac{dy}{dt} - y = t^2$.

11. $\dfrac{1}{t} \dfrac{dy}{dt} - \dfrac{y}{t^2} = \dfrac{t}{1 + t^2}$.

12. $x \dfrac{dy}{dx} + y = xe^{-10x}$.

13. $2y \ln x \, dy + 2y \, dy + \dfrac{y^2}{x} \, dx = 0$.

14. $2ye^{-5x} \, dy - x \, dy - 5y^2e^{-5x} \, dx - y \, dx - e^{-5x} \, dx = 0$.

15. $\dfrac{y^2 \, dx}{1 + x^2} + \dfrac{dx}{1 + x^2} + 2y \arc \tan x \, dy = 2x \, dx$.

16. $\dfrac{dy}{dx} = \dfrac{2y \sin 2x - \cos 2x - 2y \cos 2x}{\sin 2x + \cos 2x}$.

17. $\dfrac{dy}{dt} + 2y = \cos 5t$.

18. $\dfrac{dy}{dt} + 2y = t^2e^{-2t} \sin 5t$.

Find the exact solution to each of the following:

19. $3x^2y\,dx - dx + x^3\,dy = 0$. When $x = 1$, $y = 3$.

20. $\dfrac{dy}{y} - \dfrac{dy}{x} + \dfrac{y\,dx}{x^2} = 0$. When $x = 0.5$, $y = 1$.

21. $4\dfrac{dy}{dx} + 10y = 5$. When $x = 0$, $y = 0$.

22. $\dfrac{dy}{dx} - \dfrac{y}{x} = \dfrac{x}{5 + 4x}$. When $x = -1$, $y = 0$.

23. $\dfrac{dy}{dx} + y = \sin x$. When $x = 0$, $y = 0$.

24. $R\dfrac{dq}{dt} + \dfrac{q}{C} = E$. When $t = 0$, $q = 0$.

25. $10^{-2}\dfrac{di}{dt} + i = 2\sin 100t$. When $t = 0$, $i = 0$.

26. $10^{-3}\dfrac{di}{dt} + i = 2te^{-900t}$. When $t = 0$, $i = 0$.

27. $2xy\,dy + y^2\,dx + 4y\,dy = (4x + 6)\,dx$. When $x = 1$, $y = 3$.

28. $dy + 2y\,dx + e^{2x}\,dy + 4e^{2x}y\,dx = e^{-2x}(y\,dx + x\,dy)$. When $x = 0$, $y = 2$. (*Hint:* Each pair has the same integrating factor.)

29. $dy + y\dfrac{\sin x}{\cos x}\,dx = -4e^{-2x}\cos x\,dx$. When $x = 0$, $y = 3$.

30. $y^2\,dy + \dfrac{y^3}{x}\,dx = \dfrac{y^2}{x^2}\,dx$. When $x = 1$, $y = 3$. (*Hint:* Variables should not be mixed on right-hand side. Therefore divide by y^2.)

31. $y\,dy + y^2\,dx = e^{-2x}\,dx$. When $x = 0$, $y = 2$. (*Hint:* Substitute $y^2 = u$, express $y\,dy$ in terms of du and solve for u.)

32. $3\dfrac{dy}{dx} + \dfrac{y}{x} = \dfrac{x^2}{2y^2}$ (at $x = 2$, $y = 1$). (*Hint:* Combine the ideas suggested in Exercises 30 and 31.)

16.9 HOMOGENEOUS DIFFERENTIAL EQUATIONS

Here the word "homogeneous" refers to the dimensions of the terms in a D.E.

A constant term is dimensionless, or of zero dimensions.
Terms in x and y are one-dimensional or linear.
Terms in x^2, xy, and y^2 are two-dimensional.
Terms in x^3, x^2y, xy^2, and y^3 are three-dimensional, etc.

If, when a D.E. is put in the form $dy/dx = P/Q$ or the form $P\,dx = Q\,dy$, all terms in P and Q are of the same dimensions, the D.E. is said to be homogeneous. If this is true, by making the substitution $y = vx$, all of these terms will have the same power of x. This technique produces an equation in x and v in which the variables can be separated, y being eliminated. In general, v is a variable, since it cannot be assumed that y is proportional to x.

EXAMPLE 20.   Solve the D.E. $(x^2 - 2xy)\,dy + (x^2 - 4y^2)\,dx = 0$.
Solution: All terms are two-dimensional, indicating a homogeneous equation. Therefore substitute $y = vx$, where v and x are treated as variables. Thus $dy = v\,dx + x\,dv$, and

$$x^2(1 - 2v)(v\,dx + x\,dv) + x^2(1 - 4v^2)\,dx = 0.$$

Dividing by $x^2(1 - 2v)$,

$$v\,dx + x\,dv + (1 + 2v)\,dx = 0.$$
$$dx(v + 1 + 2v) + x\,dv = 0.$$
$$x\,dv = -(1 + 3v)\,dx.$$

Separating variables

$$\int \frac{dv}{1 + 3v} = -\int \frac{dx}{x}.$$

$$\tfrac{1}{3} \ln (1 + 3v) = -\ln x + \ln k.$$

$$\ln (1 + 3v) = 3 \ln \frac{k}{x} = \ln \frac{k^3}{x^3}.$$

Thus

$$1 + 3v = \frac{k^3}{x^3} \quad \text{and} \quad v = \frac{k^3}{3x^3} - \frac{1}{3} = \frac{A}{x^3} - \frac{1}{3}.$$

Since

$$y = vx, \quad y = \frac{A}{x^2} - \frac{x}{3}.$$

This solution should be verified.

EXAMPLE 21.   Solve the D.E.

$$\left(x + \frac{y}{x} \sqrt{x^2 + y^2} \right) dy - y\,dx = 0,$$

given that at $x = 0$, $y = 1$.

Solution: The first step is to determine that the D.E. is homogeneous. y/x is dimensionless and $\sqrt{x^2 + y^2}$ is the square root of a uniformly two-dimensional expression, so that it is one-dimensional. Hence the term $(y/x)\sqrt{x^2 + y^2}$ is one-dimensional. It is now apparent that the equation is homogeneously one-dimensional. If the substitution $y = vx$ is used, the equation becomes quite complicated, since dy, the two-termed expression $v\,dx + x\,dv$, must be multiplied by another long expression.

The substitution $x = vy$ will also lead to separation of variables, and in this instance, the result is simpler than if the substitution were $y = vx$. Let $x = vy$. Hence $dx = v\,dy + y\,dv$. Therefore

$$\left(vy + \frac{y}{vy}\,y\sqrt{v^2 + 1}\right) dy - y(v\,dy + y\,dv) = 0.$$

$$y\,dy\left(v + \frac{\sqrt{v^2 + 1}}{v} - v\right) = y^2\,dv \quad \text{and} \quad \frac{dy}{y} = \frac{v\,dv}{\sqrt{v^2 + 1}}.$$

When $y = 1$, $x = 0$, and $v = x/y = 0$. Thus

$$\int_1^y \frac{dy}{y} = \int_0^v \frac{v\,dv}{\sqrt{v^2 + 1}}.$$

$$\ln y - \ln 1 = \sqrt{v^2 + 1} - \sqrt{0 + 1}. \qquad \ln y = \sqrt{v^2 + 1} - 1.$$

$$v^2 + 1 = (\ln y + 1)^2 = (\ln y)^2 + 2\ln y + 1.$$

$$v = \sqrt{(\ln y)^2 + 2\ln y}.$$

Since $x = vy$,

$$x = y\sqrt{(\ln y)^2 + 2\ln y}.$$

To find y as a function of x is impossible here.

EXERCISES

1. Determine which of the following D.E.'s are homogeneous:

(a) $x\,dy + y\,dx = 3\,dx$.

(b) $x^2\,dy - y^2\,dx = x^2(y^2 + xy)\,dx$.

(c) $x\,dy - y\,dx = 3\sqrt{xy}\,dx$.

(d) $xy\,dy - y^2\,dx = \dfrac{x^2}{y}\sqrt{x^2 + xy}\,dx$.

(e) $\dfrac{dy}{dx} = \dfrac{\sqrt{x - y}\,\sqrt{y^3}}{x^2}$.

Solve the following D.E.'s:

2. $x\,dy - y\,dx = 0$.

3. $x\,dy - y\,dx = x\,dx$.

4. $2x^2\,dy - y^2\,dx = 2xy\,dx$.

5. $\dfrac{dy}{dx} = \dfrac{y^2}{xy - x^2}$. (*Hint:* Substitution for x is easier.)

6. $(x^2 + 2y^2)\,dx = 2xy\,dy$.

7. $\sqrt{xy}\,dy = \left(2y + y\sqrt{\dfrac{y}{x}}\right)dx$.

8. $xz\dfrac{dz}{dx} + z^2 = x^2$ (a) by the homogeneous method (b) by substituting $z^2 = 2y$ and thus $2z\dfrac{dz}{dx} = 2\dfrac{dy}{dx}$.

Find the exact solution to the following D.E.'s:

9. $2xy\,dy = (x^2 + y^2)\,dx$. When $x = 1$, $y = 0$.

10. $x\,dy = (\sqrt{x^2 - y^2} + y)\,dx$. When $x = 1$, $y = 0$.

11. $x\,dy = (\sqrt{x^2 + y^2} + y)\,dx$. When $x = 1$, $y = 0$.

12. $dy = \left(\dfrac{y}{x} + \dfrac{\sqrt{x^2 + y^2}}{2y}\right)dx$. When $x = 1$, $y = 0$.

13. $\sin\dfrac{y}{x}\,dy = \left(\dfrac{x}{y} + \dfrac{y}{x}\sin\dfrac{y}{x}\right)dx$. When $x = 1$, $y = 0$.

14. $y\,dx + (2\sqrt{x^2 + y^2} - x)\,dy = 0$. When $x = 0$, $y = 1$.

16.10   MISCELLANEOUS PROBLEMS

When differential equations of various types are to be solved, the detection of the type becomes important. The methods of solution discussed in this chapter are given here for reference:

Direct integration (Section 16.2).
Intuitive methods with exponential solutions (Section 16.3).
Method of superposition (Section 16.4).
Separation of variables (Section 16.6).
Exact differentials (Section 16.7).
Use of integrating factor to result in exact differentials (Section 16.8).
Homogeneous differential equations (Section 16.9).

The same equation may often be solved by three, or even four of these methods. This is particularly true of D.E.'s resulting from the switching of electrical circuits. In these types, one method may be used as a check on another. Superposition is often the simplest method in cases where it works.

However, when a particular solution is difficult to obtain, superposition is more complicated. The alternative methods in these instances may result in very difficult integrals. The simplest method of solution is determined only with practice.

Worded problems often give rise to differential equations. The biggest problem here often is the setting up of the differential equations. Extreme care in interpreting rates and their relationships is necessary.

EXAMPLE 22. A 3-gallon radiator containing $\frac{1}{2}$ gallon of antifreeze completely mixed with $2\frac{1}{2}$ gallons of water is strengthened by adding antifreeze and drawing off the mixture at the same constant rate, so that the radiator is always full. Find the amount of antifreeze in the mixture after one gallon of antifreeze has been poured in: (a) assuming complete mixing at all times, and (b) assuming that the increase in percent alcohol during the addition is only one half as much at the draw-off level as for the whole tank.

Solution: (a) The fact that the draw-off will contain a continuously increasing percent of alcohol gives rise to a differential equation. Let the amount of alcohol in the tank at any time be x gallons. Let the constant rate at which alcohol is added be k gallons per unit time.

With complete mixing, the ratio of alcohol to total volume for any portion of the volume is $x/3$.

Hence the rate at which alcohol is drawn off is $(x/3)k$ gallons per unit time.

The rate at which the volume of alcohol increases in the tank equals the rate at which it is added minus the rate at which it is drawn off. Thus $dx/dt = k - kx/3$. (At this stage, the consistency of units should be checked. Here each of the three terms has units of gallons per unit time.)

By the method of superposition, $x = 3 + Ae^{-kt/3}$. At $t = 0$, $x = 0.5$, and $A = -2.5$. Therefore

$$x = 3 - 2.5e^{-kt/3}.$$

The time t is not directly involved here, but if the rate of pouring is k gallons per unit time, when one gallon is poured, $1 = kt$, and $t = 1/k$. By substituting this value of t,

$$x = 3 - 2.5e^{-\frac{1}{3}} = 3 - 2.5(0.7165) = 3 - 1.79 = 1.21.$$

Hence the radiator contains 1.21 gallons of alcohol after one gallon is added.

(b) The ratio of volume of alcohol to total volume at the draw-off point for this situation equals

$\frac{1}{3}[\frac{1}{2}$ (increase in alcohol in the tank) + original alcohol in the tank]

or
$$\frac{\frac{1}{2}(x - \frac{1}{2}) + \frac{1}{2}}{3} = \frac{2x + 1}{12}.$$

Thus
$$\frac{dx}{dt} = k - \frac{k(2x + 1)}{12}, \qquad \text{or} \qquad \frac{dx}{dt} = \frac{11k}{12} - \frac{kx}{6}.$$

By the method of superposition,

$$x = 5.5 + Ae^{-kt/6}.$$

From the initial conditions, $A = -5.0$. When one gallon is added,

$$x = 5.5 - 5.0e^{-1/6} = 5.5 - 4.23 = 1.27.$$

Therefore the radiator contains 1.27 gallons of alcohol after one gallon has been added.

MISCELLANEOUS EXERCISES

Solve the following differential equations:

1. $\frac{dy}{dx} - \frac{y}{x} = 2$. When $x = 1$, $y = 2$. (Use at least two methods.)

2. $dy + 4y\, dx = 3e^{-3x}\, dx$. When $x = 0$, $y = 1$. (Use at least two methods.)

3. $\sqrt{4 - x^2}\, dy = (2x\sqrt{4 - y^2} + \sqrt{4 - y^2})\, dx$. When $x = 2$, $y = 2$.

4. $(x^2 - y^2)\frac{dy}{dx} = 2xy$. When $x = 0$, $y = 1$.

5. $\cos x \frac{dy}{dx} + y \sin x = 0$.

6. $x^2 \frac{dy}{dx} - xy = x^2 \ln x$. When $x = 1$, $y = -1$.

7. $\frac{dy}{dx} = \frac{xy^2 - xy}{x^2 - 3}$. When $x = 2$, $y = 2$.

8. $y\frac{dy}{dx} \cos^2 x = \tan x + 2$. When $x = \pi/4$, $y = 2$.

9. $\frac{1}{500}\frac{di}{dt} + i = 2e^{-200t}$. At $t = 0$, $i = 0$.

10. $\left(y^2 - xy\dfrac{dy}{dx}\right)^2 = x^2(x^2 - y^2)$. At $x = 1$, $y = 0$.

11. $R\dfrac{di}{dt} + \dfrac{i}{C} = \dfrac{d}{dt}(A \sin \omega t)$, where R, C, A, and ω are constant. When $t = 0$, $i = 0$.

12. $\dfrac{m}{k}\dfrac{dv}{dt} = \dfrac{mg}{k} - v^2$, where m, k, and g are constant. When $t = 0$, $v = 0$.

13. $(9 - x^2)\dfrac{dy}{dx} + (3 - x)y = x$.

14. Sugar in solution dissolves at a rate proportional to the amount of sugar remaining undissolved. If 2 lb of undissolved sugar reduces to 1 lb of undissolved sugar in 40 min, find how much sugar is undissolved at the end of 2 hr.

15. Using Newton's law of cooling, find the temperature of a motor in a room at 20°C after 2 hr of cooling, if it cools from 200°C to 120°C in the first hour of cooling.

16. A cylindrical tank with 8-ft depth and 2-ft radius is full of oil weighing 50 lb per cubic foot. The tank is emptied by pumping the oil to the level of the top of the tank. An element of work done in this operation is an element of weight times the vertical distance between the top of the tank and the oil level.

(a) Find the work performed in emptying the tank.

(b) When one fourth of the total work has been completed, what portion of the oil remains in the tank?

17. The end of the thread on a spool is fixed, and the spool is allowed to fall from rest, unwinding thread as it falls. Neglecting friction, find the distance it falls and the downward velocity after 3 sec of motion. Use the fact that total energy used, mgs, is equal to the sum of kinetic energy of translation $\frac{1}{2}mv^2$ plus kinetic energy of rotation $\frac{1}{2}I\omega^2$. I, the moment of inertia for this type of rotation, is equal to $\frac{1}{2}mr^2$, where m and r are the mass and radius of the spool; and ω is the angular velocity in radians per second. Assume $g = 32$ ft/sec^2.

Compare the distance of the fall at any time with the distance the spool would fall freely as one unit of mass.

18. A ventilating system moves 500 ft^3 of air per minute, pumping in air containing 0.03% carbon dioxide. If a 10,000-ft^3 room originally containing 0.10% carbon dioxide is ventilated for 40 min, find the percent carbon dioxide in the room, assuming continuous perfect mixing.

19. The velocity in cubic feet per second at which water drains from a container through a hole in the bottom is proportional to the square root of the depth of the water in the container. If water is draining through a hole in the bottom of a 2-ft high cylindrical container at such a rate that half of the water drains off in 5 min, find the time it takes to empty the container.

20. Repeat Exercise 19 for an upright conical container with its tip at the bottom and with diameter of the top equal to its height.

21. Assume that a body of mass m falling through space encounters air resistance proportional to its velocity. The net force acting on the body is its weight mg minus the air resistance, and the acceleration a of the body is controlled by the law that net force equals ma. The weight is mg pounds in Engineering Units and the mass m (in slugs) equals the weight divided by g. Assume $g = 32$.

A parachutist falling at a velocity of 100 ft per second opens his parachute and eventually attains a steady-state velocity of 20 ft per second. Determine when his velocity becomes 30 ft per second.

22. A tank contains 200 gallons of salt solution at a concentration of $\frac{1}{4}$ lb of salt per gallon. If fresh water runs into the tank at the rate of 2 gallons per minute and the mixture runs out at the rate of 4 gallons per minute, how much salt is in the tank at the end of 15 min? Assume a uniform mixture at all times.

23. In Exercise 22 assuming that a mixture containing 1 lb of salt per gallon, instead of fresh water, runs into the tank, all other conditions remaining the same, find the amount of salt in the tank at the end of 1 hr.

24. A cylindrical tank with a hole in the bottom empties 9 gallons of water in 2 hr. The rate of drain-off here is proportional to the square root of the volume of water in the tank. If, during the drain-off, water is added at a constant rate of 6 gallons per hour, find the steady-state volume of water in the tank.

25. A parachutist drops from a plane and waits 4 sec before opening his parachute. The man and his parachute weigh a total of 250 lb. Assume that the air resistance is $(1/360)v^2$ lb, where v is the velocity in feet per second.

(a) Find the time at which his velocity is 100 ft per second.

(b) Find his velocity when he opens his parachute.

(c) Find the steady velocity he would attain if he did not open his parachute, and the time at which he attains it.

(Use physical laws from Exercise 21 in setting up the differential equations.)

CHAPTER 17

Second-Order Differential Equations

17.1 DIRECT INTEGRATION AND SOME INTUITIVE METHODS

If $d^2y/dx^2 = x - 2$, since only one derivative is involved and since the right-hand side involves only the independent variable, two direct integrations will yield the general solution.

Therefore

$$\frac{dy}{dx} = \int (x - 2)\, dx = \frac{x^2}{2} - 2x + k_1.$$

Hence

$$y = \frac{x^3}{6} - x^2 + k_1 x + k_2.$$

k_1 and k_2 are the two arbitrary constants characteristic of the general solution of any second-order D.E.

If $d^2y/dx^2 = ky$, however, direct integration cannot be used. To arrive at an intuitive solution to this equation, examine carefully the following functions:

$y = Ae^{mx}.$	$y = Ae^{-mx}.$	$y = A \cos mx.$	$y = A \sin mx.$
$\dfrac{d^2y}{dx^2} = m^2 A e^{mx}$	$\dfrac{d^2y}{dx^2} = m^2 A e^{-mx}$	$\dfrac{d^2y}{dx^2} = -m^2 A \cos mx$	$\dfrac{d^2y}{dx^2} = -m^2 A \sin mx$
$= m^2 y.$	$= m^2 y.$	$= -m^2 y.$	$= -m^2 y.$

There are at least four functions whose second derivative is proportional to the original function; two of them are positively proportional and two are negatively proportional. Thus, if $d^2y/dx^2 = ky$, and k is positive, either

348

$y = Ae^{\sqrt{k}\,x}$ or $y = Ae^{-\sqrt{k}\,x}$ is a solution. However, for a general solution to a second-order differential equation, we require two arbitrary constants.

$$y = Ae^{\sqrt{k}\,x} + Be^{-\sqrt{k}\,x}$$

satisfies the original equation, and this solution contains two arbitrary constants, which can independently assume any value. It includes the two separate solutions given above, since $y = Ae^{\sqrt{k}\,x}$ when B is zero, and $y = Be^{-\sqrt{k}\,x}$ when A is zero. In general, the solution is likely to contain both a positive exponential and a negative exponential, and not just one or the other of these. Therefore the general solution is

$$y = Ae^{\sqrt{k}\,x} + Be^{-\sqrt{k}\,x}.$$

Similarly, if $d^2y/dx^2 = -ky$, (k itself being positive), the general solution is

$$y = A \cos \sqrt{k}\,x + B \sin \sqrt{k}\,y,$$

since both the sine and the cosine have a second derivative that is negatively proportional to the original function.

If two separate solutions to a second-order differential equation can be found, each containing one arbitrary constant, the general solution for y is the sum of these two separate functions, using two different arbitrary constants.

Simple harmonic motion is defined as motion in which the acceleration towards a fixed point is negatively proportional to the displacement from that point. Thus $d^2s/dt^2 = -ks$ for simple harmonic motion. Therefore

$$s = A \cos \sqrt{k}\,t + B \sin \sqrt{k}\,t.$$

If, in the most simple case, $s = 0$, when $t = 0$, a substitution yields the fact that $A = 0$. Therefore $s = B \sin \sqrt{k}\,t$. If maximum displacement from the fixed point is 6 units, $s = 6 \sin \sqrt{k}\,t$.

If, however, a maximum displacement of 6 units occurs when $t = 0$, the equation is $s = 6 \cos \sqrt{k}\,t$.

EXAMPLE 1.   If, at $t = 0$, $q = 0$ and $dq/dt = 0$, solve

$$0.10 \frac{d^2q}{dt^2} + 10q = 2.$$

Solution: Superposition may be used here. For steady-state q_1, when there is no change in q and therefore all derivatives are zero,

$$10q_1 = 2 \quad \text{and} \quad q_1 = 0.2.$$

This is consistent with our assumption, $\dfrac{d^2q_1}{dt^2} = 0$.

For transient q_2,

$$0.10\,\frac{d^2q_2}{dt^2} = -10q_2.$$

$$\frac{d^2q_2}{dt^2} = -100q_2.$$

Hence $q_2 = A \cos 10t + B \sin 10t$ (intuitively),

and $q = A \cos 10t + B \sin 10t + 0.2.$

Substituting $(0,0)$ for (t,q), $0 = A + 0 + 0.2$, and $A = -0.2$.
For the second substitution, dq/dt must be determined.

$$\frac{dq}{dt} = -10A \sin 10t + 10B \cos 10t.$$

Substituting $(0,0)$ for $(t,dq/dt)$, $0 = 0 + 10B$, and $B = 0$.
Therefore

$$q = -0.2 \cos 10t + 0.2.$$

EXERCISES

1. Find the general solution to the following differential equations:

(a) $\dfrac{d^2y}{dt^2} = 0.$ (b) $\dfrac{d^2y}{dt^2} = k.$ (c) $\dfrac{d^2y}{dx^2} = 2e^x - 3x^2.$

(d) $\dfrac{d^3y}{dt^3} = \dfrac{1}{\sqrt{t}}.$ (e) $\dfrac{d^2y}{dx^2} = \sqrt{1 + 2x}.$ (f) $\dfrac{d^2y}{dx^2} = \dfrac{1}{2\sqrt{1 - 2x}}.$

(g) $\dfrac{d^2y}{dx^2} - \dfrac{1}{(1 + x)^2} = 0.$

Find the exact solution to the following differential equations:

2. $\dfrac{d^2y}{dx^2} = -4x.$ When $x = 0$, $y = -2$, and $\dfrac{dy}{dx} = 0.$

3. $\dfrac{d^2y}{dx^2} = -4y.$ When $x = 0$, $y = 3$, and $\dfrac{dy}{dx} = 0.$

4. $\dfrac{d^2y}{dt^2} = 10e^{-5t} + 5 \sin 10t.$ When $t = 0$, $y = 1$, and $\dfrac{dy}{dt} = 1.$

5. $y'' = \dfrac{x}{(1 - x^2)^{3/2}}.$ When $x = 0$, $y = 0$, and $y' = -2.$

6. $D^2y + \dfrac{1}{\sqrt{1 - 4x}} = 0$. When $x = 0$, $y = -2$, and $Dy = 0.5$.

7. $D^2y = \dfrac{2}{(1 - 0.5x)^2}$. When $x = 0$, $y = 0$, and $Dy = 3$.

8. $\dfrac{d^2v}{dt^2} = \dfrac{2}{t}$. When $t = 1$, $v = 3$, and $\dfrac{dv}{dt} = 2$.

9. $\dfrac{d^2y}{dx^2} = \dfrac{2}{1 + 2x}$. When $x = 0$, $y = 1$, and $\dfrac{dy}{dx} = 0$.

Find the solution to the following differential equations:

10. $\dfrac{d^2i}{dt^2} - 36i = 0$. At $t = 0$, $i = -1$, and $i' = 18$.

11. $\dfrac{d^2i}{dt^2} + 36i = 0$. At $t = 0$, $i = 1$, and $i' = 6$.

12. $\dfrac{d^2s}{dt^2} = -0.25s$. At $t = 0$, s has its maximum value of 3.

13. $4\dfrac{d^2v}{dt^2} - 500t = 0$. At $t = 0$, $v = 0$, and $\dfrac{dv}{dt} = -4$.

14. $2 \times 10^{-3}\dfrac{d^2q}{dt^2} + \dfrac{2q}{10^{-3}} = 0$. At $t = 0$, $q = 5 \times 10^{-3}$, and $\dfrac{dq}{dt} = 0$.

15. $\dfrac{d^2y}{dx^2} + 16y = 8$. At $x = 0$, $y = 0$, and $y' = 2$.

16. $L\dfrac{d^2i}{dt^2} + \dfrac{i}{C} = 0$. At $t = 0$, $i = 0$, and $\dfrac{di}{dt} = \dfrac{E}{L}$.

17. $\dfrac{d^2y}{dx^2} - 10y = 12$. At $x = 0$, $y = 0$, and $\dfrac{dy}{dx} = 16$.

18. $10^{-2}D^2y - 400y = 6$. At $x = 0$, $y = 3.0 \times 10^{-2}$, and $Dy = 1.0$.

19. $4.0 \times 10^{-3}\dfrac{d^2q}{dt^2} + \dfrac{q}{4.0 \times 10^{-3}} = 3.0$. At $t = 0$, $q = 0$, and $\dfrac{dq}{dt} = 0$.

20. $2y'' + 8y = 13e^{-3x}$. At $x = 0$, $y = 0.5$, and $y' = 2.5$. (*Hint:* Find a particular solution suggested by the function on the right-hand side.)

21. $\dfrac{d^2q}{dt^2} + 90{,}000q = 8 \sin 100t$. At $t = 0$, $q = 0$, and $\dfrac{dq}{dt} = 0$. (Note that the particular solution here must be $k \sin 100t$ only. Why?)

22. $L\dfrac{d^2i}{dt^2} + \dfrac{i}{C} = A\omega \cos \omega t$. At $t = 0$, $i = 0$, and $\dfrac{di}{dt} = 0$.

23. $\dfrac{d^2y}{dx^2} - 4y = 10x + 5 \cos 4x$. At $x = 0$, $y = 1.75$, and $\dfrac{dy}{dx} = -6.5$. (Note that the particular solution cannot contain a constant term or a term in $\sin 4x$. Why?)

24. A body moves in simple harmonic motion so that its acceleration is -1.0 ft per second per second when its displacement from its point of equilibrium is 0.25 ft. If, at time $t = 0$, its velocity is 0.6 ft per second and its acceleration is 0.8 ft per second per second, find (a) its displacement at $t = 0.5$ sec, (b) its velocity at $t = 0.5$ sec, and (c) its acceleration at $t = 1.0$ sec.

25. An LC series circuit is switched into a d-c voltage source of 3.0 volts. If $L = 10^{-2}$ henry and $C = 4.0 \times 10^{-4}$ farad, find the charge q as a function of t, when at $t = 0$, q and dq/dt are both zero. [Note that $L(di/dt) = L(d^2q/dt^2)$.]

26. An LC series circuit is switched into an a-c voltage source of 2.0 sin 600t volts. If $L = 5.0 \times 10^{-3}$ henry and $C = 1.25 \times 10^{-3}$ farad, find the current in the circuit as a function of t, assuming that, at $t = 0$, i and di/dt are both zero. (*Hint:* To set up the differential equation for current, take the derivative of the D.E. $L(di/dt) + q/C$ equals the applied voltage, and use the fact that $dq/dt = i$.)

17.2   THE GENERAL SECOND-ORDER DIFFERENTIAL EQUATION WITH CONSTANT COEFFICIENTS

For the differential equation

$$a \frac{d^2y}{dx^2} + b \frac{dy}{dx} + cy = 0,$$

where a, b, and c are constant, we may use some intuitive methods to start the solution. These three terms must add to zero; therefore the basic function involved must be the same for each of the three terms and, of course, at least one of the terms must be negative. We cannot, for example, add a positive cosine function to a negative power function, and get zero for all values of the independent variable. When the sum of three terms is zero, the function must be the same for each term. There is only one function which contains the same function in all its derivatives—the exponential function. Therefore y must be an exponential function.

Therefore let $y = Ae^{mx}$ and attempt to solve for m. The following substitutions must be made in the original differential equation:

$$y = Ae^{mx}, \qquad \frac{dy}{dx} = mAe^{mx}, \qquad \text{and} \qquad \frac{d^2y}{dx^2} = m^2Ae^{mx}.$$

Hence

$$am^2Ae^{mx} + bmAe^{mx} + cAe^{mx} = 0$$

for all values of x. As predicted, the same basic function occurs in each term, and we can divide through by Ae^{mx}. (Ae^{mx} itself cannot equal zero for all values of x.)

Therefore $am^2 + bm + c = 0$. By quadratic theory, m will have two values:

$$m = \frac{-b \pm \sqrt{b^2 - 4ac}}{2a}.$$

These values of m may be real or imaginary. We must examine some of these specific cases in more detail later. However, if the values of m are real and unequal (m_1 and m_2, say), we see, from the technique of combining two separate solutions (given in Section 17.1), that

$$y = Ae^{m_1 x} + Be^{m_2 x}$$

is the general solution.

Thus, to solve the equation

$$a\frac{d^2y}{dx^2} + b\frac{dy}{dx} + cy = 0,$$

solve for m in the equation $am^2 + bm + c = 0$. (This equation is called the *auxiliary equation*, and has a recognized abbreviation, A.E.)

Values of m Real and Unequal. If the values of m are real and unequal, that is, m_1 and m_2, the general solution is

$$y = Ae^{m_1 x} + Be^{m_2 x}.$$

The technique of the auxiliary equation is not limited to second-order differential equations. For example, if

$$0.10\frac{di}{dx} + 5i = 0 \qquad \text{(a first-order D.E.)},$$

the A.E. is $0.10m + 5 = 0$, and $m = -50$. Therefore $i = Ae^{-50t}$, a solution that might have been reached by the intuitive methods of Section 16.3.

EXAMPLE 2.   Solve

$$0.10\frac{d^2i}{dt^2} + 2\frac{di}{dt} + 7.5i = 0$$

given that, at $t = 0$, $i = 0$ and $di/dt = 100$.
Solution: The A.E. is $0.10m^2 + 2m + 7.5 = 0$.

$$m^2 + 20m + 75 = 0.$$
$$(m + 5)(m + 15) = 0.$$
$$m = -5 \text{ or } -15.$$

Therefore $i = Ae^{-5t} + Be^{-15t}$.

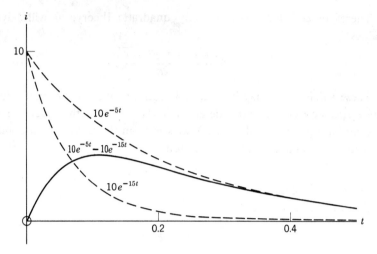

Fig. 17-1. Graph of $i = 10e^{-5t} - 10e^{-15t}$.

To solve for A and B, substitute $(0,0)$ for (t,i). Therefore $0 = A + B$ and $B = -A$.

$$\frac{di}{dt} = -5Ae^{-5t} - 15Be^{-15t}.$$

Substitute $(0,100)$ for $(t,di/dt)$.

$$100 = -5A - 15B$$
$$= -5A + 15A.$$

Hence $A = 10$ and $B = -10$.

Thus $i = 10e^{-5t} - 10e^{-15t}$ (Graphed in Fig. 17-1).

EXAMPLE 3.   If $L(di/dt) + Ri = E$, and at $t = 0$, $i = 0$, solve for i as an exact function of t, using the A.E. method.

Solution: Three other methods of solving this equation have been indicated in Example 19 of Chapter 16.

To eliminate E, both sides are differentiated with respect to t. Hence

$$L\frac{d^2i}{dt^2} + R\frac{di}{dt} = 0.$$

The A.E. is $Lm^2 + Rm = 0$. Thus

$$m(Lm + R) = 0. \qquad m = 0 \qquad \text{or} \qquad m = -\frac{R}{L}.$$

Hence $\qquad i = Ae^0 + Be^{-Rt/L} = A + Be^{-Rt/L}.$

At $t = 0$, $i = 0$; and if $i = 0$ is substituted in the original D.E., this implies that, at $t = 0$, $di/dt = E/L$.

$$\frac{di}{dt} = -\frac{RB}{L}e^{-Rt/L}.$$

Substituting $(0, 0, E/L)$ for $(t, i, di/dt)$ gives $A = -B$, and $B = -E/R$; therefore

$$A = \frac{E}{R} \quad \text{and} \quad i = \frac{E}{R}(1 - e^{-Rt/L}).$$

Values of m Real and Equal. If the roots of the auxiliary equation are real and equal, we cannot use the general solution previously given; if we did, y would be $Ae^{m_1 x} + Be^{m_1 x}$ or

$$e^{m_1 x}(A + B) = ke^{m_1 x},$$

which has only one independent arbitrary constant. There must be two arbitrary constants. Therefore, to develop the general solution in this situation, more rigorous methods are used, as follows:

With equal roots, the auxiliary equation may always be written in the form $\qquad m^2 - 2m_1 m + m_1^2 = 0,$

the left-hand side being the perfect square of $m - m_1$ giving two equal roots, $m = m_1$. This implies that the original differential equation is

$$\frac{d^2 y}{dx^2} - 2m_1 \frac{dy}{dx} + m_1^2 y = 0.$$

This may also be written as

$$\frac{d^2 y}{dx^2} - m_1 \frac{dy}{dx} = m_1 \frac{dy}{dx} - m_1^2 y.$$

The left-hand side is now the exact derivative of $dy/dx - m_1 y$. Therefore

$$\frac{d}{dx}\left(\frac{dy}{dx} - m_1 y\right) = m_1\left(\frac{dy}{dx} - m_1 y\right).$$

The structure here may be more obvious if the expression in parentheses is replaced by u. Thus

$$\frac{du}{dx} = m_1 u, \quad \text{and} \quad u = Ae^{m_1 x}.$$

$$\frac{dy}{dx} - m_1 y = Ae^{m_1 x} \quad \text{(a first-order D.E.).}$$

The left-hand side can now be transformed to an exact derivative by using an integrating factor of $e^{-m_1 x}$. Therefore

$$e^{-m_1 x} \frac{dy}{dx} - m_1 y e^{-m_1 x} = A.$$

$$\frac{d}{dx}(y e^{-m_1 x}) = A.$$

$$y e^{-m_1 x} = \int A\, dx = Ax + B \quad \text{and} \quad y = (Ax + B)e^{m_1 x}.$$

Thus, if the A.E. has two equal roots (both equal to m_1, say), the general solution is

$$y = (Ax + B)e^{m_1 x}.$$

EXAMPLE 4.   Find q as an exact function of t if, when $t = 0$, $q = 0.050$ and $dq/dt = 2$, and for all values of t

$$0.50 \frac{d^2 q}{dt^2} + 10 \frac{dq}{dt} + 50q = 0.$$

Solution: The A.E. is $0.50 m^2 + 10m + 50 = 0.$

or $m^2 + 20m + 100 = 0.$

Thus $(m + 10)^2 = 0.$ $m = -10$ or -10 (real and equal).

Therefore $q = (At + B)e^{-10t}.$

By substituting $(0, 0.050)$ for (t, q), $0.050 = B$.

$$\frac{dq}{dt} = -10e^{-10t}(At + B) + Ae^{-10t}.$$

By substituting $(0, 2)$ for $(t, dq/dt)$ and 0.050 for B,

$$2 = -10 \times 0.050 + A. \qquad A = 2.5.$$

Therefore $q = (2.5t + 0.050)e^{-10t}.$ (See Fig. 17-2.)

Values of m Imaginary. Imaginary roots occur in conjugate pairs. If one root is $p + jq$, the other is $p - jq$. Moreover, the sum of these conjugates is real, and their difference involves only j terms.

If the roots of the auxiliary equation are imaginary ($p + jq$ and $p - jq$), they cannot be equal and therefore a correct mathematical solution to the differential equation is $y = Ae^{(p+jq)x} + Be^{(p-jq)x}.$

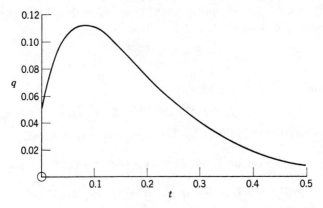

Fig. 17-2. Graph of $q = (2.5t + 0.050)e^{-10t}$.

However, it is difficult to interpret the right-hand side and to envisage a graphical representation of y. It appears to contain some imaginary factors, and yet we know that it must be a real function. Example 18 of Chapter 15 is a similar problem.

To eliminate the imaginary factors:

$$y = Ae^{px}e^{jqx} + Be^{px}e^{-jqx}$$
$$= e^{px}(Ae^{jqx} + Be^{-jqx})$$
$$= e^{px}[A(\cos qx + j \sin qx) + B(\cos qx - j \sin qx)]$$
$$= e^{px}[(A + B) \cos qx + j(A - B) \sin qx].$$

This expression must be real; this means that A and B themselves must be two complex conjugates, so that $(A + B)$ is real and $(A - B)$ contains only j terms, which, multiplied by the j coefficient, give a real constant. Since this is so, a single real constant could replace $(A + B)$ and another single real constant could replace $j(A + B)$.

Therefore the general solution to the differential equation for this type of equation is $y = e^{px}(A \cos qx + B \sin qx)$.
Here A and B are two independent arbitrary constants, but not the original A and B of this solution.

EXAMPLE 5.   Find an exact solution to the following differential equation:

$$\frac{d^2y}{dx^2} + 2\frac{dy}{dx} + 10y = 0;$$

at $x = 0$, $y = 2$, and $dy/dx = 1$.

Verify that the same solution results from using the *j* form of solution.
Solution: The A.E. is $m^2 + 2m + 10 = 0$. Therefore

$$m = \frac{-2 \pm \sqrt{4 - 40}}{2} = -1 \pm j3.$$

Therefore the general solution is

$$y = e^{-x}(A \cos 3x + B \sin 3x).$$

By substituting (0,2) for (x,y), $2 = 1(A) + 0$. Hence $A = 2$.

$$\frac{dy}{dx} = e^{-x}(-3A \sin 3x + 3B \cos 3x) - e^{-x}(A \cos 3x + B \sin 3x).$$

Thus $1 = 1(0 + 3B) - 1(A + 0)$, $A = 2$, and $B = 1$. Therefore the exact solution is

$$y = e^{-x}(2 \cos 3x + \sin 3x).$$

If the *j* form of solution is used,

$$y = Ae^{(-1+j3)x} + Be^{(-1-j3)x}.$$

By substituting (0,2) for (x,y), $2 = Ae^0 + Be^0$. Thus

$$A + B = 2. \qquad (1)$$

$$\frac{dy}{dx} = (-1 + j3)Ae^{(-1+j3)x} + (-1 - j3)Be^{(-1-j3)x}.$$

At $x = 0$, $\qquad \dfrac{dy}{dx} = (-1 + j3)A - (1 + j3)B.$

Hence $\qquad\qquad 1 = -(A + B) + j3(A - B)$

$$= -2 + j3(A - B).$$

Therefore $\qquad j3(A - B) = 3 \qquad A - B = -j. \qquad (2)$

By using (1) and (2), $A = (2 - j)/2$ and $B = (2 + j)/2$, two complex conjugates, as predicted above. Thus

$$y = \frac{2 - j}{2} e^{(-1+j3)x} + \frac{2 + j}{2} e^{(-1-j3)x}$$

$$= \frac{2 - j}{2} e^{-x}e^{j3x} + \frac{2 + j}{2} e^{-x}e^{-j3x}$$

$$= e^{-x}\left[(e^{j3x} + e^{-j3x}) + \frac{j}{2}(-e^{j3x} + e^{-j3x})\right]$$

$$= e^{-x}\left[2 \cos 3x + \frac{j}{2}(-2j \sin 3x)\right]$$

$$= e^{-x}(2 \cos 3x + \sin 3x), \qquad \text{as above.}$$

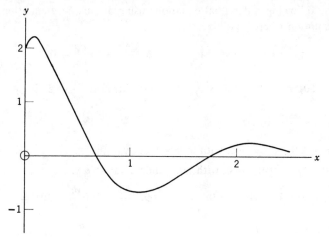

Fig. 17-3. Graph of $y = e^{-x}(2 \cos 3x + \sin 3x) = \sqrt{5}\, e^{-x} \sin (3x + 63.4°)$.

By using the method indicated in Section 9.1, this solution may be put in the form $y = \sqrt{5}\, e^{-x} \sin (3x + 63.4°)$, which is a sine function with a decaying amplitude (Fig. 17-3). This is usually called a damped oscillation.

Summary: To solve an equation of the form $a(d^2y/dx^2) + b(dy/dx) + cy = 0$, use the auxiliary equation $am^2 + bm + c = 0$. Three possible types of roots and the solution in each instance are shown below.

Type of Roots	Real and Unequal (m_1 and m_2)	Real and Equal (both equal to m_1)	Imaginary $(p + jq)$ and $(p - jq)$
Solution	$y = Ae^{m_1 x} + Be^{m_2 x}$.	$y = (Ax + B)e^{m_1 x}$.	$y = e^{px}(A \cos qx + B \sin qx)$.

EXERCISES

1. Determine the auxiliary equation for each of the following differential equations:

(a) $\dfrac{d^2y}{dx^2} - 5\dfrac{dy}{dx} + 4y = 0.$

(b) $L\dfrac{d^2q}{dt^2} + R\dfrac{dq}{dt} + \dfrac{q}{C} = 0.$

(c) $L\dfrac{d^2i}{dt^2} = -\dfrac{i}{C}.$

(d) $L\dfrac{d^2i}{dt^2} + R\dfrac{di}{dt} = 0.$

(e) $\dfrac{dv}{dt} = -50v.$

(f) $D^3y + 4D^2y - 3y = 0.$

Solve the following differential equations using an auxiliary equation, giving the exact solution where possible:

2. $\dfrac{d^2y}{dx^2} - 25y = 0.$

3. $\dfrac{d^2y}{dx^2} - 25\dfrac{dy}{dx} = 0.$

4. $\dfrac{dy}{dx} + 50y = 5.$ When $x = 0$, $y = 0$ and therefore $\dfrac{dy}{dx} = 5.$

5. $\dfrac{d^3y}{dx^3} - 9\dfrac{dy}{dx} = 0.$

6. $D^2y + 5Dy + 6y = 0$, with independent variable x.

7. $(2D^2 - D - 10)y = 0$, with independent variable x.

8. $\dfrac{d^2i}{dt^2} + 200\dfrac{di}{dt} + 6400i = 0.$

9. $\dfrac{d^2i}{dt^2} + 200\dfrac{di}{dt} + 10^4i = 0.$

10. $\dfrac{d^2y}{dt^2} - 8\dfrac{dy}{dt} - 4y = 0.$

11. $\dfrac{d^2y}{dt^2} + 4\dfrac{dy}{dt} + 8y = 0.$

12. $4y'' - 20y' + 25y = 0$, with independent variable x.

13. $2\dfrac{d^2y}{dx^2} - 6\dfrac{dy}{dx} + 7y = 0.$

14. $L\dfrac{d^2i}{dt^2} + R\dfrac{di}{dt} + \dfrac{i}{C} = 0$, where $R = 3\sqrt{\dfrac{L}{C}}.$

15. $L\dfrac{d^2i}{dt^2} + R\dfrac{di}{dt} + \dfrac{i}{C} = 0$, where $R = 2\sqrt{\dfrac{L}{C}}.$

16. $L\dfrac{d^2i}{dt^2} + R\dfrac{di}{dt} + \dfrac{i}{C} = 0$, where $R = \sqrt{\dfrac{L}{C}}.$

17. $(0.5D^2 - 30D + 400)y = 0.$ When $x = 0$, $y = 0$, and $Dy = 40.$

18. $4.0 \times 10^{-3}\dfrac{d^2v}{dt^2} + 10^3v = 0.$ When $t = 0$, $v = 0$, and $\dfrac{dv}{dt} = 200.$

19. $\dfrac{d^2y}{dx^2} = 50y.$ When $x = 0$, $y = 0$, and $\dfrac{dy}{dx} = -20\sqrt{2}.$

20. $\dfrac{d^2q}{dt^2} + 300\dfrac{dq}{dt} = 0.$ When $t = 0$, $q = 0$, and $\dfrac{dq}{dt} = 6.0.$

21. $0.5\dfrac{d^2y}{dx^2} - 30\dfrac{dy}{dx} + 450y = 0.$ When $x = 0$, $\dfrac{dy}{dx} = 0$, and $y = 0.$

22. $10^{-3}\dfrac{d^2i}{dt^2} + 2.0\dfrac{di}{dt} + 10^3i = 0.$ When $t = 0$, $i = 0$, and $\dfrac{di}{dt} = 415.$

23. $10^{-3}\dfrac{d^2i}{dt^2} + 2.0\dfrac{di}{dt} + 2.0 \times 10^3i = 0.$ When $t = 0$, $i = 0$, and $\dfrac{di}{dt} = 800.$

24. $10^{-2}\dfrac{d^2i}{dt^2} + 5.0\dfrac{di}{dt} + 1850i = 0.$ When $t = 0$, $i = 0$, and $\dfrac{di}{dt} = 600.$

25. $2\dfrac{d^2s}{dt^2} + \dfrac{ds}{dt} + 10.25s = 0.$ When $t = 0$, $s = 0.5$, and $\dfrac{ds}{dt} = -\dfrac{1}{8}.$

26. $8.0 \times 10^{-3}\dfrac{d^2q}{dt^2} + 3.2\dfrac{dq}{dt} + 192q = 0.$ When $t = 0$, $q = 10^{-2}$, and $\dfrac{dq}{dt} = 0.$

Solve the differential equation

$$L\dfrac{d^2i}{dt^2} + R\dfrac{di}{dt} + \dfrac{i}{C} = 0,$$

if, at $t = 0$, $i = 0$, and $di/dt = E/L$, for each of the following conditions. (L, R, C, and E are positive.)

27. If $R = 0$.   **28.** If C is infinite.

29. If $R^2 = \dfrac{4L}{C}.$ **30.** If $R^2 > \dfrac{4L}{C}.$

31. If $R^2 < \dfrac{4L}{C}.$

32. Find the exact solution to the differential equation, $d^3y/dx^3 + 8y = 0$, if, at $x = 0$, $y = 0$, and $dy/dx = 3$, and at $x = \pi/\sqrt{3}$, $y = 0$.

17.3   EQUATIONS OF THE FORM $a\dfrac{d^2y}{dx^2} + b\dfrac{dy}{dx} + cy = k$ or $f(x)$

The auxiliary equation method depends on the right-hand side being zero. To solve equations in which right-hand side is not zero, the method of superposition may be used.

EXAMPLE 6.   Solve the differential equation

$$0.01\dfrac{d^2q}{dt^2} + 2\dfrac{dq}{dt} + 1000q = 6\cos 400t,$$

given that at $t = 0$, $q = 0$, and $dq/dt = 0$.

Solution: The sustained value of q (that is a particular solution for q, not in this case a constant or steady-state value) could contain the function on the right-hand side plus all its derivatives. Therefore, let

$$q_1 = k\cos 400t + n\sin 400t.$$

Then $\dfrac{dq_1}{dt} = -400k\sin 400t + 400n\cos 400t.$

and $\dfrac{d^2q_1}{dt^2} = -160{,}000k\cos 400t - 160{,}000n\sin 400t.$

By substituting in the differential equation to find the appropriate k and n,

$$-1600k \cos 400t - 1600n \sin 400t - 800k \sin 400t$$

$$+ 800n \cos 400t + 1000k \cos 400t + 1000n \sin 400t = 6 \cos 400t.$$

$$(-600k + 800n) \cos 400t - (600n + 800k) \sin 400t = 6 \cos 400t.$$

Hence
$$-600k + 800n = 6.$$

$$-800k - 600n = 0.$$

$$k = -\frac{9}{2500} \quad \text{and} \quad n = \frac{12}{2500}.$$

Therefore

$$q_1 = -\frac{9}{2500} \cos 400t + \frac{12}{2500} \sin 400t.$$

The transient q_2 satisfies the equation

$$0.01 \frac{d^2 q_2}{dt^2} + 2 \frac{d^2 q_2}{dt} + 1000 q_2 = 0.$$

The A.E. is $0.01m^2 + 2m + 1000 = 0$.

$$m = \frac{-2 \pm \sqrt{4 - 40}}{0.02} = -100 \pm j300.$$

Therefore $q_2 = e^{-100t}(A \cos 300t + B \sin 300t)$.

$$q = -\frac{9}{2500} \cos 400t + \frac{12}{2500} \sin 400t + e^{-100t}(A \cos 300t + B \sin 300t).$$

By using the fact that, at $t = 0$, $q = 0$, and $dq/dt = 0$, we find $A = 9/2500$ and $B = -13/2500$. The details of this calculation are left as an exercise for the student. Thus

$$q = -\frac{9}{2500} \cos 400t + \frac{12}{2500} \sin 400t$$

$$+ e^{-100t}\left(\frac{9}{2500} \cos 300t - \frac{13}{2500} \sin 300t\right).$$

This function is essentially an undamped oscillation of frequency $200/\pi$ superimposed on a damped oscillation of frequency $150/\pi$. When the damped oscillation has vanished, a simple sinusoidal function remains.

EXERCISES

1. Find the particular solution for each of the following differential equations:

(a) $\dfrac{d^2y}{dx^2} - 5\dfrac{dy}{dx} + 8y = 10.$

(b) $L\dfrac{d^2q}{dt^2} + R\dfrac{dq}{dt} + \dfrac{q}{C} = E,$ where $L, R, C,$ and E are constant.

(c) $2\dfrac{d^2y}{dx^2} + 5\dfrac{dy}{dx} - y = 2e^{-x}.$

(d) $\dfrac{d^2y}{dt^2} + \dfrac{dy}{dt} + y = \cos t.$

Solve each of the following differential equations giving the general solution:

2. $\dfrac{d^2y}{dx^2} - 10\dfrac{dy}{dx} + 16y = 24.$

3. $2.5 \times 10^{-3}\dfrac{d^2q}{dt^2} + 50\dfrac{dq}{dt} + \dfrac{q}{6.25 \times 10^{-6}} = 2.0.$

4. $2.5 \times 10^{-3}\dfrac{d^2q}{dt^2} + 40\dfrac{dq}{dt} + \dfrac{q}{4.0 \times 10^{-6}} = 3.5.$

5. $2.5 \times 10^{-3}\dfrac{d^2q}{dt^2} + 50\dfrac{dq}{dt} + \dfrac{q}{4.0 \times 10^{-6}} = 2.5e^{-5000t}.$

6. $5.0 \times 10^{-3}\dfrac{d^2q}{dt^2} + \dfrac{q}{8.0 \times 10^{-6}} = 6.3 \sin 2000t.$

7. $0.5\dfrac{d^2y}{dx^2} + 4\dfrac{dy}{dx} + 5y = 6e^{-5x}.$

8. $5\dfrac{d^2y}{dx^2} + 2\dfrac{dy}{dx} + y = 10 \cos 2x.$

9. If q satisfies the differential equation $L\dfrac{d^2q}{dt^2} + R\dfrac{dq}{dt} + \dfrac{q}{C} = E,$ where $L, R,$ $C,$ and E are constant for any one problem, solve the following problems.

(a) If R is doubled, what multiples of $L, C,$ and E must be used to leave q unchanged at all times?

(b) If C is one-fourth its original value, R is twice its original value and L remains unchanged, assuming the oscillating situation, what is the effect on the frequency of the oscillation?

Find the exact solution to each of the following differential equations:

10. $\dfrac{d^2y}{dx^2} + 10\dfrac{dy}{dx} - 144y = 6.$ When $x = 0, y = 1,$ and $\dfrac{dy}{dx} = -\dfrac{1}{3}.$

11. $\dfrac{d^2y}{dx^2} + 10\dfrac{dy}{dx} + 34y = 10.$ When $x = 0, y = -1,$ and $\dfrac{dy}{dx} = 2.$

12. $0.5\dfrac{d^2y}{dt^2} + 6\dfrac{dy}{dt} + 18y = 27.$ When $t = 0, y = 0,$ and $\dfrac{dy}{dt} = 8.$

13. $\dfrac{d^2y}{dx^2} + 3\dfrac{dy}{dx} = 6$. When $x = 0$, $y = 1$, $\dfrac{dy}{dx} = -5$, and therefore $\dfrac{d^2y}{dx^2} = 21$.

(*Hint:* Differentiate both sides.)

14. $0.002\dfrac{d^2q}{dt^2} + 3.2\dfrac{dq}{dt} + \dfrac{q}{5 \times 10^{-4}} = 2.4 \sin 1000t$. When $t = 0$, both q and $\dfrac{dq}{dt}$ are zero.

15. $2.0 \times 10^{-3}\dfrac{d^2i}{dt^2} + 4.0\dfrac{di}{dt} + \dfrac{i}{4.0 \times 10^{-4}} = \dfrac{d}{dt}\,6.5 \sin 1000t$.   When $t = 0$, both i and $\dfrac{di}{dt}$ are zero.

16. $D^2y + 2Dy + y = xe^{-2x}$. When $x = 0$, $y = 3$, and $Dy = -1$.

17. $0.2y'' - y' + 1.2y = e^{2x}$. When $x = 0$, $y = 2$, and $y' = 0$. (*Note:* The particular solution here cannot be ke^{2x}, since the transient solution has a term in e^{2x}. See the note with Exercise 18 following Section 16.4 for the method of finding the particular solution.)

18. $y'' + y = 2 \sin x$. When $x = 0$, $y = 1$, and $y' = 0$. (*Note:* The particular solution cannot be $k \cos x + n \sin x$. See the note with Exercise 18 following Section 16.4.)

19. $(D^3 + D^2 - D - 1)y = 2$. At $x = 0$, $y = -2$, $Dy = 3$, and $D^2y = 2$.

17.4   THE *RLC* CIRCUIT

Consider a circuit consisting of a resistor, an inductor, and a capacitor in series with a voltage source E (Fig. 17-4) and a switching arrangement. At all times when the switch is closed the sum of all voltage drops equals the voltage E.

Therefore
$$L\dfrac{di}{dt} + Ri + \dfrac{q}{C} = E.$$

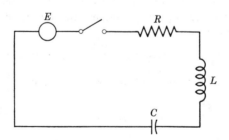

Fig. 17-4.

At the instant the switch is closed ($t = 0$, say) there can be no current, since the inductor prevents any instantaneous jump in current. Therefore, at $t = 0$, the voltage across the resistor is zero. Assuming the capacitor has been previously discharged at $t = 0$, $q = 0$; hence there can be no voltage across the capacitor. Thus at $t = 0$, the only voltage drop is across the inductor, and $L(di/dt) = E$. An instant later, however, the current has increased and charge has flowed to the capacitor, so that there is some voltage across each of the three elements.

First, assume a constant E. From the graph of current against time (Fig. 17-5) we find that at point A, $di/dt = 0$; therefore the voltage across $L [= L(di/dt)]$ has dropped to zero, and the sum of the two voltages across R and C equals E. At point B, the voltage drop across L is negative and the voltage drop across $R (\propto i)$ is less than that at A; therefore the capacitor C is still charging and the voltage drop across C is increasing. If the capacitor is still far from being fully charged, the voltage drop across C increases still more, and under certain conditions the current will follow the graph from B to C. At C, the capacitor is fully charged, and

$$\frac{q}{C} = E, \qquad L\frac{di}{dt} = 0, \qquad Ri = 0.$$

This is a one-surge type of current.

However, if the capacitor is already fully charged at the point B, at this point

$$\frac{q}{C} = E, \qquad \text{and} \quad L\frac{di}{dt} = -Ri.$$

In this case, the shaded area is

$$\int_0^t i\, dt = q = CE,$$

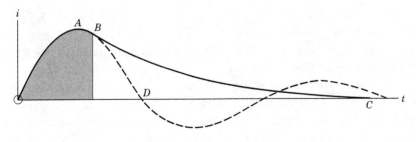

Fig. 17-5.

the full charge on the capacitor. The current (a continuous function, since there is an inductor in the circuit) will become zero at D, and at D the capacitor is overcharged by an amount equivalent to the area under BD. The current now becomes negative, and an oscillating condition is set up, in which the capacitor becomes successively overcharged and undercharged, but the oscillations are damped, so that the current finally drops to zero, with the capacitor fully charged.

We have, then, two possible types of transient currents: the one-surge current and the damped oscillation, and our mathematical techniques should be sufficient to determine which type of transient current will flow in any specific instance.

By analyzing the circuit mathematically,

$$L\frac{di}{dt} + Ri + \frac{q}{C} = E.$$

This is a differential equation, but there are three variables, t, i, and q. However, since $dq/dt = i$, the equation is transformed into an equation with two variables by simply taking a derivative of each side with respect to t.

Therefore

$$L\frac{d^2i}{dt^2} + R\frac{di}{dt} + \frac{i}{C} = 0.$$

The right-hand side has become zero, and an auxiliary equation may be used directly.

The A.E. is $Lm^2 + Rm + i/C = 0$, and

$$m = \frac{-R \pm \sqrt{R^2 - 4L/C}}{2L};$$

if $4L/C > R^2$,

$$m = \frac{-R \pm j\sqrt{4L/C - R^2}}{2L}.$$

We now see mathematically how the current can at times be a single surge and at others a damped oscillation. If $R^2 \geq 4L/C$, the roots are real, resulting in a single-surge current. If $R^2 < 4L/C$, the roots are imaginary, resulting in a damped oscillation.

For real roots and positive values of R, L, and C, all values of m are negative, since $\sqrt{R^2 - 4L/C}$ is less in magnitude than R; the current will therefore consist of decay components. If $R^2 = 4L/C$, we have the borderline condition between a one-surge type of current and a damped oscillation. This is known as *critical damping*. If R is decreased from this critical value, the result is a

damped oscillation. The following tabulation is a summary of the effects
of changing R, all of which may be deduced from a knowledge of the three
types of solution of the differential equation.

R	Type of Transient Current	Effect on Damping	Effect on Frequency
0	Undamped oscillation		
Increasing to critical value	Damped oscillation	Increasing	Decreasing
Critical: $R = \sqrt{4L/C}$	Single surge		Frequency has become 0; (infinite period)
Increasing beyond critical R	Single surge	Increasing	

Thus, mathematics can predict what type of transient current will result
from any specific combination of R, L, and C.

Appendix II gives a summary showing how solvable differential equations
arise from RLC circuits, including charge and discharge, with current, charge,
and voltage equations. Note that for all equations referring to the RLC
series circuit, the pattern of constants on the left-hand side is identical.
Therefore the transient charge q and voltage v will be the same type of
function as the transient current i. For that matter, even with a variable
voltage source, the transient components can still be analyzed as they were
in the table, since the variable voltage source controls only the *sustained*
values of current, charge, and voltage. The solution here is achieved by
superposition.

The initial conditions at the instant of switching ($t = 0$) are controlled by
electrical restrictions such as the following.

(a) An inductor prevents any sudden jump in its current.

(b) A capacitor prevents any sudden jump in voltage across it, etc.

These initial conditions allow us to solve for arbitrary constants and
get an exact solution.

There is a mechanical analogy to the RLC circuit; that is, simple harmonic
motion with the effect of air resistance.

For simple harmonic motion

$$\frac{d^2s}{dt^2} = -ks,$$

which gives an undamped oscillation. This equation can be solved either intuitively or by using an A.E. which is $m^2 = -k$, $m = \pm j\sqrt{k}$,   etc.

Taking the resistance of the spring and air resistance into account, the oscillation is damped. In most cases, the combined spring and air resistance is proportional to velocity and, of course, in an opposite direction.

Now acceleration is always proportional to net force. Here, since acceleration is proportional to net force and the force of spring and air resistance is proportional to velocity, the acceleration will be reduced by an amount proportional to velocity. Therefore

$$\frac{d^2s}{dt^2} = -ks - k_1 \frac{ds}{dt}, \quad \text{or} \quad \frac{d^2s}{dt^2} + k_1 \frac{ds}{dt} + ks = 0.$$

Unless the damping is extremely heavy (that is, k_1 is very high), the result is a damped oscillation.

In both the electrical and mechanical cases, the damping is caused by the resistive element.

EXERCISES

1. Assuming a series RLC circuit, find the critical resistance for each of the following values of L and C:

    (a) $L = 2.0 \times 10^{-3}$; $C = 5.0 \times 10^{-4}$.   (b) $C = L/10$.
    (c) $L = 3.2 \times 10^{-4}$; $C = 1.8 \times 10^{-3}$.

2. For a series RLC circuit, determine in each of the following cases whether the transient electrical quantities are a single surge, a damped oscillation, or an undamped oscillation:

    (a) $R = 50$, $L = 0.008$, and $C = 0.000012$.
    (b) $R = 0$, $L = 2 \times 10^{-2}$, and $C = 4 \times 10^{-4}$.
    (c) $R = 3.2$, $L = 1.28 \times 10^{-3}$, and $C = 5.0 \times 10^{-4}$.
    (d) $R = 12$, $L = 5.1 \times 10^{-4}$, and $C = 9.4 \times 10^{-5}$.

3. Determine the frequency of the oscillation of the transient electrical quantities for each of the following values of R, L, and C in a series RLC circuit:

    (a) $R = 0$, $L = 5.4 \times 10^{-2}$, and $C = 1.5 \times 10^{-3}$.
    (b) $R = 0.4$, and $L = C/16$.
    (c) $R = \sqrt{8.0}$, $L = 3.4 \times 10^{-4}$, and $C = 8.0 \times 10^{-5}$.
    (d) $R = 24$, $L = 1.8 \times 10^{-2}$, and $C = 5.6 \times 10^{-5}$.

By using Appendix II as a guide in setting up the differential equations and initial conditions, find (a) the charge q, and (b) the current i, as functions of time t

measured from the moment of switch opening in a series RLC circuit for each of the following cases:

4. $R_T = 0$, $L = 10^{-3}$, and $C = 10^{-3}$; E is constant at 2.0 volts.

5. $R_T = 10$, $L = 4.0 \times 10^{-3}$, and $C = 2.5 \times 10^{-4}$; E is constant at 7.5 volts.

6. $R_T = 20$, $L = 2.0 \times 10^{-3}$, and $C = 2.0 \times 10^{-5}$; E is constant at 1.8 volts.

7. $R_T = 6.0$, $L = 2.5 \times 10^{-2}$, and $C = 10^{-3}$; E is constant at 6.0 volts.

By using Appendix II for the differential equations and initial conditions, find (a) the current i, and (b) the charge q, as functions of time t measured from the moment of switch closing in a series RLC circuit for each of the following cases:

8. $R = 17$, $L = 2.0 \times 10^{-4}$, and $C = 1.25 \times 10^{-5}$; E is constant at 6.0 volts.

9. $R = 4.0$, $L = 10^{-2}$, and $C = 5.0 \times 10^{-4}$; E is constant at 4.0 volts.

10. $R = 0$, $L = 10^{-2}$, and $C = 4.0 \times 10^{-4}$; $E = 7.5 \sin 1000t$. (*Note:* At $t = 0$, $di/dt = E/L = 0$.)

11. $R = 5.0$, $L = 10^{-2}$, and $C = 2.5 \times 10^{-3}$; $E = 10 \cos 200t$. (*Note:* At $t = 0$, $di/dt = E/L = 10/10^{-2} = 1000$.)

12. $R = 2.0$, $L = 10^{-3}$, and $C = 2.0 \times 10^{-4}$; $E = 8 \cos 1000t - 4 \sin 1000t$.

13. $R = 1.0$, $L = 10^{-3}$, and $C = 8.0 \times 10^{-5}$; $E = 14.5 \sin 5000t$.

When a switch is closed to apply a voltage of E volts to a circuit consisting of a resistance R in series with a parallel combination of an inductance L and a capacitance C, the current i in the inductance is governed by the equation

$$RLC\frac{d^2i}{dt^2} + L\frac{di}{dt} + Ri = E.$$

At $t = 0$, $i = 0$, and $di/dt = 0$. Find i as a function of t for the following conditions:

14. $R = 1.0$, $L = 2.0 \times 10^{-2}$, and $C = 5.0 \times 10^{-3}$; $E = 8.0$.

15. $R = 5.0$, $L = 10^{-2}$, and $C = 10^{-3}$; $E = 13 \cos 200t$.

16. What is the critical resistance for the values of L and C used in Exercise 14? Determine the effect on the transient current of a resistance (a) less than the critical resistance and (b) greater than the critical resistance.

17. In a constant current circuit consisting of a parallel combination of a resistance R, an inductance L, and a capacitance C, the differential equation governing the current i in the inductance on closing the switch is

$$CL\frac{d^2i}{dt^2} + \frac{L}{R}\frac{di}{dt} + i = I,$$

and, at $t = 0$, both i and di/dt are equal to zero. If $R = 2.0$, $L = 4.0 \times 10^{-3}$, $C = 7.5 \times 10^{-4}$, and $I = 2.5$, find i as a function of t.

18. Neglecting resistance, the free end of a spring moves in simple harmonic motion, and when its displacement s from its neutral position is 2.0 cm, its acceleration towards its neutral position is 8.0 cm per second per second. Taking

into account the resistive force, which reduces the acceleration by twice the velocity, set up the differential equation of displacement s, and find s as a function of t, if, at $t = 0$, $s = 7.0$, and the velocity is zero.

17.5   OTHER TECHNIQUES

There are endless examples of differential equations with special techniques for solution that we have not previously covered. However, we will limit the discussion here to two more types, in which the key to the solution is to reduce the second-order differential equations to a first-order differential equation and use methods discussed in Chapter 16.

Term in Dependent Variable Missing

EXAMPLE 7.   Solve the differential equation

$$\frac{d^2y}{dx^2} + 10\frac{dy}{dx} = 2x.$$

Solution: In this case it is easily seen that one straight integration of both sides would be feasible, producing the equation

$$\frac{dy}{dx} + 10y = x^2 + k_1.$$

This may be solved by superposition or by an integrating factor.

$$\text{I.F.} = e^{\int 10\,dx} = e^{10x}.$$

Therefore $e^{10x}\,dy + 10e^{10x}y\,dx = e^{10x}(x^2 + k_1)\,dx.$

$$\int d(ye^{10x}) = \int e^{10x}(x^2 + k_1)\,dx.$$

The right-hand side can be integrated by using integration by parts through two steps. This gives

$$ye^{10x} = e^{10x}\left(\frac{x^2 + k_1}{10} - \frac{x}{50} + \frac{1}{500}\right) + k_2.$$

$$y = \frac{x^2}{10} - \frac{x}{50} + k_2 e^{-10x} + k_3.$$

An easier solution would be to substitute a single letter p (say) for (dy/dx). Therefore

$$\frac{d^2y}{dx^2} = \frac{dp}{dx} \quad \text{and} \quad \frac{dp}{dx} + 10p = 2x;$$

by superposition or an I.F.

$$p = \frac{x}{5} - \frac{1}{50} + Ae^{-10x}.$$

Since $p = dy/dx$, one integration gives

$$y = \frac{x^2}{10} - \frac{x}{50} + k_2 e^{-10x} + k_3.$$

The second method involves less calculation than the first, because the integration is left until the end.

EXAMPLE 8.   A falling body of mass m starting from rest encounters air resistance which is proportional to its velocity. When this velocity is 300 ft per second its acceleration is 30 ft per second per second. Using $g = 32$ ft per second per second, find its velocity and displacement as functions of time.

Solution: Acceleration $\dfrac{d^2s}{dt^2} = \dfrac{\text{net force}}{m}$,

and gravitational force is $32m$. Therefore

$$\frac{d^2s}{dt^2} = \frac{32m - k_1 \dfrac{ds}{dt}}{m} \qquad \text{or} \qquad \frac{d^2s}{dt^2} + k_2 \frac{ds}{dt} = 32.$$

When $ds/dt = 300$, $d^2s/dt^2 = 30$. Therefore $k_2 = 1/150$. Since $ds/dt = v$,

$$\frac{dv}{dt} + \frac{1}{150} v = 32. \qquad \therefore \ v = 4800 + Ae^{-t/150}.$$

Since the body starts from rest, at $t = 0$, $v = 0$. Hence $A = -4800$ and $v = 4800 - 4800e^{-t/150}$.

Note that the body reaches a sustained velocity of 4800 ft per second, at which time the air resistance matches the force of gravity.

$$s = \int v \, dt = 4800 \int (1 - e^{-t/150}) \, dt = 4800(t + 150e^{t/150}) + k_3.$$

Assuming $s = 0$ when $t = 0$, $s = 4800(t + 150e^{-t/150} - 150)$.

Term in Independent Variable Missing

EXAMPLE 9.   Solve the differential equation

$$\frac{d^2y}{dx^2} + 10 = 4y.$$

Solution: As an exercise, solve this using an auxiliary equation. Another technique is to begin as in Example 7 by putting $dy/dx = p$. However, if we continued in the same way as before,

$$\frac{dp}{dx} + 10 = 4y,$$

and three variables appear in the equation, making it impossible to solve directly. To get around this difficulty, we note that

$$\frac{dp}{dx} = \frac{dp}{dy}\frac{dy}{dx} = \frac{dp}{dy}p.$$

Therefore $p\frac{dp}{dy} + 10 = 4y$ (a D.E. with only two variables).

Thus $\int p\,dp = \int (4y - 10)\,dy.$ $\dfrac{p^2}{2} = 2y^2 - 10y + k.$

$$p = 2\sqrt{y^2 - 5y + k_1}.\qquad \frac{dy}{dx} = 2\sqrt{y^2 - 5y + k_1}.$$

$$\int \frac{dy}{\sqrt{(y - \frac{5}{2})^2 + k_2}} = 2\int dx.\qquad \sinh^{-1}\frac{y - \frac{5}{2}}{k_3} = 2x + k_4.$$

Hence $y - \frac{5}{2} = k_3 \sinh(2x + k_4).$ $\therefore\ y = \frac{5}{2} + k_3 \sinh(2x + k_4).$

By using the logarithmic form of the answer to the last integral on the right-hand side, this solution may be put in the form

$$y = \tfrac{5}{2} + Ae^{2x} + Be^{-2x},$$

the solution obtained by using an auxiliary equation originally.

EXAMPLE 10.   Solve the D.E.

$$y\frac{d^2y}{dx^2} = \left(\frac{dy}{dx}\right)^2 - 4.$$

When $x = 1$, $y = 0$; and when $y = 1$, $dy/dx = \sqrt{5}$.

Solution: Since dy/dx is raised to a power, it is reasonable to substitute a single letter for it. Therefore let $dy/dx = p$. Hence $d^2y/dx^2 = dp/dx$.

$$y\frac{dp}{dx} = p^2 - 4\qquad \text{(an equation with 3 variables)}.$$

$$\frac{dp}{dx} = \frac{dp}{dy}\frac{dy}{dx} = p\frac{dp}{dy}.$$

Therefore $yp \dfrac{dp}{dy} = p^2 - 4$ and $\displaystyle\int_{\sqrt{5}}^{p} \dfrac{p \, dp}{p^2 - 4} = \int_{1}^{y} \dfrac{dy}{y}.$

Hence $\tfrac{1}{2} \ln (p^2 - 4) - \tfrac{1}{2} \ln 1 = \ln y - \ln 1;$

$$\tfrac{1}{2} \ln (p^2 - 4) = \ln y.$$

Thus $\ln (p^2 - 4) = 2 \ln y = \ln y^2.$ $p^2 - 4 = y^2.$

$$\left(\frac{dy}{dx}\right)^2 = y^2 + 4 \quad \text{and} \quad \frac{dy}{dx} = \sqrt{y^2 + 4}.$$

Therefore

$$\int_{0}^{y} \frac{dy}{\sqrt{y^2 + 4}} = \int_{1}^{x} dx.$$

$$\sinh^{-1} \frac{y}{2} - \sinh^{-1} 0 = x - 1.$$

$$\frac{y}{2} = \sinh (x - 1) \qquad \text{or} \quad y = 2 \sinh (x - 1) = e^{x-1} - e^{-x+1}.$$

This solution should be checked.

EXERCISES

Reduce the following differential equations to first order and find their solution. Wherever the D.E. may be solved by another method, use this other method as a check.

1. $\dfrac{d^2y}{dx^2} + \dfrac{dy}{dx} = 0.$

2. $\dfrac{d^2y}{dx^2} + \dfrac{dy}{dx} = 10.$

3. $\left(\dfrac{d^2y}{dx^2}\right)^2 + \left(\dfrac{dy}{dx}\right)^2 = 1.$

4. $y\dfrac{d^2y}{dx^2} - \left(\dfrac{dy}{dx}\right)^2 = 0.$

5. $\dfrac{d^2y}{dx^2} - \left(\dfrac{dy}{dx}\right)^2 = 1.$

6. $\dfrac{d^2s}{dt^2} = g - \dfrac{ds}{dt}.$

7. $\dfrac{d^2y}{dx^2} + \dfrac{dy}{dx} = e^x.$

8. $\dfrac{d^2y}{dx^2} + \left(\dfrac{dy}{dx}\right)^2 = \dfrac{dy}{dx}.$ When $x = -\infty$, $y = 0$, and when $x = 0$, $dy/dx = \tfrac{1}{2}.$

9. $\dfrac{d^2y}{dx^2} - 4y = 0.$ When $x = 0$, $y = 1$, and $\dfrac{dy}{dx} = 2.$

10. $\dfrac{d^2y}{dx^2} = e^y \dfrac{dy}{dx}.$ When $x = 2$, $y = 0$, and $\dfrac{dy}{dx} = 1.$

11. $\dfrac{d^2y}{dx^2} + 2\dfrac{dy}{dx} = 4x.$

12. $\dfrac{d^2y}{dt^2} + 2\dfrac{dy}{dt} = 8 \sin 2t.$

13. $\dfrac{d^2y}{dx^2} + 3\dfrac{dy}{dx} = y\dfrac{dy}{dx}$. When $x = -2$, $y = 5$, and $\dfrac{dy}{dx} = 2$.

14. $D^2y = Dy\left(1 + \dfrac{1}{x}\right)$. When $x = 1$, $y = 1$, and $Dy = e$.

15. $xy'' - y' = 3x^2$. When $x = 1$, y and y' are both zero.

16. $xy'' - y' = 3x$. When $x = 1$, y and y' are both zero.

17. $y\dfrac{d^2y}{dx^2} = 2\left(\dfrac{dy}{dx}\right)^2$.   **18.** $\dfrac{d^2y}{dx^2} = 4\dfrac{dy}{dx} + \left(\dfrac{dy}{dx}\right)^3$.

19. $y\dfrac{d^2y}{dx^2} + (\ln y)^{-2}\dfrac{dy}{dx} = 0$. When $x = 0$, $y = e$, and $\dfrac{dy}{dx} = 1$.

20. $x\dfrac{d^2y}{dx^2} = \dfrac{dy}{dx} + \left(\dfrac{dy}{dx}\right)^3$.

21. $y\dfrac{d^2y}{dx^2} = \left(\dfrac{dy}{dx}\right)^2 + \dfrac{y^2}{2}$. When $x = 0$, $y = 1$, and $\dfrac{dy}{dx} = 0$.

MISCELLANEOUS EXERCISES

Solve the following differential equations:

1. $\dfrac{d^2y}{dx^2} = 8$.   **2.** $\dfrac{d^2y}{dx^2} + 4\dfrac{dy}{dx} = 8$.

3. $\dfrac{d^2y}{dx^2} + 4\dfrac{dy}{dx} = 8y$.   **4.** $\dfrac{d^2y}{dx^2} + 4\dfrac{dy}{dx} + 8y = 0$.

5. $\dfrac{d^2y}{dx^2} + 4\dfrac{dy}{dx} + 8y = 8$.   **6.** $\dfrac{d^2y}{dx^2} + 4\dfrac{dy}{dx} = e^{-3x}$.

7. $\dfrac{d^2y}{dx^2} + 2y\dfrac{dy}{dx} = 0$. When $x = 0$, $y = 1$, and $\dfrac{dy}{dx} = -1$.

8. $\dfrac{d^2y}{dx^2} + 4\dfrac{dy}{dx} = e^{-4x}$.   **9.** $\dfrac{d^2y}{dx^2} + 4\dfrac{dy}{dx} = 4x$.

10. $\dfrac{d^2y}{dx^2} - 4y = 10e^{3x}$.   **11.** $\dfrac{d^2y}{dx^2} + 4y = 6.5e^{3x}$.

12. $\dfrac{d^2y}{dx^2} + 9y = 8\sin 2x$.

13. $\dfrac{d^2y}{dx^2} + \sin 2y\dfrac{dy}{dx} = 0$. When $x = 0$, $y = 0$, and $\dfrac{dy}{dx} = 1$.

14. $10^{-1}\dfrac{d^2q}{dt^2} + 200\dfrac{dq}{dt} + 10^5q = 40$. When $t = 0$, q and $\dfrac{dq}{dt}$ are both zero.

15. $0.50\dfrac{d^2i}{dt^2} + 600\dfrac{di}{dt} + \dfrac{i}{2.0 \times 10^{-6}} = \dfrac{d}{dt}60\sin 1000t$. When $t = 0$, $i = 0$, and $\dfrac{di}{dt} = 0$.

16. A bullet comes to rest after penetrating a piece of wood to a depth of 4 in. in 0.002 sec. Assuming constant negative acceleration, find the speed of the bullet on entering the wood.

17. A body of mass m falling from rest encounters air resistance proportional to its velocity and finally reaches a steady-state velocity of 500 ft per second. Assuming gravitational acceleration $g = 32$ ft per second per second find (a) the velocity and (b) the distance of the fall, when the body has fallen for 10 sec.

18. The negative acceleration due to air resistance on a body of mass m falling from rest is $1/800$ times the square of the velocity. Assuming $g = 32$, find the velocity (a) after 10 sec, and (b) after an infinite time.

19. The angular acceleration of a pendulum of small amplitude, neglecting resistance, is negatively proportional to the angular displacement from the vertical. The period of the pendulum governed by this law is 1.0 sec. Resistance adds another component to this acceleration, which, measured in degrees per second per second, is 0.08 times the angular velocity in degrees per second and in a direction opposed to this velocity. If the pendulum is drawn 2.0° from the vertical and released, find its amplitude in degrees after 20 sec.

20. A bullet strikes a block of wood at a velocity of 1600 in. per second and penetrates 6.0 in. before coming to rest. Assuming it is retarded at a rate proportional to the square root of its velocity, find the time elapsed before it comes to rest.

21. A cylindrical log h feet high and with specific gravity 0.50 is held upright, so that its base rests on the surface of a body of water, and then released. Assuming it remains upright, and neglecting friction:

(a) Find the two positions of the log when its velocity is zero.

(b) Find the acceleration at these two positions.

(c) Find the length of log submerged in water as a function of t. Assume $g = 32$.

(*Hint:* The buoyant force is equal to the weight of the water displaced.)

22. If the retardation due to friction in feet per second per second in Exercise 21 is four times the velocity and the height of the log is 4.0 ft:

(a) Find the first time after $t = 0$ when the velocity is zero.

(b) Hence find the length of log submerged when the log is at its lowest point.

CHAPTER 18

Laplace Transforms

18.1 DEFINITION AND USE IN DIFFERENTIAL EQUATIONS

If y is any function of t, the *Laplace transform* of y, designated as $\bar{y}$ or $\overline{f(t)}$ or $\mathscr{L}(y)$, is defined as follows:

$$\bar{y} = \int_0^\infty ye^{-st}\, dt,$$

where s represents any number, the value of which is never specified, for which the definite integral may be calculated as a finite expression. We may consider s as a parameter. Of course x may be used instead of t to represent the independent variable; t is used here because the independent variable for most of the applied problems is time. It is seen that the Laplace transform $\bar{y}$, being a definite integral, will not be a function of t but will be expressed in terms of the parameter s.

As a simple example, consider the function $y = c$, where c is a constant. Then

$$\bar{y} = \int_0^\infty ce^{-st}\, dt = -\frac{c}{s}\left[e^{-st}\right]_0^\infty = -\frac{c}{s}[0 - 1] = \frac{c}{s}.$$

In this case s may be any number greater than zero.

Before we proceed to calculate the Laplace transforms of other functions, it may be well to investigate how a transformation of this type may be of some use. One of the main uses of Laplace transforms is in the solution of differential equations with constant coefficients where certain initial conditions, at $t = 0$, are known. The analysis of transient conditions in electrical circuits produces exactly this type of differential equation, as we have seen

376

in Chapters 16 and 17. The initial conditions are imposed by electrical laws, and they always specify the value of the dependent variable and, where necessary, the values of its successive derivatives at $t = 0$. By using Laplace transforms in the solution of this type of differential equation, the calculations for the solution are almost entirely algebraic. We may see the principles involved from the following analysis.

In general, the design of the integral representing the Laplace transform is chosen so that we may represent the transform of any derivative of a dependent variable in terms of the transform of the dependent variable itself and the initial conditions at $t = 0$. Thus, if the transform of a complete differential equation is taken, a relationship is established involving the transform of the dependent variable, the parameter s, and numerical terms, which may be solved for the transform of the dependent variable. If the transform of the dependent variable is adapted to a combination of standard transforms by algebraic methods, the dependent variable may be expressed as a function of t.

We will now examine in detail the method by which the transform of any derivative of y is expressed in terms of the transform of y itself. In the following calculations and in some of the problems to follow, the symbol D is sometimes used to represent d/dt; similarly, D^2 means d^2/dt^2, D^3y means d^3y/dt^3, etc. We should note that the square, cube, and higher "powers" used in these symbols do not represent powers, but rather indicate the order of the derivative. Other symbols used are:

y_0 to represent the value of y when $t = 0$,
D_0 to represent the value of dy/dt when $t = 0$,
$D_0{}^2$ to represent the value of d^2y/dt^2 when $t = 0$, etc.

From our definition of $\bar{y}$, it follows that

$$\overline{Dy} = \int_0^\infty \frac{dy}{dt} e^{-st} \, dt = \int_0^\infty e^{-st} \, dy.$$

This integral may be calculated by parts, letting $e^{-st} = u$ and $dy = dv$. Hence $-se^{-st} \, dt = du$ and $y = v$.

Thus

$$\overline{Dy} = \left[ye^{-st} \right]_0^\infty + s \int_0^\infty ye^{-st} \, dt.$$

The value of s must be considered limited to a value for which $[ye^{-st}]_0^\infty$ is finite; for example, if $y = e^{4t}$,

$$\left[ye^{-st} \right]_0^\infty = \left[e^{-(s-4)t} \right]_0^\infty,$$

and, unless $s > 4$, the calculation for the upper limit will produce an infinite result, e^∞; however, if $s > 4$, the calculation for the upper limit gives $e^{-\infty} = 0$. In practice, the student need not be concerned about this point of rigor, since there are always *some* values of s for which finite answers are obtained, and these values of s do not affect the results in any way. Thus, for appropriate values of s, (ye^{-st}) always approaches zero as t approaches ∞. The value of ye^{-st} when $t = 0$ is 1 times the value of y when $t = 0$. Moreover, by definition,

$$\int_0^\infty ye^{-st}\, dt = \bar{y}.$$

Therefore $\overline{Dy} = 0 - y_0 + s\bar{y}$, or $\overline{Dy} = s\bar{y} - y_0$. This expresses the transform of dy/dt in terms of the transform of y and the value of y when $t = 0$.

Similarly,
$$\overline{D^2y} = \int_0^\infty D^2ye^{-st}\, dt.$$

Letting $e^{-st} = u$ and $D^2y\, dt = dv$, then

$$-se^{-st}\, dt = du \quad \text{and} \quad Dy = v.$$

Hence

$$\overline{D^2y} = [Dye^{-st}]_0^\infty + s\int_0^\infty Dye^{-st}\, dt.$$

By the same reasoning as that used in the calculation of $\overline{Dy}$, we see that

$$\overline{D^2y} = 0 - D_0 + s\overline{Dy} = s(s\bar{y} - y_0) - D_0,$$

substituting the previously calculated expression for $\overline{Dy}$. Thus

$$\overline{D^2y} = s^2\bar{y} - sy_0 - D_0.$$

From the calculations of $\overline{Dy}$ and $\overline{D^2y}$, it becomes apparent that the transform of a higher derivative may be expressed in terms of the transform of the derivative one order lower, and thus by a step-by-step process it may be expressed in terms of the transform of the original y. In general,

$$\overline{D^ny} = s^n\bar{y} - s^{n-1}y_0 - s^{n-2}D_0 \cdots - sD_0{}^{n-2} - D_0{}^{n-1}.$$

EXAMPLE 1.   Find the Laplace transform of the differential equation

$$D^2y + 8Dy + 25y = 0$$

in terms of $\bar{y}$, if, at $t = 0$, $y = 2$, and $Dy = 1$. Thus find $\bar{y}$.
Solution: The Laplace transform of each separate term may be found, and the transform of the complete left-hand side will be the sum of the

separate transforms. This reasoning is sound, since a transform is an integral, and we may combine transforms of separate terms as we combine integrals of separate terms. For the same reason, the transform of a constant multiple of a variable equals the constant multiple times the transform of the variable; moreover, the transform of zero is zero. Any rules of combination applying to integration may be applied to Laplace transforms.

The transform of the differential equation is therefore

$$\overline{D^2y} + 8\overline{Dy} + 25\bar{y} = 0;$$

$$s^2\bar{y} - sy_0 - D_0 + 8(s\bar{y} - y_0) + 25\bar{y} = 0.$$

Substituting the given values of y_0 and D_0,

$$s^2\bar{y} - 2s - 1 + 8(s\bar{y} - 2) + 25\bar{y} = 0,$$

or

$$\bar{y}(s^2 + 8s + 25) = 2s + 17 \quad \text{and} \quad \bar{y} = \frac{2s + 17}{s^2 + 8s + 25}.$$

The parameter s appears at least once in all Laplace transforms.

EXERCISES

1. (a) If $dy/dt = -8y$, and, at $t = 0$, $y = 3$, find $\bar{y}$.

(b) Solve for y as an exact function of t by using the differential equation and initial condition given in (a).

(c) Use (a) and (b) to deduce the Laplace transform of $y = e^{-8t}$.

(d) Verify your answer to (c) by using the definition of a Laplace transform.

2. If $\dfrac{d^2y}{dx^2} = 4y$, and, at $x = 0$, $y = 0$, and $\dfrac{dy}{dx} = 8$, find $\mathcal{L}(y)$.

3. If $\dfrac{d^2i}{dt^2} = -\omega^2 i$, and, at $t = 0$, $i = 0$, and $\dfrac{di}{dt} = 5\omega$, find $\mathcal{L}(i)$.

4. If, for $y = f(t)$, $(D^2 + 6D + 9)y = 0$, $y_0 = 0$, and $D_0 = 20$, find $\bar{y}$.

5. If $L\dfrac{dv}{dt} + Rv = 0$, and $v_0 = -E$, find $\mathcal{L}(v)$.

6. If $L\dfrac{d^2i}{dt^2} + R\dfrac{di}{dt} + \dfrac{i}{C} = 0$, and, at $t = 0$, $i = 0$, and $\dfrac{di}{dt} = \dfrac{E}{L}$, find $\bar{i}$.

7. If $L\dfrac{d^2q}{dt^2} + R\dfrac{dq}{dt} + \dfrac{q}{C} = 0$, and, at $t = 0$, $q = CV$, and $\dfrac{dq}{dt} = 0$, find $\bar{q}$.

18.2 CALCULATION OF LAPLACE TRANSFORMS

For every function of t (or x, or u ...), there is a corresponding unique function of s representing the Laplace transform. Conversely, for every Laplace transform, there is a corresponding unique original function. Thus, to solve a differential equation with constant coefficients and with specified initial conditions, it is possible to transform every term in the equation and then solve for the transform of the dependent variable, as in Example 1; the remaining step will be to find the original function corresponding to this transform. It is therefore necessary first to calculate the transforms of as many functions as we may need and use the results for a standard reference in the solution of particular problems.

Some of the commonly occurring functions are transformed from first principles in the following paragraphs. A fairly complete table of transforms is included at the end of this section. The calculation of $\bar{y}$, when $y = c$, has already been established as $\bar{y} = c/s$.

The Transform of a Single Power Function. First consider $y = t$.

$$\bar{y} = \int_0^\infty t e^{-st} \, dt$$

$$= -\frac{1}{s} \left[t e^{-st} \right]_0^\infty + \frac{1}{s} \int_0^\infty e^{-st} \, dt \qquad \text{(by parts).}$$

We can show that $\lim_{t \to \infty} [t e^{-st}] = 0$, provided $s > 0$. This is an indeterminate form $\infty \, 0$, but the limit in this case is zero. Hence

$$\bar{y} = 0 - \frac{1}{s^2} \left[e^{-st} \right]_0^\infty = -\frac{1}{s^2} [0 - 1] = \frac{1}{s^2}.$$

Next consider $y = t^2$.

$$\bar{y} = \int_0^\infty t^2 e^{-st} \, dt.$$

Let $t^2 = u$ and $e^{-st} \, dt = dv$. $2t \, dt = du$, $\dfrac{-1}{s} e^{-st} = v$.

Therefore

$$\bar{y} = -\frac{1}{s} \left[t^2 e^{-st} \right]_0^\infty + \frac{2}{s} \int_0^\infty t e^{-st} \, dt.$$

Here also, $\lim_{t \to \infty} [t^2 e^{-st}] = 0$ provided $s > 0$. Thus $\bar{y} = 0 + (2/s)\bar{t}$. From above, $\bar{t} = 1/s^2$. Hence $\bar{y} = 2/s^3$. Similarly, if $y = t^3$, we find that

$$\bar{y} = \frac{3}{s} \bar{t^2} = \frac{3 \times 2}{s^4} = \frac{3!}{s^4}.$$

In general, if $y = t^n$, $\bar{y} = n!/s^{n+1}$. Here s is any positive number.

The Transform of an Exponential. Given $y = e^{-ct}$.

$$\bar{y} = \int_0^\infty e^{-ct}e^{-st}\,dt = \int_0^\infty e^{-(s+c)t}\,dt = -\frac{1}{s+c}(0-1) = \frac{1}{s+c}.$$

This assumes $s + c$ is positive; that is $s > -c$.

The Transforms of cos ωt and sin ωt. Given $y = \cos \omega t$.

$$\bar{y} = \int_0^\infty \cos \omega t e^{-st}\,dt.$$

This may be calculated by parts in two steps but $\bar{y}$ may easily be calculated in terms of the transform of an exponential if the exponential form of cos ωt is used.

$$y = \cos \omega t = \tfrac{1}{2}(e^{j\omega t} + e^{-j\omega t}).$$

By using the fact that the transform of e^{-ct} is $1/(s+c)$,

$$\bar{y} = \frac{1}{2}\left(\frac{1}{s-j\omega} + \frac{1}{s+j\omega}\right) = \frac{\tfrac{1}{2}(s+j\omega+s-j\omega)}{s^2+\omega^2} = \frac{s}{s^2+\omega^2}.$$

Given
$$y = \sin \omega t = \frac{1}{2j}(e^{j\omega t} - e^{-j\omega t}),$$

$$\bar{y} = \frac{1}{2j}\left(\frac{1}{s-j\omega} - \frac{1}{s+j\omega}\right) = \frac{1}{2j}\left(\frac{s+j\omega-s+j\omega}{s^2+\omega^2}\right) = \frac{\omega}{s^2+\omega^2}.$$

The Transform of a Damped Oscillation. Given $y = e^{-ct}\cos \omega t$.

$$y = e^{-ct}\frac{e^{j\omega t} + e^{-j\omega t}}{2} = \tfrac{1}{2}[e^{-(c-j\omega)t} + e^{-(c+j\omega)t}].$$

Thus

$$\bar{y} = \frac{1}{2}\left(\frac{1}{s+c-j\omega} + \frac{1}{s+c+j\omega}\right)$$

$$= \frac{1}{2}\left(\frac{s+c+j\omega+s+c-j\omega}{(s+c)^2+\omega^2}\right)$$

$$= \frac{s+c}{(s+c)^2+\omega^2}.$$

In much the same way we may prove that, if

$$y = e^{-ct}\sin \omega t, \qquad \bar{y} = \frac{\omega}{(s+c)^2+\omega^2}.$$

TABLE OF LAPLACE TRANSFORMS

$$y = f(t). \qquad \bar{y} = \int_0^\infty y e^{-st}\, dt.$$

y	$\bar{y}$	Note
1. c	$\dfrac{c}{s}$	
2. t^n	$\dfrac{n!}{s^{n+1}}$	$n = 0, 1, 2, 3, \ldots$ $0! = 1.$
3. e^{-ct}	$\dfrac{1}{s+c}$	
4. $t^n e^{-ct}$	$\dfrac{n!}{(s+c)^{n+1}}$	$n = 0, 1, 2, 3, \ldots$ $0! = 1.$
5. $\cos \omega t$	$\dfrac{s}{s^2 + \omega^2}$	
6. $\sin \omega t$	$\dfrac{\omega}{s^2 + \omega^2}$	
7. $e^{-ct} \cos \omega t$	$\dfrac{s+c}{(s+c)^2 + \omega^2}$	
8. $e^{-ct} \sin \omega t$	$\dfrac{\omega}{(s+c)^2 + \omega^2}$	
9. $e^{-ct} \cosh at$	$\dfrac{s+c}{(s+c)^2 - a^2}$	
10. $e^{-ct} \sinh at$	$\dfrac{a}{(s+c)^2 - a^2}$	
11. $t \cos \omega t$	$\dfrac{s^2 - \omega^2}{(s^2 + \omega^2)^2}$	
12. $t \sin \omega t$	$\dfrac{2\omega s}{(s^2 + \omega^2)^2}$	
13. $\cosh at$	$\dfrac{s}{s^2 - a^2}$	
14. $\sinh at$	$\dfrac{a}{s^2 - a^2}$	
15. Dy or $\dfrac{dy}{dt}$	$s\bar{y} - y_0$	y_0 is the value of y at $t = 0.$
16. $D^2 y$ or $\dfrac{d^2 y}{dt^2}$	$s^2 \bar{y} - s y_0 - D_0$	D_0 is the value of $\dfrac{dy}{dt}$ at $t = 0.$
17. $D^n y$ or $\dfrac{d^n y}{dt^n}$	$s^n \bar{y} - s^{n-1} y_0 - s^{n-2} D_0 \cdots - D_0^{\,n-1}$	

EXERCISES

1. Establish the validity of transforms 9, 10, 13 and 14 in the table of Laplace transforms.

2. Find the Laplace transform of $y = \sin(\omega t + \pi/6)$. (*Hint:* Express y as the sum of a sine and a cosine.)

For the following differential equations, find the transform of the whole equation, and then calculate the transform of the dependent variable. Make use of the table where necessary.

3. $10^{-2}\dfrac{di}{dt} + 3i = 8$; at $t = 0$, $i = 0$.

4. $CR\dfrac{dv}{dt} + v = E\sin\omega t$; at $t = 0$, $v = 0$.

5. $\dfrac{dy}{dx} = 20y + x^4$; $y_0 = 2$.

6. $\dfrac{dy}{dx} + 3y = e^{-3x} + 2e^{-5x}x^2$; $y_0 = 0$.

7. $L\dfrac{d^2q}{dt^2} + R\dfrac{dq}{dt} + \dfrac{q}{C} = 3\sin\omega t + 4\cos\omega t$; $q_0 = 0$ and $D_0 = 0$.

18.3 DETERMINATION OF A FUNCTION FROM ITS TRANSFORM

The problem of finding $\bar{y}$ from a given differential equation is usually not too difficult. However, knowing $\bar{y}$, the problem of finding y usually presents difficulties. In general, $\bar{y}$ has to be manipulated to put it into one or more of the standard forms listed in the table. The problem is entirely algebraic, and it may be illustrated by a few examples.

EXAMPLE 2.   Find y, if

$$\bar{y} = \frac{3}{(s + 20)^4} + \frac{2s + 5}{3s^2 + 300}.$$

Solution: The table of Laplace transforms will be used throughout.
The first fraction corresponds closely to transform 4 in the table, with $c = 20$ and $n + 1 = 4$; the numerator must be $n! = 3! = 6$ to make it completely standard. It may therefore be expressed as

$$\frac{1}{2}\left[\frac{3!}{(s + 20)^4}\right]$$

where the expression within the large brackets is a standard transform.

The second fraction is a combination of transforms 5 and 6, and it must be split into the form

$$\frac{2s}{3s^2 + 300} + \frac{5}{3s^2 + 300}.$$

In the standard forms of the transforms each denominator must begin with s^2, and each numerator is dependent on one of the terms in the denominator. These two fractions are therefore made to fit the standard forms by making numerical factors into coefficients of the fractions as follows:

$$\frac{2s}{3s^2 + 300} + \frac{5}{3s^2 + 300} = \frac{2}{3}\left[\frac{s}{s^2 + 10^2}\right] + \frac{5}{3}\left[\frac{1}{s^2 + 10^2}\right].$$

Then the last fraction is once again manipulated into

$$\frac{5}{30}\left[\frac{10}{s^2 + 10^2}\right].$$

The quantities within square parentheses are exact transforms. Therefore

$$\bar{y} = \frac{1}{2}\left[\frac{3!}{(s + 20)^4}\right] + \frac{2}{3}\left[\frac{s}{s^2 + 10^2}\right] + \frac{1}{6}\left[\frac{10}{s^2 + 10^2}\right].$$

Now y is determined from the table by finding an original function for each transform. Assuming $y = f(t)$,

$$y = \tfrac{1}{2}t^3e^{-20t} + \tfrac{2}{3}\cos 10t + \tfrac{1}{6}\sin 10t.$$

EXAMPLE 3.   Find q if

$$\bar{q} = \frac{500}{s(s^2 + 100s + 3400)}.$$

Solution: No standard transform has more than one factor in its denominator, other than a power of one unique factor. Therefore the method of splitting into partial fractions may often have to be used.

$$\bar{q} = \frac{5/34}{s} + \frac{As + B}{s^2 + 100s + 3400},$$

the first numerator being determined by cover-up rule. Hence

$$\bar{q} = \frac{(5/34)s^2 + (500/34)s + 500 + As^2 + Bs}{s(s^2 + 100s + 3400)}.$$

Equating coefficients of powers of s in the numerators of the two expressions for $\bar{q}$, $\frac{5}{34} + A = 0$; $A = -\frac{5}{34}$. $500/34 + B = 0$;

$B = -500/34$. The numerical terms are each 500, which is a check on the accuracy of the first calculation by cover-up rule. Thus

$$\bar{q} = \frac{5/34}{s} - \frac{5}{34} \frac{s + 100}{s^2 + 100s + 3400}.$$

The final fraction here is now made to fit standard transforms 7 and 8 in the table by using the method of completing the square and splitting the fraction appropriately. Therefore

$$\bar{q} = \frac{5/34}{s} - \frac{5}{34} \frac{s + 100}{(s + 50)^2 + 30^2}$$

$$= \frac{5/34}{s} - \frac{5}{34} \frac{s + 50}{(s + 50)^2 + 30^2} - \frac{5}{34} \frac{50}{(s + 50)^2 + 30^2}$$

$$= \frac{5/34}{s} - \frac{5}{34} \frac{s + 50}{(s + 50)^2 + 30^2} - \frac{5}{34} \frac{5}{3} \frac{30}{(s + 50)^2 + 30^2}.$$

By using the table to reverse the transforms,

$$q = \frac{5}{34} \left(1 - e^{-50t} \cos 30t - \frac{5e^{-50t}}{3} \sin 30t \right).$$

EXAMPLE 4.   Find i, if

$$\bar{i} = \frac{3}{(s + 5)^4} + \frac{2s}{(s + 5)^2} + \frac{3s + 16}{s(s^2 + 64)}.$$

Solution: The student may be tempted to take a common denominator. If this is done, it will be necessary to undo it later by splitting into partial fractions. The quickest procedure is therefore to treat each of the separate fractions individually. The first fraction may be expressed as

$$\frac{1}{2} \left[\frac{3!}{(s + 5)^4} \right],$$

fitting transform 4. The second fraction is

$$\frac{A}{(s + 5)^2} + \frac{B}{s + 5} = \frac{As + 5A + B}{(s + 5)^2};$$

$$A = 2 \quad \text{and} \quad 5A + B = 0. \quad B = -10.$$

The third fraction is

$$\frac{\frac{1}{4}}{s} + \frac{Cs + D}{s^2 + 64} = \frac{s^2/4 + 16 + Cs^2 + Ds}{s(s^2 + 64)};$$

$$\tfrac{1}{4} + C = 0. \quad C = -\tfrac{1}{4}. \quad D = 3.$$

Therefore $\bar{i}$ may be expressed as

$$\frac{1}{2}\left[\frac{3!}{(s+5)^4}\right] + \frac{2}{(s+5)^2} - \frac{10}{s+5} + \frac{\frac{1}{4}}{s} - \frac{1}{4}\left(\frac{s}{s^2+8^2}\right) + \frac{3}{8}\left(\frac{8}{s^2+8^2}\right).$$

$$i = \tfrac{1}{2}t^3 e^{-5t} + 2te^{-5t} - 10e^{-5t} + \tfrac{1}{4} - \tfrac{1}{4}\cos 8t + \tfrac{3}{8}\sin 8t.$$

EXERCISES

Assuming that the dependent variable is a function of t, find the expression for the dependent variable, being given its transform, in the following exercises.

1. $\bar{y} = \dfrac{1}{s^3}.$

2. $\bar{q} = \dfrac{1}{(2s+5)^2}.$

3. $\bar{i} = \dfrac{s+6}{2s^2+10}.$

4. $\bar{y} = \dfrac{2s-5}{s(s+2)}.$

5. $\bar{v} = \dfrac{3s+4}{s^3+2s}.$

6. $\bar{v} = \dfrac{3s+4}{s^3+2s^2}.$

7. $\bar{y} = \dfrac{s+5}{s(2s+7)(s+1)}.$

8. $\bar{y} = \dfrac{5}{(s+1)^3} + \dfrac{s-2}{(s+1)^2}.$

9. $\bar{i} = \dfrac{2s}{s^2+6s+25}.$

10. $\bar{y} = \dfrac{s+2}{s^2+6s+6}.$

11. $\bar{v} = \dfrac{s^2}{(s+1)^2(s+2)}.$

12. $\bar{y} = \dfrac{s+1}{s^2(s^2+4)}.$

13. $\bar{y} = \dfrac{3}{s^2(s^2+4s+13)}.$

14. $\bar{i} = \dfrac{2}{s^4+s}.$

15. $\bar{y} = \dfrac{4}{(s^2+25)(s^2+100)}.$

16. $\bar{y} = \dfrac{4}{(s^2+25)(s^2+100)} + \dfrac{10}{s^2+25}.$

17. $\bar{q} = \dfrac{2s}{(s^2+25)(s^2+100)}.$

18. $\bar{v} = \dfrac{4}{(s^2+25)(s^2+4s+13)} + \dfrac{2s}{s^2+4s+13}.$

18.4   SOLUTION OF DIFFERENTIAL EQUATIONS BY LAPLACE TRANSFORMS

We may now combine the facts in the foregoing sections and obtain a complete picture. The Laplace transform method of solving differential equations may be used if all the derivatives and the dependent variable in the differential equation have constant coefficients and also if the number of specified conditions at t (or x, or $u \ldots$) $= 0$ matches the order of the differential equation, the specified conditions being the value(s) of the dependent variable and its successive derivatives at $t = 0$.

The method of solution by Laplace transforms may then be summarized as follows:

1. Transform the whole equation, term by term, by using the table.
2. Solve for the transform of the dependent variable.
3. Put this expression into a combination of standard transforms.
4. Use the table to find the original function for each transform successively.

EXAMPLE 5.   Solve the differential equation

$$10^{-2} \frac{di}{dt} + 2i = 3 \sin 200t; \quad \text{at } t = 0, \, i = 0.$$

Solution: Transforming the whole equation,

$$10^{-2}(s\bar{i} - 0) + 2\bar{i} = \frac{3 \times 200}{s^2 + 200^2}.$$

$$\bar{i} = \frac{6 \times 10^4}{(s + 200)(s^2 + 200^2)}$$

$$= \frac{\frac{3}{4}}{s + 200} + \frac{As + B}{s^2 + 200^2}.$$

Solve for A and B by conventional methods, then

$$\bar{i} = \frac{3/4}{s + 200} + \frac{-3s/4 + 150}{s^2 + 200^2},$$

or

$$\bar{i} = \frac{\frac{3}{4}}{s + 200} - \frac{3}{4}\left(\frac{s}{s^2 + 200^2}\right) + \frac{3}{4}\left(\frac{200}{s^2 + 200^2}\right).$$

$$i = \tfrac{3}{4}(e^{-200t} - \cos 200t + \sin 200t).$$

EXAMPLE 6.   Solve the differential equation

$$\frac{d^2q}{dt^2} + 100 \frac{dq}{dt} + 2500q = 50; \quad \text{at } t = 0, \, q = 0 \text{ and } dq/dt = 0.$$

Solution: Taking transforms,

$$s^2\bar{q} - 0 + 100s\bar{q} - 0 + 2500\bar{q} = \frac{50}{s}.$$

$$\bar{q} = \frac{50}{s(s^2 + 100s + 2500)} = \frac{50}{s(s + 50)^2}$$

$$= \frac{\frac{1}{50}}{s} + \frac{A}{(s + 50)^2} + \frac{B}{s + 50}$$

$$= \frac{\frac{1}{50}(s^2 + 100s + 2500) + As + Bs^2 + 50Bs}{s(s + 50)^2}.$$

$$B = -\tfrac{1}{50}; \qquad A = -1.$$

$$\bar{q} = \frac{1}{50}\left[\frac{1}{s} - \frac{50}{(s + 50)^2} - \frac{1}{s + 50}\right].$$

$$q = \frac{1}{50}(1 - 50te^{-50t} - e^{-50t}).$$

The transform of

$$(aD^2 + bD + c)y = f(t),$$

with the terms involving initial values transposed, is

$$(as^2 + bs + c)\bar{y} = \overline{f(t)} + a(sy_0 + D_0) + by_0.$$

This equation, known as the subsidiary equation, may be used directly, instead of transforming each separate derivative. Its left-hand side has a close similarity to the left-hand side of the differential equation.

EXERCISES

1. Which of the following differential equations can be solved directly by using Laplace transforms? Give reasons where answers are negative.

(a) $\dfrac{dy}{dt} - 8y = e^{-2t}$.

(b) $\dfrac{dy}{dt} - 8y = e^{-2t}$; at $t = 1$, $y = -3$.

(c) $\dfrac{d^2y}{dt^2} + 3ty = 2\sin 4t$; $y_0 = 0$ and $D_0 = 0$.

(d) $t\dfrac{d^2y}{dt^2} + 3ty = t^3e^{-5t}$; at $t = 0$, y and $\dfrac{dy}{dt}$ are both 1.

Solve the following differential equations by Laplace transforms:

2. $2 \times 10^{-3} \dfrac{di}{dt} + 4i = 0$; at $t = 0$, $i = 20$.

3. $L \dfrac{di}{dt} + Ri = 0$; at $t = 0$, $i = \dfrac{E}{R}$.

4. $\dfrac{dq}{dt} + 500q = 10$; at $t = 0$, $q = 0$.

5. $\dfrac{dy}{dx} - 10y = 3e^{-4x}$; at $x = 0$, $y = 2$.

6. $\dfrac{di}{dt} + 100i = 200e^{-100t}$; at $t = 0$, $i = 0$.

7. $4 \dfrac{di}{dt} + 600i = 1000 (\sin 50t + \cos 50t)$; at $t = 0$, $i = 0$.

8. $10^{-2} \dfrac{dq}{dt} + 5q = 2 \sin (400t + \pi/3)$; at $t = 0$, $q = 0$.

9. $4 \dfrac{dy}{dx} + 20y = 180xe^{-20x}$; at $x = 0$, $y = 2$.

10. $4 \dfrac{dy}{dx} + 20y = 40 + 5e^{-5x}$; at $x = 0$, $y = 2$.

11. $3 \dfrac{d^2y}{dt^2} + 20 \dfrac{dy}{dt} + 40y = 0$; at $t = 0$, y and $\dfrac{dy}{dt}$ are both zero.

12. $\dfrac{d^2y}{dt^2} + 5 \dfrac{dy}{dt} - 14y = 0$; at $t = 0$, $y = 0$, and $\dfrac{dy}{dt} = 6$.

13. $\dfrac{d^2q}{dt^2} + 50 \dfrac{dq}{dt} + 600q = 20$; at $t = 0$, q and $\dfrac{dq}{dt}$ are both zero.

14. $\dfrac{d^2q}{dt^2} + 50 \dfrac{dq}{dt} + 625q = 20$; at $t = 0$, q and $\dfrac{dq}{dt}$ are both zero.

15. $(D^2 + 80D + 2000)i = 0$; at $t = 0$, $i = 0$, and $\dfrac{di}{dt} = 1000$.

16. $(D^2 + 80D + 1400)i = 0$; at $t = 0$, $i = 0$, and $\dfrac{di}{dt} = 1000$.

17. $\dfrac{d^2\theta}{dt^2} + 2\pi \dfrac{d\theta}{dt} + 5\pi^2\theta = 0$; at $t = 0$, $\theta = 0.10$, and $\dfrac{d\theta}{dt} = 0$.

18. $L \dfrac{d^2q}{dt^2} + \dfrac{q}{C} = 5 \sin 100t$; at $t = 0$, q and $\dfrac{dq}{dt}$ are both zero. $L = 5 \times 10^{-3}$,

and $C = 5 \times 10^{-3}$.

19. $L\dfrac{d^2i}{dt^2} + R\dfrac{di}{dt} + \dfrac{i}{C} = -6000 \sin 2000t$; at $t = 0$, $i = 0$, and $\dfrac{di}{dt} = 750$.
$L = 4 \times 10^{-3}$, $R = 5$, and $C = 10^{-3}$.

20. $RLC\dfrac{d^2i}{dt^2} + L\dfrac{di}{dt} + Ri = 2\cos 10{,}000t + 4$; at $t = 0$, i and $\dfrac{di}{dt}$ are both zero.
$R = 5$, $L = 2 \times 10^{-3}$, and $C = 2 \times 10^{-5}$.

21. $(D^3 + 200D^2 + 50{,}000D)v = 0$; $v_0 = 0$, $D_0 = 0$, $D_0{}^2 = 400$.

CHAPTER 19

Fourier Analysis

19.1 WAVE ANALYSIS

The mathematical series known as the *Fourier series* deals with periodic functions. A *periodic function* is one in which the graph for one interval of time known as the *period* repeats itself identically in successive equal intervals of time. The most familiar periodic functions are the sine and cosine functions. Thus, $\sin 100\pi t$ completes a cycle in $\frac{1}{50}$ sec (from $100\pi t = 0$ radians to $100\pi t = 2\pi$ radians) and then repeats this cycle in each successive period of $\frac{1}{50}$ sec.

Currents and voltages are very often periodic, but seldom are they perfect sine or cosine functions. The most perfectly designed a-c generators may generate a voltage that is very close to being sinusoidal, but some harmonics will always be present to give slight distortions. The resulting voltage will be periodic, but not exactly sinusoidal. There are, of course, voltages generated in square waves, sawtooth waves, half-rectified sine waves, etc., all of which are periodic but not even approximately sinusoidal.

Figure 19-1 represents the graph of a voltage consisting of a fundamental of 10 sin θ volts and a third harmonic (that is, a voltage whose frequency is three times the frequency of the fundamental) with 2-volt amplitude. The graphs of the fundamental and the third harmonic are shown separately, and the composite graph is plotted simply by superimposing the ordinate of the graph of the third harmonic on the ordinate of the graph of the fundamental for all values of θ. We see that the resulting graph is periodic with a period equal to that of the fundamental, and that it roughly approximates a sine wave.

This process of combining two or more periodic waves into a single wave form is referred to as harmonic synthesis. The main concern of the work in this chapter is a reversal of this process, namely the sorting out of the various components of a single periodic wave into a d-c component, a fundamental, and the harmonics, a process which is known as harmonic analysis, or wave analysis.

Before we proceed to the mathematics involved in harmonic analysis, we must evaluate several definite integrals that are involved in the proof of the rules of harmonic analysis. These definite integrals follow.

$$\int_k^{k+2\pi} \sin m\theta \, d\theta = 0.$$

$$\int_k^{k+2\pi} \cos m\theta \, d\theta = 0.$$

$$\int_k^{k+2\pi} \sin m\theta \sin n\theta = 0.$$

$$\int_k^{k+2\pi} \cos m\theta \cos n\theta \, d\theta = 0.$$

$$\int_k^{k+2\pi} \sin m\theta \cos n\theta \, d\theta = 0.$$

m and n are positive integers, $m \neq n$, and k is a constant (positive, negative, or zero). Thus the interval between lower and upper limits is 2π radians.

$$\int_k^{k+2\pi} \sin m\theta \cos m\theta \, d\theta = 0.$$

$$\int_k^{k+2\pi} \sin^2 m\theta \, d\theta = \pi.$$

$$\int_k^{k+2\pi} \cos^2 m\theta \, d\theta = \pi.$$

To summarize, the definite integral over an interval of $\theta = 2\pi$ of the sine or the cosine of a whole multiple of θ or the product of the sine and cosine in any two-factor combination is always equal to zero, except for the integral of the $\sin^2$ or $\cos^2$ of the multiple angle, both of which have a value of π.

These integrals are readily evaluated. The proof of one of the more complicated of these is as follows:

$$\int_k^{k+2\pi} \sin m\theta \sin n\theta \, d\theta = \frac{1}{2}\int_k^{k+2\pi} [\cos(m-n)\theta - \cos(m+n)\theta] \, d\theta$$

$$= \frac{1}{2}\left[\frac{\sin(m-n)\theta}{m-n} - \frac{\sin(m+n)\theta}{m+n}\right]_k^{k+2\pi}$$

$$= \frac{1}{2}\left\{\frac{\sin[(m-n)(k+2\pi)] - \sin[(m-n)k]}{m-n}\right.$$

$$\left. - \frac{\sin[(m+n)(k+2\pi)] - \sin[(m+n)k]}{m+n}\right\}.$$

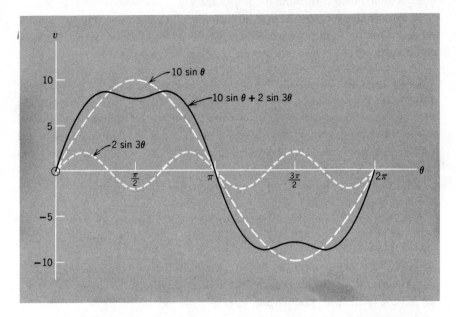

Fig. 19-1. Periodic voltage consisting of a fundamental and a third harmonic.

Since m and n both represent positive integers, $m - n$ is a positive or negative integer and $m + n$ is a positive integer. Thus

$$\sin (m - n)(k + 2\pi) = \sin [(m - n)k + (m - n)2\pi],$$

and since $(m - n)2\pi$ is some whole multiple of one complete cycle,

$$\sin [(m - n)k + (m - n)2\pi] = \sin (m - n)k.$$

Thus the numerator of the first fraction in the braces equals zero. Similarly, the numerator of the second fraction equals zero. Hence

$$\int_{k}^{k + 2\pi} \sin m\theta \sin n\theta \, d\theta = 0.$$

The evaluation of the other six definite integrals in this list is left to the student.

19.2 THE FOURIER SERIES

Any single-valued periodic function of θ with no discontinuity or only a finite number of finite discontinuities may be expressed as the sum of the following components:

1. A constant component, usually called the d-c component, which may be any positive or negative constant or zero.

2. A fundamental sinusoidal component.

3. An infinite series of sinusoidal harmonics (that is, a second, third, ..., ∞th harmonic). For certain functions, some or all of these harmonics may not be present (e.g. the sine function itself has no harmonics).

Since a single sine function with a phase displacement may always be expressed as the sum of a sine function and a cosine function with no phase displacement, it is usually more convenient to find the fundamental and all harmonic components as the sum of a sine and a cosine function with no phase displacement; for example, $K \sin (3\theta + \phi)$ would be calculated originally as $k_1 \sin 3\theta + k_2 \cos 3\theta$.

Thus, a periodic function $f(\theta)$ may be expressed in the form

$$a_0 + a_1 \cos \theta + a_2 \cos 2\theta + a_3 \cos 3\theta + \cdots$$
$$+ b_1 \sin \theta + b_2 \sin 2\theta + b_3 \sin 3\theta \cdots.$$

a_0 is the d-c component, a_1 is the amplitude of the fundamental cosine component, a_2 is the amplitude of the second cosine harmonic, b_m is the amplitude of the mth sine harmonic, etc.

We now proceed to find $a_0, a_1, a_2, \ldots$ and $b_1, b_2, \ldots$. The method used to solve for these coefficients is somewhat similar to that used in the Maclaurin series of Section 15-3, in that, by certain mathematical manipulations, we successively isolate each term on the right-hand side, making all other terms vanish. These manipulations depend on the values of the eight definite integrals that have previously been established in this section.

To isolate the term in a_0, it is only necessary to integrate both sides of the equation between limits of k and $k + 2\pi$. Thus

$$\int_k^{k+2\pi} f(\theta)\, d\theta = \int_k^{k+2\pi} a_0\, d\theta + \int_k^{k+2\pi} a_1 \cos \theta\, d\theta$$

$$+ \int_k^{k+2\pi} a_2 \cos 2\theta\, d\theta + \cdots + \int_k^{k+2\pi} b_1 \sin \theta\, d\theta + \int_k^{k+2\pi} b_2 \sin 2\theta\, d\theta \cdots.$$

All definite integrals after the first term on the right-hand side are zero, and therefore

$$\int_k^{k+2\pi} f(\theta)\, d\theta = a_0 \int_k^{k+2\pi} d\theta = a_0[\theta]_k^{k+2\pi} = a_0(k + 2\pi - k) = 2\pi a_0.$$

$$a_0 = \frac{\int_k^{k+2\pi} f(\theta)\, d\theta}{2\pi}.$$

We see that the value of the d-c component a_0 is the mean value of the function over one cycle starting at any point in the cycle.

To isolate any one of the succeeding terms on the right-hand side, we note that, in the preceding list of eight integrals, only those involving the integration of $\sin^2 m\theta$ and $\cos^2 m\theta$ have a numerical value other than zero. Hence, to isolate the term in $\cos m\theta$, we may multiply both sides by $\cos m\theta$ and then integrate both sides over a complete cycle.

Thus, to solve for a_1,

$$\int_k^{k+2\pi} f(\theta) \cos\theta\, d\theta = a_0 \int_k^{k+2\pi} \cos\theta\, d\theta + a_1 \int_k^{k+2\pi} \cos^2\theta\, d\theta$$

$$+ a_2 \int_k^{k+2\pi} \cos\theta \cos 2\theta\, d\theta + \cdots$$

$$+ b_1 \int_k^{k+2\pi} \cos\theta \sin\theta\, d\theta + b_2 \int_k^{k+2\pi} \cos\theta \sin 2\theta\, d\theta \cdots .$$

All definite integrals on the right-hand side are zero except the term in a_1, which equals $a_1\pi$. Therefore

$$a_1\pi = \int_k^{k+2\pi} f(\theta) \cos\theta\, d\theta \quad \text{and} \quad a_1 = \frac{1}{\pi} \int_k^{k+2\pi} f(\theta) \cos\theta\, d\theta.$$

Similarly, to solve for a_m, the coefficient of $\cos m\theta$, we should multiply both sides of the equation by $\cos m\theta$ and then integrate both sides over a complete cycle; the only term on the right-hand side having a value other than zero is $a_m \int_k^{k+2\pi} \cos^2 m\theta\, d\theta$, which equals $a_m\pi$. Hence

$$a_m = \frac{1}{\pi} \int_k^{k+\pi} f(\theta) \cos m\theta\, d\theta.$$

To solve for b_m, we should multiply both sides of the original equation by $\sin m\theta$ and then take the integral of both sides over a complete cycle; the resulting equation is

$$\int_k^{k+2\pi} f(\theta) \sin m\theta\, d\theta = b_m\pi,$$

$$b_m = \frac{1}{\pi} \int_k^{k+2\pi} f(\theta) \sin m\theta\, d\theta.$$

In summary, any single-valued periodic function of θ with only a finite number of finite discontinuities, may be expressed by the Fourier series

$$f(\theta) = a_0 + a_1 \cos\theta + a_2 \cos 2\theta + \cdots + a_m \cos m\theta \cdots \text{ to } \infty$$
$$+ b_1 \sin\theta + b_2 \sin 2\theta + \cdots + b_m \sin m\theta \cdots \text{ to } \infty.$$

where

$$a_0 = \frac{1}{2\pi} \int_k^{k+2\pi} f(\theta) \, d\theta,$$

$$a_m = \frac{1}{\pi} \int_k^{k+2\pi} f(\theta) \cos m\theta \, d\theta,$$

and

$$b_m = \frac{1}{\pi} \int_k^{k+2\pi} f(\theta) \sin m\theta \, d\theta.$$

The corresponding cosine and sine terms, such as $a_m \cos m\theta + b_m \sin m\theta$ may be combined into a single sine function of the form $k \sin (m\theta + \phi)$ by the methods explained in Section 9.1.

The question arises whether the function is completely represented by this series. May there not be some terms in the cosine and sine of fractional multiples of θ? It may be assumed that the Fourier series represents the function completely, provided there are no discontinuities or only a finite number of finite discontinuities per cycle.

19.3   ANALYSIS OF SQUARE WAVES

A square wave of voltage or current may be considered as one in which the current or voltage over separate intervals of time is constant, changing instantaneously from one constant value to another in the course of the cycle.

EXAMPLE 1.   Find the d-c, fundamental, and harmonic components of the square wave of Fig. 19-2.
Solution: In order to apply the foregoing methods for determining the coefficients, it is always necessary to establish the equation of the graph, or, in this instance, to express i as a function of θ. For the current of Fig. 19-2, there is a finite discontinuity at $\theta = \pi$, and the equation from $\theta = 0$ to $\theta = \pi$ is $i = A$; from $\theta = \pi$ to $\theta = 2\pi$ the equation is $i = -A$. When it is necessary to integrate this function over a complete cycle from $\theta = 0$ to $\theta = 2\pi$, the integral must be expressed as the sum of two integrals, $\theta = \pi$ being the upper limit of the first integral and the lower limit of the second integral.

The d-c component, a_0, is therefore

$$\frac{1}{2\pi} \left(\int_0^\pi A \, d\theta + \int_\pi^{2\pi} -A \, d\theta \right) = \frac{A}{2\pi} (\pi - 0 - 2\pi + \pi)$$

$$= 0.$$

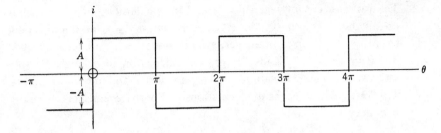

Fig. 19-2. A square wave $i = A$ for one half-cycle and $i = -A$ for the second half-cycle.

More directly, the d-c component is the mean value of i over the complete cycle, and since $i = A$ for one half-cycle and $i = -A$ for the other half-cycle, the mean value $= 0$.

To evaluate the amplitude of the fundamental cosine component by using the rules, we see that the function to be integrated is not i, but $i \cos \theta$; hence

$$a_1 = \frac{1}{\pi} \left(\int_0^\pi A \cos \theta \, d\theta - \int_\pi^{2\pi} A \cos \theta \, d\theta \right)$$

$$= \frac{A}{\pi} \{ [\sin \theta]_0^\pi - [\sin \theta]_\pi^{2\pi} \}$$

$$= \frac{A}{\pi} (0 - 0 - 0 + 0) = 0.$$

$$a_m = \frac{1}{\pi} \left(\int_0^\pi A \cos m\theta \, d\theta - \int_\pi^{2\pi} A \cos m\theta \, d\theta \right)$$

$$= \frac{A}{\pi} \left\{ \left[\frac{\sin m\theta}{m} \right]_0^\pi - \left[\frac{\sin m\theta}{m} \right]_\pi^{2\pi} \right\}$$

$$= \frac{A}{\pi} (0) = 0. \text{ (Since } m \text{ is a positive integer.)}$$

Hence the wave contains no cosine components.

$$b_1 = \frac{1}{\pi} \left(\int_0^\pi A \sin \theta \, d\theta - \int_\pi^{2\pi} A \sin \theta \, d\theta \right)$$

$$= \frac{A}{\pi} \{ [-\cos \theta]_0^\pi + [\cos \theta]_\pi^{2\pi} \}$$

$$= \frac{A}{\pi} [-(-1 - 1) + (1 - (-1))] = \frac{4A}{\pi}.$$

$$b_m = \frac{A}{\pi} \left\{ \left[-\frac{\cos m\theta}{m} \right]_0^\pi + \left[\frac{\cos m\theta}{m} \right]_\pi^{2\pi} \right\}.$$

The value of cos $m\theta$ when $\theta = \pi$ and m is a positive integer depends on whether m is even or odd. If m is even, cos $m\pi = 1$, and, if m is odd, cos $m\pi = -1$. If m is an integer, even or odd, the value of cos $m\theta$ when $\theta = 0$ or 2π is $+1$. Thus the general calculation of b_m must be split into two separate calculations, one for an even m and one for an odd m. This very often happens in calculations of Fourier coefficients. If m is even,

$$b_m = \frac{A}{m\pi}[-(1-1) + (1-1)] = 0.$$

If m is odd,

$$b_m = \frac{A}{m\pi}\{-[-1-1] + [1-(-1)]\} = \frac{4A}{m\pi}.$$

This calculation applies if $m = 1$, and b_1 does not have to be calculated independently, as shown above, but may be treated as the value of b_m when $m = 1$. Now we see that $b_3 = 4A/3\pi$, $b_5 = 4A/5\pi \cdots$. The Fourier series is therefore

$$i = \frac{4A}{\pi}\left(\sin\theta + \frac{\sin 3\theta}{3} + \frac{\sin 5\theta}{5} + \frac{\sin 7\theta}{7}\cdots\right).$$

It may seem incredible that the sum of an infinite number of functions whose graphs are continuous curves can result in a function whose graph consists of straight lines. It may therefore be worthwhile to calculate a harmonic synthesis of the series in order to be convinced of its validity. At any value of θ within the range from 0 to π radians, the sum of the series should equal A, the value of the original function within this range. At $\theta = \pi/6$, for instance, the series is

$$\frac{4A}{\pi}\left(\sin 30° + \frac{\sin 90°}{3} + \frac{\sin 150°}{5} + \frac{\sin 210°}{7}\cdots\right)$$

$$= \frac{4A}{\pi}\left(\frac{1}{2} + \frac{1}{3} + \frac{\frac{1}{2}}{5} - \frac{\frac{1}{2}}{7} - \frac{1}{9} - \frac{\frac{1}{2}}{11} + \frac{\frac{1}{2}}{13} + \frac{1}{15} + \frac{\frac{1}{2}}{17}\right.$$

$$\left. - \frac{\frac{1}{2}}{19} - \frac{1}{21} - \frac{\frac{1}{2}}{23} + \frac{\frac{1}{2}}{25} + \frac{1}{27}\cdots\right).$$

This series is converging slowly, but if it is taken as far as the 27th harmonic as indicated, the result, using slide rule throughout, is

$$\frac{4A}{\pi}(0.500 + 0.333 + 0.100 - 0.0714 - 0.1111 - 0.0454$$

$$+ 0.0384 + 0.0667 + 0.0294 - 0.0263 - 0.0477$$

$$- 0.0217 + 0.0200 + 0.0370) = 1.02A,$$

which is close enough to A to be convincing.

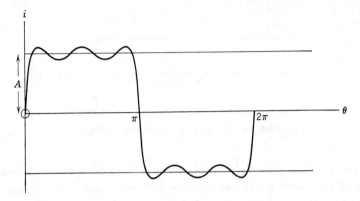

Fig. 19-3. Synthesis of the first three terms of Example 1.

At a point of finite discontinuity, such as $\theta = 0$ or $\theta = \pi$, we see that all terms of this series are zero, and the series is equal to zero. In general, at a point of finite discontinuity this type of calculation yields the mean value of the two alternative values at that point.

Figure 19-3 shows the graph of the function resulting from the synthesis of the first three terms of the foregoing Fourier series (that is, as far as the fifth harmonic). With only three terms considered, the graph already approaches the shape of the square wave. If the sum of all terms after the fifth harmonic is superimposed on this graph, the square wave is exactly duplicated, except for the points of discontinuity.

The function of Example 1, graphed in Fig. 19-2 has odd symmetry; that is, $f(-\theta) = -f(\theta)$, which means that the value of the dependent variable for any value of the independent variable to the left of the origin is the negative of its value for the same absolute value of the independent variable to the right of the origin. This same odd symmetry is a characteristic of the

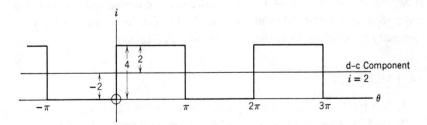

Fig. 19-4. Square wave with a d-c component of 2.

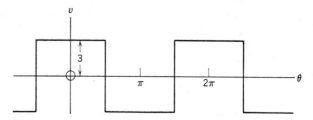

Fig. 19-5. A voltage wave with even symmetry.

sine function. Any periodic function that has this odd symmetry when the values of the independent variable are measured from a horizontal axis representing the d-c component will have no cosine components in its Fourier series. The stipulation regarding the horizontal axis is necessary to cover such wave forms as that of Fig. 19-4, which has a d-c component (equal to the mean value of the function over one cycle) of 2. Since this d-c component is the first term of the Fourier series, the given rule analyzes the symmetry of the part of the wave to be superimposed on $i = 2$, that is, the wave in which the values of the dependent variable are measured from the displaced horizontal axis $i = 2$. We see that, using this displaced horizontal axis, the graph has odd symmetry, and therefore there will be no cosine harmonics in the Fourier series.

For a wave with even symmetry [in which $f(-\theta) = f(\theta)$] there will be no sine harmonics in the Fourier series, since even symmetry is a characteristic of the cosine function. The characteristic of even symmetry does not depend on the vertical position of the horizontal axis, and even symmetry will be apparent whether there is a d-c component or not, without having to displace the horizontal axis. The wave of Fig. 19-5 has even symmetry, and therefore there will be no sine components in the Fourier series.

These rules of symmetry may be used to shorten the number of calculations in finding a Fourier series for a wave with either of these types of symmetry. However, the student should at first calculate the cosine harmonics for a wave with odd symmetry and the sine harmonics for a wave with even symmetry in order to verify that they are all equal to zero.

EXERCISES

1. Is it possible to find a Fourier series for $y = \tan \theta$? Explain.
2. Find the Fourier series for the periodic current wave of Fig. 19-4.
3. Find the Fourier series for the voltage wave of Fig. 19-5.

Find the Fourier series for the following periodic functions, and sketch each wave:

   4. $i = 0$ from $\theta = 0$ to $\theta = \pi/2$ and from $\theta = 3\pi/2$ to $\theta = 2\pi$; $i = 2$ from $\theta = \pi/2$ to $\theta = 3\pi/2$.

   5. $i = 2$ from $\theta = 0$ to $\theta = \pi/2$ and from $\theta = 3\pi/2$ to $\theta = 2\pi$; $i = 0$ from $\theta = \pi/2$ to $\theta = 3\pi/2$.

   6. $v = -1$ from $\theta = 0$ to $\theta = \pi$; $v = 3$ from $\theta = \pi$ to $\theta = 2\pi$.

   7. $i = A$ from $\theta = 0$ to $\theta = 3\pi/2$; $i = 0$ from $\theta = 3\pi/2$ to $\theta = 2\pi$.

   8. $i = 3$ from $\theta = 0$ to $\theta = \pi/2$; $i = -1$ from $\theta = \pi/2$ to $\theta = 2\pi$.

   9. $v = 2$ from $\theta = 0$ to $\theta = \pi$; $v = -1$ from $\theta = \pi$ to $\theta = 3\pi/2$; $v = 0$ from $\theta = 3\pi/2$ to $\theta = 2\pi$.

   10. $i = 2$ from $\theta = -\pi$ to $\theta = 0$ and from $\theta = \pi/2$ to $\theta = \pi$; $i = -3$ from $\theta = 0$ to $\theta = \pi/2$.

19.4   WAVES INVOLVING DISCONTINUOUS POWER FUNCTIONS

The methods established in Section 19.2 may be used to find the amplitude of any of the components of the Fourier series for any periodic function for which a Fourier series exists, provided the function can be expressed in the form of an equation. The only part of the calculation that may differ for different wave forms is the technique used in calculating the integrals.

For wave forms described as triangular or sawtooth, the equation will contain some multiple of the independent variable, and the calculation of the constant coefficients in the Fourier series will require the technique of integration by parts.

EXAMPLE 2.   Find the Fourier series for the sawtooth wave of Fig. 19-6. *Solution: i* must first be expressed as a function of θ. The slope of the straight line is $2/2\pi = 1/\pi$, and, since the line passes through the origin, the equation is $i = \theta/\pi$, a continuous function from $\theta = 0$ to $\theta = 2\pi$.

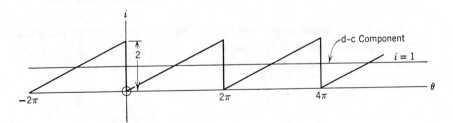

Fig. 19-6. A sawtooth wave.

The d-c component is

$$\frac{1}{2\pi} \int_0^{2\pi} \frac{\theta}{\pi} \, d\theta = \frac{1}{2\pi^2} \left[\frac{\theta^2}{2} \right]_0^{2\pi} = \frac{1}{4\pi^2} [4\pi^2] = 1.$$

This result might have been seen intuitively as the average value of a current which increases linearly from 0 to 2.

The wave has odd symmetry when values of i are measured from the displaced axis $i = 1$; therefore there are no cosine components. The amplitude of the mth sine component is

$$b_m = \frac{1}{\pi} \int_0^{2\pi} \frac{\theta}{\pi} \sin m\theta \, d\theta.$$

Let $\qquad\qquad \theta = u \qquad$ and $\quad \sin m\theta \, d\theta = dv.$

Then $\qquad\qquad d\theta = du \qquad$ and $\quad \dfrac{-\cos m\theta}{m} = v.$

Hence

$$b_m = \frac{1}{\pi^2} \left[\frac{-\theta \cos m\theta}{m} + \int \frac{\cos m\theta \, d\theta}{m} \right]_0^{2\pi}$$

$$= \frac{1}{\pi^2} \left[\frac{-\theta \cos m\theta}{m} + \frac{\sin m\theta}{m^2} \right]_0^{2\pi}$$

$$= \frac{1}{\pi^2} \left[\frac{-2\pi(1) + 0(1)}{m} + \frac{0 - 0}{m^2} \right] = \frac{-2}{m\pi}.$$

Therefore $\qquad b_1 = \dfrac{-2}{\pi}, \qquad b_2 = \dfrac{-2}{2\pi}, \qquad b_3 = \dfrac{-2}{3\pi} \cdots$

Thus $\qquad i = 1 - \dfrac{2}{\pi} \left(\sin \theta + \dfrac{\sin 2\theta}{2} + \dfrac{\sin 3\theta}{3} + \dfrac{\sin 4\theta}{4} \cdots \right).$

EXAMPLE 3. Find the Fourier series for the voltage wave of Fig. 19-7, assuming that the curved portion is parabolic with an equation of the form $v = k\theta^2$.

Solution: The variation of v with θ must be expressed as an equation for each separate graph. The equation of the parabolic graph is of the form $v = k\theta^2$, and it passes through the point $(\pi, 4)$. Hence

$$4 = k\pi^2 \qquad \text{and} \quad k = \frac{4}{\pi^2}.$$

$$v = \frac{4}{\pi^2} \theta^2 \qquad \text{from} \quad \theta = 0 \text{ to } \theta = \pi.$$

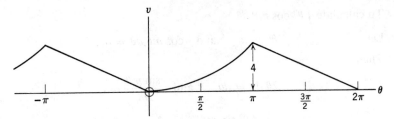

Fig. 19-7.

The complete cycle of the wave used to calculate the coefficients in the Fourier series may commence at any point in the cycle as long as the period is taken as 2π rad. Thus the calculations assuming a cycle from $\theta = -\pi$ to $\theta = \pi$ will yield coefficients identical to the calculations assuming a cycle from $\theta = 0$ to $\theta = 2\pi$. The first calculations will be simpler, since the straight line in the interval from $-\pi$ to zero passes through the origin and its equation will not contain a constant term. However, the equation of the straight line over the interval $\theta = \pi$ to $\theta = 2\pi$ would contain an extra constant term. The cycle will therefore be considered as commencing at $\theta = -\pi$. The straight line has a slope of $-4/\pi$ and its equation is $v = -4\theta/\pi$.

Amplitude of d-c component is

$$\frac{1}{2\pi} \left(\int_{-\pi}^{0} -\frac{4\theta}{\pi} \, d\theta + \int_{0}^{\pi} \frac{4\theta^2}{\pi^2} \, d\theta \right) = \frac{1}{2\pi} \left\{ -\frac{4}{\pi} \left[\frac{\theta^2}{2} \right]_{-\pi}^{0} + \frac{4}{\pi^2} \left[\frac{\theta^3}{3} \right]_{0}^{\pi} \right\}$$

$$= \frac{1}{2\pi} \left(0 + \frac{4}{\pi} \frac{\pi^2}{2} + \frac{4}{\pi^2} \frac{\pi^3}{3} - 0 \right)$$

$$= \frac{1}{2\pi} \left(2\pi + \frac{4\pi}{3} \right) = \frac{5}{3}.$$

Amplitude of the mth cosine component is

$$\frac{1}{\pi} \left(-\frac{4}{\pi} \int_{-\pi}^{0} \theta \cos m\theta \, d\theta + \frac{4}{\pi^2} \int_{0}^{\pi} \theta^2 \cos m\theta \, d\theta \right).$$

To calculate $\int \theta \cos m\theta \, d\theta$:

Let $\qquad\qquad \theta = u \qquad$ and $\quad \cos m\theta \, d\theta = dv.$

Then

$$d\theta = du \quad \text{and} \quad \frac{\sin m\theta}{m} = v.$$

$$I_1 = \frac{\theta \sin m\theta}{m} - \int \frac{\sin m\theta}{m} \, d\theta = \frac{\theta \sin m\theta}{m} + \frac{\cos m\theta}{m^2}.$$

To calculate $\int \theta^2 \cos m\theta \, d\theta$:

Let
$$\theta^2 = u \qquad \text{and} \quad \cos m\theta \, d\theta = dv.$$

Then
$$2\theta \, d\theta = du \qquad \text{and} \quad \frac{\sin m\theta}{m} = v.$$

$$I_2 = \frac{\theta^2 \sin m\theta}{m} - \frac{2}{m} \int \theta \sin m\theta \, d\theta.$$

Let
$$\theta = u_1 \qquad \text{and} \quad \sin m\theta \, d\theta = dv.$$

Then
$$d\theta = du_1 \qquad \text{and} \quad -\frac{\cos m\theta}{m} = v.$$

$$I_2 = \frac{\theta^2 \sin m\theta}{m} - \frac{2}{m} \left(-\frac{\theta \cos m\theta}{m} + \int \frac{\cos m\theta}{m} \, d\theta \right)$$

$$= \frac{\theta^2 \sin m\theta}{m} + \frac{2\theta \cos m\theta}{m^2} - \frac{2 \sin m\theta}{m^3}.$$

Therefore the amplitude of the mth cosine component is

$$\frac{1}{\pi} \left\{ -\frac{4}{\pi} \left[\frac{\theta \sin m\theta}{m} + \frac{\cos m\theta}{m^2} \right]_{-\pi}^{0} \right.$$

$$\left. + \frac{4}{\pi^2} \left[\frac{\theta^2 \sin m\theta}{m} + \frac{2\theta \cos m\theta}{m^2} - \frac{2 \sin m\theta}{m^3} \right]_{0}^{\pi} \right\}.$$

In this calculation, since m is an integer, $m\theta$ is always a whole multiple of π; hence $\sin m\theta = 0$ for all limits. The value of $\cos m\theta$ for a limit of $\theta = \pi$ or $\theta = -\pi$ will be -1 if m is odd and $+1$ if m is even. It is therefore necessary to make two separate calculations at this point. In some cases it may be necessary to make more than two separate calculations, and it may be simpler in the long run to drop the generalization of the mth coefficient a_m and calculate a_1, a_2, a_3, ... individually. For this particular problem it is not too difficult to maintain the generalization using two separate calculations.

If m is odd:

$$a_m = \frac{1}{\pi} \left[-\frac{4}{\pi} \left(0 + \frac{1}{m^2} - \frac{-1}{m^2} \right) + \frac{4}{\pi^2} \left(0 + \frac{2\pi(-1)}{m^2} - 0 - 0 \right) \right]$$

$$= -\frac{16}{m^2 \pi^2}.$$

If m is even:

$$a_m = \frac{1}{\pi}\left[-\frac{4}{\pi}\left(\frac{1}{m^2} - \frac{1}{m^2}\right) + \frac{4}{\pi^2}\left(\frac{2\pi}{m^2} - 0\right)\right] = +\frac{8}{m^2\pi^2}.$$

$$b_m = \frac{1}{\pi}\left(-\frac{4}{\pi}\int_{-\pi}^{0}\theta \sin m\theta \, d\theta + \frac{4}{\pi^2}\int_{0}^{\pi}\theta \sin m\theta \, d\theta\right).$$

The calculation of these two integrals, using integration by parts, is very similar to that of the two integrals involved in a_m, which were previously calculated. It is left to the student to verify that

$$\int \theta \sin m\theta \, d\theta = \frac{-\theta \cos m\theta}{m} + \frac{\sin m\theta}{m^2},$$

and

$$\int \theta^2 \sin m\theta \, d\theta = -\frac{\theta^2 \cos m\theta}{m} + \frac{2\theta \sin m\theta}{m^2} + \frac{2\cos m\theta}{m^3}.$$

If m is odd:

$$b_m = \frac{1}{\pi}\left\{ -\frac{4}{\pi}\left[0 - \frac{\pi(-1)}{m}\right] + \frac{4}{\pi^2}\left[-\frac{\pi^2(-1)}{m} + 0 + \frac{2(-1)}{m^3} - \frac{2}{m^3}\right]\right\}$$

$$= \frac{1}{\pi}\left\{-\frac{4}{m} + \frac{4}{m} - \frac{16}{\pi^2 m^3}\right\} = -\frac{16}{\pi^3 m^3}.$$

If m is even:

$$b_m = \frac{1}{\pi}\left\{ -\frac{4}{\pi}\left[-\frac{\pi(1)}{m}\right] + \frac{4}{\pi^2}\left[-\frac{\pi^2(1)}{m} + \frac{2(1)}{m^3} - \frac{2(1)}{m^3}\right]\right\}$$

$$= \frac{1}{\pi}\left\{\frac{4}{m} - \frac{4}{m}\right\} = 0.$$

Hence

$$v = \frac{5}{3} + \frac{8}{\pi^2}\left(-2\cos\theta + \frac{\cos 2\theta}{2^2} - \frac{2}{3^2}\cos 3\theta + \frac{\cos 4\theta}{4^2}\cdots\right)$$

$$- \frac{16}{\pi^3}\left(\sin\theta + \frac{\sin 3\theta}{3^3} + \frac{\sin 5\theta}{5^3} + \frac{\sin 7\theta}{7^3}\cdots\right).$$

EXERCISES

Sketch one cycle of each of the following periodic wave forms and establish the Fourier series for each:

1. *i* varies directly with θ from $\theta = -\pi$ to $\theta = +\pi$, reaching a peak value of 2.
2. $v = 3\theta/\pi$ from $\theta = 0$ to $\theta = \pi$, and $v = 0$ from $\theta = \pi$ to $\theta = 2\pi$.

3. $v = -\theta$ from $\theta = -2\pi$ to $\theta = 0$.

4. i has a triangular wave form, in which $i = A\theta/\pi$ from $\theta = 0$ to $\theta = \pi$ and i decreases to zero at a uniform rate from $\theta = \pi$ to $\theta = 2\pi$.

5. i increases steadily from 0 to 4 between $\theta = 0$ and $\theta = \pi$, and $i = 4$ for the balance of the cycle.

6. i increases steadily from 2 to 4 between $\theta = 0$ and $\theta = \pi$, and $i = 2$ from $\theta = \pi$ to $\theta = 2\pi$.

7. $i = \theta^2/2$ from $\theta = -\pi$ to $\theta = \pi$.

8. $v = 2\theta^2/\pi^2$ from $\theta = 0$ to $\theta = \pi$, and $v = 2$ from $\theta = \pi$ to $\theta = 2\pi$.

9. $i = 4\theta/\pi - \theta^2/\pi^2$ from $\theta = 0$ to $\theta = \pi$, and $i = 3$ from $\theta = \pi$ to $\theta = 2\pi$.

10. By combining sine and cosine components of equal frequency in the answer to Exercise 5, express this answer as the sum of the d-c component and sine components only, with phase displacement where necessary, taking the series as far as the fifth harmonic.

19.5 OTHER TYPES OF WAVES

The equation representing the whole cycle or any portion of the whole cycle of a periodic wave may contain other functions; for example, a trigonometric or exponential function. The method of finding the coefficients of the component terms is still basically the same, although the actual technique of integration varies.

If a portion of the wave may be represented as a sine function, for example, it will be necessary in the calculations to determine integrals of the form $\int \sin k\theta \sin m\theta \, d\theta$ and $\int \sin k\theta \cos m\theta \, d\theta$, respectively. This may be done by changing the product of the two trigonometric functions to a sum or difference by a standard formula. If part of the wave is represented by an exponential function, an integrand of the form $e^{k\theta} \sin m\theta$ or $e^{k\theta} \cos m\theta$ will appear in the calculations; these are typical examples of the cyclic case of integration by parts.

EXAMPLE 4. Find a Fourier series for the half-rectified sine wave, plotted in Fig. 19.8, in which $i = A \sin \theta$ from $\theta = 0$ to $\theta = \pi$, and $i = 0$ from $\theta = \pi$ to $\theta = 2\pi$.

Solution: The d-c component is

$$a_0 = \frac{1}{2\pi} \left(\int_0^\pi A \sin \theta \, d\theta + 0 \right) = -\frac{A}{2\pi} [\cos \theta]_0^\pi = \frac{A}{\pi}.$$

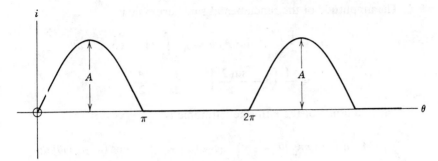

Fig. 19-8. The half-rectified sine wave.

The amplitude of the fundamental cosine component is

$$a_1 = \frac{1}{\pi} \int_0^\pi A \sin \theta \cos \theta \, d\theta = \frac{A}{2\pi} \int_0^\pi \sin 2\theta \, d\theta$$

$$= -\frac{A}{4\pi} [\cos 2\theta]_0^\pi = -\frac{A}{4\pi} [1 - 1] = 0.$$

The amplitude of the mth cosine harmonic,

$$a_m = \frac{1}{\pi} \int_0^\pi A \sin \theta \cos m\theta \, d\theta = \frac{A}{2\pi} \int_0^\pi [\sin (m + 1)\theta - \sin (m - 1)\theta] \, d\theta$$

$$= \frac{A}{2\pi} \left[-\frac{\cos (m + 1)\theta}{m + 1} + \frac{\cos (m - 1)\theta}{m - 1} \right]_0^\pi.$$

If m is even:

$$a_m = \frac{A}{2\pi} \left(\frac{1}{m + 1} - \frac{1}{m - 1} + \frac{1}{m + 1} - \frac{1}{m - 1} \right)$$

$$= \frac{A}{\pi} \left(\frac{1}{m + 1} - \frac{1}{m - 1} \right).$$

If m is odd and m ≠ 1:

$$\cos (m + 1)\pi = \cos (m - 1)\pi = 1 \quad \text{and} \quad a_m = 0.$$

Note that if $m = 1$, the expression for a_m is indeterminate, since the fraction with denominator $m - 1$ becomes infinite for both upper and lower limits. This is the reason for the separate calculation of a_1 previously.

The amplitude of the fundamental sine harmonic is

$$b_1 = \frac{1}{\pi} \int_0^\pi A \sin^2 \theta \, d\theta = \frac{A}{2\pi} \int_0^\pi (1 - \cos 2\theta) \, d\theta$$

$$= \frac{A}{2\pi} \left[\theta - \frac{\sin 2\theta}{2} \right]_0^\pi = \frac{A\pi}{2\pi} = \frac{A}{2}.$$

The amplitude of the mth sine harmonic is

$$\frac{1}{\pi} \int_0^\pi A \sin \theta \sin m\theta \, d\theta = \frac{A}{2\pi} \int_0^\pi [\cos (m - 1)\theta - \cos (m + 1)\theta] \, d\theta$$

$$= \frac{A}{2\pi} \left[\frac{\sin (m - 1)\theta}{m - 1} - \frac{\sin (m + 1)\theta}{m + 1} \right]_0^\pi = 0$$

for all integral values of m.

Therefore

$$i = \frac{A}{\pi} \left[1 - \left(\frac{1}{1} - \frac{1}{3} \right) \cos 2\theta - \left(\frac{1}{3} - \frac{1}{5} \right) \cos 4\theta \right.$$

$$\left. - \left(\frac{1}{5} - \frac{1}{7} \right) \cos 6\theta \cdots \right] + \frac{A}{2} \sin \theta$$

$$= \frac{A}{\pi} \left(1 + \frac{\pi}{2} \sin \theta - \frac{2}{3 \times 1} \cos 2\theta - \frac{2}{5 \times 3} \cos 4\theta - \frac{2}{7 \times 5} \cos 6\theta \cdots \right).$$

EXERCISES

Establish the Fourier series for the following periodic waves and sketch each graph:

1. $i = 6 \cos \theta$ from $\theta = 0$ to $\theta = \pi/2$; $i = 0$ from $\theta = \pi/2$ to $\theta = 3\pi/2$; $i = 6 \cos \theta$ from $\theta = 3\pi/2$ to $\theta = 2\pi$.

2. $v = \sin (\theta/2)$ from $\theta = 0$ to $\theta = 2\pi$.

3. $v = \sin \theta$ from $\theta = 0$ to $\theta = \pi$; $v = -\sin \theta$ from $\theta = \pi$ to $\theta = 2\pi$: (a) by making use of the answer to Exercise 2, and (b) as a separate calculation.

4. $i = 4 \cos \theta$ from $\theta = 0$ to $\theta = \pi/2$; i increases at a constant rate from 0 to 4 between $\theta = \pi/2$ and $\theta = 2\pi$.

5. $i = 2e^{\theta/\pi}$ from $\theta = 0$ to $\theta = \pi$; $i = 2$ from $\theta = \pi$ to $\theta = 2\pi$.

6. $i = 3e^{-\theta/2\pi}$ from $\theta = -\pi$ to $\theta = 0$; $i = 3$ from $\theta = 0$ to $\theta = \pi$.

19.6   WAVE ANALYSIS BY APPROXIMATE GRAPHICAL METHODS

In all periodic waves considered up to this point, it has been assumed that the equation of all portions of the wave is given or that this equation may be

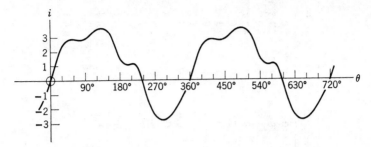

Fig. 19-9.

established by analytical geometry. If this is not true, approximate methods must be used to find an approximate Fourier series.

Each definite integral to be calculated in the exact method explained previously represents an area. The amplitude of the d-c component is the mean value of the function over a complete cycle. We see that

$$a_m = \frac{1}{\pi} \int_k^{k+2\pi} f(\theta) \cos m\theta \, d\theta$$

is twice the mean value of the function $f(\theta) \cos m\theta$ over a complete cycle. Similarly b_m is twice the mean value of the function $f(\theta) \sin m\theta$ over a complete cycle. Any organized method of finding approximate mean values will therefore yield the approximate Fourier coefficients. The following example illustrates one such method applied to a given wave.

EXAMPLE 5.   Find an approximate Fourier series for the current wave of Fig. 19-9 as far as the third harmonic components.
Solution: The value of the current i at 30° intervals is read from the graph, and entered in the data tabulation shown here. Assuming that the average current over the cycle is approximately equal to the average of the twelve readings taken, the d-c component a_0 equals the algebraic sum of the readings of current divided by 12. In this case the algebraic sum of the twelve readings of i is 10.2, and a_0 is approximately equal to $10.2/12 = 0.85$, as indicated in the tabulation. The values of the cosine and sine of θ, 2θ, and 3θ for each of the twelve values of θ in the first column are recorded in columns. The fundamental cosine component a_1 is calculated by first calculating the value of $i \cos \theta$ for all twelve values of θ and then taking twice the average of these values. The average of the values being the algebraic sum divided by 12, a_1 equals (approximately) the algebraic sum divided by 6, as indicated in the tabulation.

TABULATION OF DATA FOR EXAMPLE 5

θ	i	$\cos\theta$	$i\cos\theta$	$\sin\theta$	$i\sin\theta$	$\cos 2\theta$	$i\cos 2\theta$	$\sin 2\theta$	$i\sin 2\theta$	$\cos 3\theta$	$i\cos 3\theta$	$\sin 3\theta$	$i\sin 3\theta$
0°	0	1	0	0	0	1	0	0	0	1	0	0	0
30°	2.5	0.866	2.165	0.5	1.25	0.5	1.25	0.866	2.165	0	0	1	2.5
60°	2.8	0.5	1.4	0.866	2.425	−0.5	−1.7	0.866	2.425	−1	−2.8	0	0
90°	2.7	0	0	1	2.7	−1	−2.7	0	0	0	0	−1	−2.7
120°	3.5	−0.5	−1.75	0.866	3.031	−0.5	−1.75	−0.866	−3.031	1	3.5	0	0
150°	3.3	−0.866	−2.858	0.5	1.65	0.5	1.65	−0.866	−2.858	0	0	1	3.3
180°	1.5	−1	−1.5	0	0	1	1.5	0	0	−1	−1.5	0	0
210°	1.1	−0.866	−0.953	−0.5	−0.65	0.5	0.65	0.866	0.953	0	0	−1	−1.1
240°	−0.6	−0.5	0.3	−0.866	0.520	−0.5	0.3	0.866	−0.520	1	−0.6	0	0
270°	−2.5	0	0	−1	2.5	−1	2.5	0	0	0	0	1	−2.5
300°	−2.7	0.5	−1.35	−0.866	2.338	−0.5	1.35	−0.866	2.338	−1	2.7	0	0
330°	−1.4	0.866	−1.212	−0.5	0.7	0.5	−0.7	−0.866	1.212	0	0	−1	1.4
Sum	10.2		−5.758		16.464		2.35		2.684		1.3		0.9

$$a_0 = \frac{10.2}{12}\qquad a_1 = \frac{-5.758}{6}\qquad b_1 = \frac{16.464}{6}\qquad a_2 = \frac{2.35}{6}\qquad b_2 = \frac{2.684}{6}\qquad a_3 = \frac{1.3}{6}\qquad b_3 = \frac{0.9}{6}$$

$$= 0.85 \qquad\quad = -0.96 \qquad\quad = 2.74 \qquad\quad = 0.39 \qquad\quad = 0.45 \qquad\quad = 0.22 \qquad\quad = 0.15$$

$$i \doteq 0.85 - 0.96\cos\theta + 2.74\sin\theta + 0.39\cos 2\theta + 0.45\sin 2\theta + 0.22\cos 3\theta + 0.15\sin 3\theta\cdots.$$

This general process is repeated for the succeeding sine and cosine components as indicated. For higher harmonics the calculations tend to become less lengthy since some of the products are identical to products previously calculated, and also when the 30° interval is chosen, more zeros appear in the values of the sine and cosine. Of course, if more readings of current are taken at smaller intervals, the resulting series will be more accurate.

EXERCISES

Use the approximate method outlined above to find the Fourier series for:

1. The square wave of Example 1 as far as the fifth harmonic noting that the wave has odd symmetry and therefore there will be no cosine harmonics. (The value of i at a point of discontinuity must be taken as the mean value of the top and bottom points of the vertical line.)

2. The half-rectified sine wave of Example 4, as far as the fourth harmonic.

APPENDIX I

Transients in RL and RC Circuits

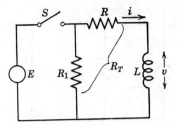

Fig. A-1. *RL* circuit.

SWITCH CLOSING

At $t = 0$, $i = 0$, and v is the value of E at $t = 0$.

$$L\frac{di}{dt} + Ri = E. \tag{1}$$

$$v = E - Ri.$$

Therefore

$$\frac{dv}{dt} = \frac{dE}{dt} - R\frac{di}{dt}$$

or

$$\frac{dv}{dt} = \frac{dE}{dt} - \frac{Rv}{L}. \tag{2}$$

SWITCH OPENING

At $t = 0$, $i = I_0$, and $v = -I_0 R_T$. If E is constant, $I_0 = E/R$.

$$L\frac{di}{dt} + R_T i = 0. \tag{3}$$

$$v + R_T i = 0.$$

Therefore

$$\frac{dv}{dt} + R_T\frac{di}{dt} = 0$$

or

$$\frac{dv}{dt} = -\frac{R_T v}{L}. \tag{4}$$

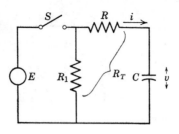

Fig. A-2. *RC* circuit.

SWITCH CLOSING WITH UNCHARGED CAPACITOR

At $t = 0$, $i = E_0/R$, $v = 0$, and $q = 0$, where E_0 is the value of E at $t = 0$,

$$Ri + \frac{q}{C} = E.$$

Therefore

$$R\frac{di}{dt} + \frac{i}{C} = \frac{dE}{dt}. \tag{5}$$

$$R\frac{dq}{dt} + \frac{q}{C} = E. \tag{6}$$

$$RC\frac{dv}{dt} + v = E. \tag{7}$$

SWITCH OPENING

At $t = 0$, $i = -V_0/R_T$, $v = V_0$, and $q = CV_0$. If E is constant, $V_0 = E$.

$$R_Ti + \frac{q}{C} = 0.$$

Therefore

$$R_T\frac{di}{dt} + \frac{i}{C} = 0. \tag{8}$$

$$R_T\frac{dq}{dt} + \frac{q}{C} = 0. \tag{9}$$

$$R_TC\frac{dv}{dt} + v = 0. \tag{10}$$

APPENDIX II

Transients in RLC Series Circuits

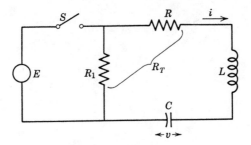

Fig. A-3. *RLC* circuit.

SWITCH CLOSING WITH UNCHARGED CAPACITOR

At $t = 0$, $i = 0$, $di/dt = E_0/L$, $q = 0$, $dq/dt = 0$, $v = 0$, and $dv/dt = 0$, where E_0 is the value of E when $t = 0$.

$$L\frac{di}{dt} + Ri + \frac{q}{C} = E.$$

Therefore

$$L\frac{d^2i}{dt^2} + R\frac{di}{dt} + \frac{i}{C} = \frac{dE}{dt}. \qquad (1)$$

$$L\frac{d^2q}{dt^2} + R\frac{dq}{dt} + \frac{q}{C} = E. \qquad (2)$$

$$LC\frac{d^2v}{dt^2} + RC\frac{dv}{dt} + v = E. \qquad (3)$$

SWITCH OPENING

At $t = 0$, if E is variable, values of i, di/dt, q, dq/dt, v, and dv/dt need to be specified; if E is constant, $i = 0$, $di/dt = -E/L$, $q = CE$, $dq/dt = 0$, $v = E$, and $dv/dt = 0$.

$$L\frac{di}{dt} + R_Ti + \frac{q}{C} = 0.$$

Therefore

$$L\frac{d^2i}{dt^2} + R_T\frac{di}{dt} + \frac{i}{C} = 0. \qquad (4)$$

$$L\frac{d^2q}{dt^2} + R_T\frac{dq}{dt} + \frac{q}{C} = 0. \qquad (5)$$

$$LC\frac{d^2v}{dt^2} + R_TC\frac{dv}{dt} + v = 0. \qquad (6)$$

Answers to Odd-Numbered Exercises

Page 5, Sec. 1.1

	No. 1	No. 3
Independent Variable	T	A
Dependent Variable	H	R
Constant(s)	V	L, T
Functional Notation	$H = f(T)$	$R = f(A)$

5. (a) $2x - 1 - 6x^2$, (b) $2x^2 - 1$, (c) $3(x - 2)^2$, (d) -7, (e) $6x^2 - 1$,

(f) $3(2x - 1)^2$. **7.** $X = \sqrt{\left(\dfrac{E}{4}\right)^2 - 9}$. **9.** $X = 5 - \sqrt{\dfrac{v^2}{9} - 4}$.

Page 11, Sec. 1.2

1. Power function (a), (b), (c), (e), (f), (h), (j). Direct variation (b). Inverse variation (h). Linear variation (b), (c). Quadratic function (e), (j). Odd symmetry (b), (f), (g), (h). Even symmetry (a), (d), (j). **3.** Power function, quadratic function, even symmetry. **5.** Power function, quadratic. **7.** Power function, odd symmetry. **9.** (a) $I = \frac{2}{5}E$, (b) 7.5. **11.** (a) 13.3, (b) 13.3. **13.** (a) $v = -4t + 20$, (b) 0. **15.** (a) $y = 3x^2 - 3$, (b) 24, (c) $\pm\sqrt{3}$. **17.** (a) $w = \dfrac{165 \times 3960^2}{(h + 3960)^2}$, (b) lb-mile². **19.** (a) $P = 2i - \frac{4}{3}i^2$, (b) $P = \dfrac{36r}{(4 + 3r)^2}$.

Page 17, Sec. 2.2

1. (a) 1.5, (b) -0.4, (c) ∞, (d) 0, (e) ∞. **3.** (a) 0.5, (b) -1, (c) 0.4, (d) -0.4. **5.** (a) 45°, (b) 8.1°, (c) 0, (d) 90°, (e) 45°, (f) 26.6°, (g) 90°, (h) 63.4°. **7.** (a) 33.7°, (b) 90°, (c) 26.6°, (d) 14.0°. **9.** Coordinates of P are (3,4). Slope of $PA = \frac{1}{2}$. Slope of $PB = -2$.

Page 22, Sec. 2.3

1. Slope is always 2. **3.** At (1,0). **5.** $y = 0.5$. **7.** $I = 2E$. **9.** (a) m, (b) -7, (c) $-\frac{4}{3}$. **11.** $2x - 3y + 12 = 0$.

421

Page 27, Sec. 2.4

1. (a) Parabola, (b) circle, (c) ellipse, (d) rectangular hyperbola, (e) parabola, (f) rectangular hyperbola.   **3.** (a) Hyperbola, (b) circle, (c) parabola, (d) hyperbola, (e) ellipse, (f) parabola.   **5.** $x^2 + y^2 = 9$. Circle.   **7.** $x^2/a^2 + y^2/b^2 = 1$. Ellipse.   **9.** Tangent at (2,1).   **11.** $x^2/25 + y^2/9 = 1$. Ellipse.

Page 29, Sec. 2.5

1. 5, 4.47, and 5.   **3.** $(1, -8)$.   **7.** 9 square units.   **13.** $y^2 = 4x$. Parabola. **15.** $9x^2 - 16y^2 = 144$. Hyperbola.

Page 34, Sec. 3.1

1. -2.   **3.** 6.   **5.** 600.   **7.** -32.   **9.** $20 - 32t - 16\,\Delta t$; -44. **11.** $3t^2 + 3t(\Delta t) + (\Delta t)^2 - 6$; 7.   **13.** (a) -1.98, (b) 1.98.   **15.** (a) -0.00875, (b) -0.50.

Page 36, Sec. 3.1, second section

1. (a) 6 ft/sec, (b) -2 ft/sec, (c) 2 ft/sec, (d) 0.   **3.** 2 ft/sec^3.   **5.** 6 ft/sec^2. **7.** -3.06 lbs/in.^{2}/ft^3.   **9.** 450 amp/sec.   **11.** $4t + 2\,\Delta t$.

Page 41, Sec. 3.2

1. 3.   **3.** $3x$.   **5.** $-2/x^2$.   **7.** $-1/(3 + x)^2$.   **9.** 3, 3, 4, -4, $-\frac{1}{2}$, 12, -0.04, -0.016.   **11.** 27 ft/sec.   **13.** 9.8 m/sec.   **15.** (a) 5.5 amp, (b) -6 amp.

Page 46, Sec. 3.3

1. $6x$.   **3.** $-3/x^2$.   **5.** $8x$.   **7.** $3 + 5/x^2$.   **9.** $-5/2x^{3/2} - 2/x^2$.   **11.** $3 - 2/t^3$. **13.** $\sqrt{2} + 1/\sqrt{2t}$.   **15.** $3/4t^{3/4} - 1/t^{3/2}$.   **17.** $3 + 1/x^2 - 1/x^{3/2}$.   **19.** 3.   **21.** -18. **23.** $1/t^2$.   **25.** 1.   **27.** $2t$.   **29.** $1/2x^{1/2} - 1/2x^{3/2}$.   **31.** 2.25.   **33.** 30.5.

Page 51, Sec. 3.4

1. (a) 240 ft/sec, (b) 960 ft/sec.   **3.** 1.   **5.** -1.5 ft/sec^2.   **7.** 13 rad/sec^2. **9.** $y = 6x$.   **11.** $(3, -34)$ and $(-2,96)$.   **13.** (a) 100 ft/sec, (b) -4 ft/sec^2, (c) 1250 ft.   **15.** (1,1); 90°.   **17.** $2y + x = 4$.   **21.** (a) $i = 3t^{1/3} + 2$, (b) $v = 0.06t + 0.1$, (c) $i = 2.25\sqrt{t} + 0.15/\sqrt{t}$.   **23.** (a) 0.023 volt, (b) 0.

Page 57, Sec. 4.1

1. (a) 6, (b) -10, (c) $0.5t^{-1.5}$, (d) $6/t^3 - 1.5/t^{2.5}$.   **3.** (a) (i) -32 ft/sec^2, (ii) 0; (b) (i) $-\frac{1}{8}$ ft/sec^2, (ii) $\frac{3}{4}$ ft/sec^3; (c) (i) 1 ft/sec^2, (ii) 1 ft/sec^3.   **5.** (a) 3 and -1, (b) negative, (c) 6, (d) $x > 1$, (e) $x < 1$, (f) 6.   **7.** (a) $0 < t < 2$, (b) $0 < t < 1$.

Page 66, Sec. 4.2

1. (a) Increasing, (b) increasing, (c) decreasing, (d) decreasing.   **3.** (a) increasing, (b) decreasing, (c) decreasing, (d) increasing.   **5.** (a) dy/dx is negative, (b) d^2y/dx^2 is positive, (c) dy/dx is zero, (d) d^2y/dx^2 is zero   **7.** At A, dy/dx is zero; d^2y/dx^2 is positive. At B, dy/dx is positive; d^2y/dx^2 is zero. At C, dy/dx is zero;

d^2y/dx^2 is negative. At D, dy/dx is negative; d^2y/dx^2 is zero. At E, dy/dx is zero; d^2y/dx^2 is positive. **9.** (a) False, (b) false, (c) false, (d) true. **11.** None. **13.** A minimum at $(0,0)$. **15.** (a) 16 ft, (b) 0, (c) -32 ft/sec^2, (d) 2 sec. **17.** Point of inflection at $(2,8)$. **19.** (a) Maximum at $(0,4)$, minima at $(\sqrt{3}, -5)$ and $(-\sqrt{3}, -5)$, points of inflection at $(1, -1)$ and $(-1, -1)$, (b) $\pm\sqrt{6}$. **21.** (a) Maximum at $(-3,32)$, minimum at $(1,0)$, point of inflection at $(-1,16)$, (b) 3, (c) -5. **23.** (a) Maximum value is -4; minimum value is 4, (b) discontinuous function. **25.** 12. **27.** Point of inflection at $(1,4)$.

Page 76, Sec. 4.3

1. $L/2$. **3.** (a) 2, (b) $\sqrt{3}$. **5.** (a) 1/256 coulomb, (b) maximum at $\frac{1}{6}$ sec., minimum at 0 sec. **7.** 3 amps, 1 amp, 24 watts. **9.** r. **11.** Length and width of 2.52 ft, height of 1.26 ft. **13.** 2.36 in., 228 cu in. **15.** 8 ohms. **17.** $\frac{1}{2}$. **19.** (a) $2t/9 - t^2/3$ amp, (b) $2t/9 - t^3/3$ amp, (c) 0.700 amp at 0.471 sec. **21.** (b) 8 square units. **23.** $\frac{10}{3}$ ohms. **25.** (a) 352,000 cir. mil, (b) 336,400 c.m. **27.** 2.26 in., 3.18 in. **29.** 200. **31.** $\frac{17}{16}$ in.

Page 82, Sec. 5.1

1. (a) $(10x + 3/2\sqrt{x})\,dx$, (b) $(7 + 5/x^2)\,dx$, (c) dE/R, (d) $-E\,dR/R^2$. **3.** (a) -0.25, (b) -2.25. **5.** (a) $\frac{97}{16}$, 399/44, (b) $\frac{14}{3}$, 6.94, (c) -0.0766, -0.088.

Page 84, Sec. 5.2

1. Straight line. **3.** 3.00. **5.** -4.2%. **7.** -2%. **9.** $+2.5\%$. **11.** 3000 cu mile. **13.** 9%.

Page 87, Sec. 5.3

1. $-4x/y$. **3.** $(6x - 1)/(2y - 2)$. **5.** $\pm\frac{34}{27}$. **7.** $-\frac{1}{32}$. **9.** (a) R/Z, (b) $R/\sqrt{R^2 + X^2}$, (c) $\sqrt{Z^2 - X^2}/Z$. **11.** -0.06. **13.** $\frac{1}{4}$. **15.** 3 ohms. **17.** (a) $\frac{5}{24}$ ft, (b) 1.2 ft.

Page 94, Sec. 6.1

1. $2x$. **3.** $6x^2 + 2x - 4$. **5.** $-10v^4 - 20v^3 + 36v^2$. **7.** $-2/t^3 - 3/t^4$. **9.** $6x^5 - 10x^4 - 8x^3 + 6x^2 + 8x + 4$. **11.** $[v(du/dx) - u(dv/dx)]/v^2$. **13.** $2/(\theta + 1)^2$. **15.** $(8w^2 - 24w - 54)/(2w - 3)^2$. **17.** 3. **19.** 13/8. **21.** $y + 2x = 8$. **23.** $66/(3x + 1)^3$. **25.** $6/(1 + x)^3$. **27.** Max at $(5, 0.1)$, min at $(-1, -0.5)$. **31.** (a) 0.4 ohm, (b) 0.1 ohm.

Page 98, Sec. 6.2

1. $6(4t + 3t^2)$; $4 + 6t$; $12(4t + 3t^2)(2 + 3t)$. **3.** $(3t^2 + 1)/(t^2 + 1)$; $t/\sqrt{t^2 + 1}$; $(3t^3 + t)/(t^2 + 1)^{3/2}$. **5.** $2t - 2$; $6t$; $(t - 1)/3t$. **7.** $-2(t - 1)^2$; $-1/(t - 1)^2$; 2. **9.** $12t(t - 1)/(3t^2 + 2t - 1)^3$. **11.** 8 ft/sec. **13.** -3.56 cubic units/min. **15.** 1.28 ohm/sec. **17.** $+1.79$ ohm/sec. **19.** -16 ft/sec. **21.** (a) 12%/hr, (b) 8%/hr. **23.** 0.15 in./sec.

Page 103, Sec. 6.3

1. $4x^3$.   **3.** $6(2x - 1)^2$.   **5.** $(1 - x)/\sqrt{2x - x^2}$.
7. $1.5\sqrt{2v^{3/4} - v^{-1/4}}\,(1.5v^{-1/4} + 0.25v^{-3/4})$.   **9.** $-8\pi\theta(2\theta^2 - 1)/3(2\theta^2 + 1)^3$.
11. $3(t + 10)/2(t + 5)^{3/2}$.   **13.** 1.   **15.** -31.   **17.** $X^2/(R^2 + X^2)^{3/2}$.
19. $8R(3t^{1/2} + 4t^{-1/2} - 8)$.   **21.** $(-15t^3 + 12t^2 - 14t + 8)/(1 - 2t)^{3/2}$.   **23.** $\sqrt{2}$.
25. Where $Z = \sqrt{R^2 + 4\pi^2 f^2 L^2}$: (a) $1/Z$, (b) $-ER/Z^3$, (c) $-4E\pi^2 f L^2/Z^3$.
27. $1/2\sqrt{t^2 - 4t}$.

Page 107, Sec. 6.4

1. 0; E/R.   **3.** $\frac{1}{3}$ sec.   **5.** ωL; $E^2/2\omega L$.   **7.** (a) 0, (b) $-0.671E_0$.   **9.** (a) $\sqrt{2t^3 - 1}/3$,
(b) $\sqrt{2t^3 - 1}/2$, (c) $t^2/\sqrt{2t^3 - 1}$, (d) $(2t^3 + 2t^2 - 1)/2\sqrt{2t^3 - 1}$.   **11.** R_p.
13. $1/2\pi fC$; πfCE^2.   **15.** ± 3.56.   **17.** ± 6.3.   **19.** 2.08 ft/sec.   **21.** The airport.
23. X.   **25.** RC/G.

Page 113, Sec. 7.1

1. $2x^3 + x^2 - 7x + C$.   **3.** $3.5v^2 - 4^{1/2} - 3v^{2/3} + C$.   **5.** $i = 7t + C$.
7. $i = t^3/3 - 8t^{1/2}/5 + 2t^2 + C$.   **9.** (a) $4t^{3/2}/3 + 3t + C$,
(b) $8t^{5/2}/15 + 3t^2/2 + Ct + C_1$.   **11.** $y = 3x - 20x^{3/4} - 4$.

Page 116, Sec. 7.2

1. (a) $u + C$, (b) $x^2 - 3x + C$, (c) du, (d) $2t + 5/t^2$.   **3.** $x^2 - 3/x + C$.
5. $6t^{1/2} - 4t^{3/2}/3 + C$.   **7.** $vu^2/2 + v^2u^3/3 - u^3/3v^2 + C$.   **9.** $4x^{7/4} - 4x^{5/4} + C$.
11. $4x^{5/2}/15 + 4x^{3/2}/3 + Cx + C_1$.   **13.** $x^2 + 1/x - 1$.
15. $y = p^2/2 - 4p^{3/2}/3 - 3p + 74/3$.   **17.** $y = 1.5x^2 - 5x - 6.5$.
19. $y = 2x^3/3 + 2\sqrt{x} - 493$.   **21.** (a) $at + v_0$, (b) $at^2/2 + v_0t$.   **23.** $1.5t^3$.
25. (a) -8 ft/sec², (b) 20 ft/sec, (c) $-8t + 20$, (d) $-4t^2 + 20t$.   **27.** 12.5 ft/sec².
29. (a) $0.5x^3 - x^2 + Cx + C_1$, (b) $0.5x^3 - x^2 + 2x - 1$.   **31.** (a) 10.8 rev,
(b) 12.6 rev.

Page 119, Sec. 7.3

1. 0.01 coulomb.   **3.** 139 volts.   **5.** $8t + 16t^{1/2} - 20$.   **7.** 2 sec.
9. (a) $5t^{1/2} + 5t^{-1/2} - 4$, (b) 4.85 amp.   **11.** (a) 7 dyne-cm, (b) 12 dyne-cm.
13. 2.2 volts.

Page 126, Sec. 8.1

1. 9 square units.   **3.** (a) $x^3/3$, (b) $\frac{1}{3}$.   **5.** (a) $\frac{22}{3}$, (b) 9, (c) 1.5, (d) 10^4.   **7.** $\frac{32}{3}$.
9. 4.

Page 130, Sec. 8.2

1. $\frac{8}{3}$.   **3.** $\frac{16}{3}$.   **5.** $\frac{2}{3}$.   **7.** 0.   **9.** 2.   **11.** 5.2.   **13.** $\frac{4}{3}$.   **15.** 2.   **17.** ∞.   **19.** 20/3.
21. (a) $\frac{4}{3}$, (b) $\frac{20}{3}$, (c) $x = 3$.   **23.** (a) Displacement, (b) velocity, (c) charge.

Page 135, Sec. 8.3

5. (b) 3 sec, (c) 4.5 ft, (d) -3 ft/sec², (e) 4.5 sec, (f) at $t = 4.5$, area below
t-axis equals previous area above t-axis, (g) first time at $t = 1$ sec.   **7.** 10.8 erg.

9. (a) 2 erg, (b) 2 erg.   **11.** $4\pi a^2 \sigma/c^2$.   **13.** (a) $t = \frac{1}{4}$ sec, (b) $\frac{1}{4}$ sec. to 1 sec.
15. (a) 40/21 joule, (b) $-800/21$ joule, (c) 7/3 sec.   **17.** 1.164 joule.   **19.** (a) No,
(b) displacement from initial position, (c) above.   **21.** (a) ∞, (b) 0, (c) 0.01 sec,
(d) maximum.

Page 144, Sec. 8.5

1. (a) 6, 6.4; (b) 0, 4.55; (c) -0.5, 3.54; (d) 5, 5.   **5.** (a) 16π, (b) 24π.
7. (a) $15\pi/128$, (b) 0.88π.   **9.** 7.24 cu in.   **11.** (a) 0.45, (b) 0, (c) 2.80, (d) -0.237,
(e) 0.79.   **13.** 0.10 watt.

Page 154, Sec. 9.1

1.

	Amplitude	Angular Velocity (rad/sec)	Frequency (cps)	Period (sec)
(a)	7.5	200π	100	0.01
(b)	2.3	400	$200/\pi$	$\pi/200$
(c)	5.2	377	60	0.0167
(d)	5.2	10^4	$5 \times 10^3/\pi$	$2\pi \times 10^{-4}$

3. (a) $3\cos(\theta - 60°)$, (b) $5\sin(\theta + 68°)$, (c) $4\cos(300\pi t - 5\pi/8)$,
(d) $2\sin 120\pi t$.   **5.** (a) e_1 lags e_2 by $\pi/4$, (b) i_1 leads i_2 by $\pi/2$, (c) v_1 and v_2 are
180° out of phase, (d) i_1 leads i_2 by $5\pi/6$.   **9.** $6.1\sin(\theta + 34.7°)$.
11. $8.7\sin(\theta + 46.7°)$.   **13.** $9.1\sin(377t + 26°)$.   **15.** Impossible.
17. $5.0\cos(500t - 6.9°)$.   **19.** $1.7\sin(\theta + 61.9°)$.   **21.** $-0.252\sin(800t + 7.5°)$.
23. $2.4\sin(2000t + 67.5°)$.

Page 161, Sec. 9.3

1. 16.   **3.** 1.4.   **5.** (a) $\log_b 2 + 3\log_b x$, (b) $\frac{2}{3}\log_b x$, (c) 2; 7; 3.
9. $u/(\ln R - \frac{1}{3}\ln T)$.   **11.** (a) 4.3070, (b) 2.5823, (c) 2.05.   **13.** 5.2.
15. (a) $(x^2 + 3)/(2x + 1)^3 = b^k$, (b) $y - 2 = b^{\sqrt{2x+5}}$.   **17.** (a) $\ln e^a$, (b) $e^{\ln a}$.
21. (a) Natural growth, (b) natural decay, (c) a horizontal straight line, (d) faster
growth, (e) faster decay, (f) height is tripled throughout.   **23.** $q = CE\,e^{-t/RC}$.
25. (a) $q = CE(1 - e^{-t/RC})$.

Page 170, Sec. 9.5

7. (a) -1.175, (b) 3.762, (c) -0.987.   **9.** $\sinh x$.   **11.** 1.   **13.** 1.   **15.** 37.6.
17. 1.444.   **19.** $\ln(x \pm \sqrt{x^2 - 1})$, or $\pm\ln(x + \sqrt{x^2 - 1})$.   **21.** $\pi/4$.   **23.** $\pi/2$.
25. $\sqrt{1 - 4x^2}/2x$.   **27.** $\pi/6$.   **29.** $x - y$.   **31.** $6x/(x^2 + 9)$.   **33.** $\arcsin(1/x)$.
35. $\arctan[5/(3x - 1)]$.

Page 179, Sec. 10.2

1. (a) $6\cos 6x$, (b) $-2\sin(2\theta - \pi/3)$, (c) $200\pi\sec^2(200\pi t - \pi/4)$.
3. $-3\cos^2\theta\sin\theta$.   **5.** $3\sec^2 2x/\sqrt{\tan 2x}$.   **7.** 0.   **9.** $1.5(1 + \cos^2 3x)^{-3/2}\sin 6x$.
11. $-\operatorname{cosec} x \cot x$.   **13.** $-\operatorname{cosec}^2 x$.   **15.** -299.   **17.** 326.   **21.** $y' = 2\cos 2x$,
$y'' = -4\sin 2x$, $y''' = -8\cos 2x$, $y'''' = 16\sin 2x$.   **23.** $y' = -A\sin x$,
$y'' = -A\cos x$, $y''' = A\sin x$, $y'''' = A\cos x$.   **25.** $y = A\cos x$ or $A\sin x$.

27. $y = A \cos 2x$ or $A \sin 2x$.   **29.** $3(\sec^2 3\theta + 1) \sin 3\theta$.
31. $\theta^2 \cos \theta(3 \cos \theta - 2\theta \sin \theta)$.   **33.** $\sin x/(\cos 2x)^{3/2}$.
35. $x^2[2x \sec^3 2x(1 + \sin^2 2x) + 3 \sin 2x \sec^2 2x]$.
37. $[\cos (x + y)]/[1 - \cos (x + y)]$.   **39.** $y = 5x - 0.285$.
41. Maximum $= 2.236$; minimum $= -2.236$.

Page 183, Sec. 10.3

1. (a) 1, (b) 1, (c) 1, (d) 0.707.   **3.** 0.   **5.** $-0.4/\pi f$.   **7.** $-1/40\pi$.   **9.** 0.5.   **11.** $\pi/2$.
13. $\sin \theta + C$.   **15.** $\frac{1}{2}\theta + \frac{1}{8} \sin 4\theta + C$.   **17.** $-\frac{1}{4} \cos 2x + C$.
19. $(\sin 100\pi t)/100\pi + C$.   **21.** $0.25(1.5\theta - \sin 2\theta + 0.125 \sin 4\theta) + C$.
23. $1/2f$.   **25.** $1/50\pi$.   **27.** 0.   **29.** 0.0122.

Page 191, Sec. 10.4

1. (a) $d^2y/dx^2 = -9y$, (b) $d^2y/dx^2 = -9y$, (c) $d^2y/dx^2 = -k^2y$,
(d) $d^2y/dx^2 = -k^2y$, (e) $d^2y/dx^2 = -ky$.   **3.** (a) 2.5 in., (b) $v = -25\pi \sin 10\pi t$;
$a = -250\pi^2 \cos 10\pi t$, (c) $v = 0$; $a = -250\pi^2$, (d) $s = 0$; $a = 0$.   **5.** $a = -16s$.
7. -7.96 amp.   **9.** (a) $P = 12 \sin^2 200\pi t$, (c) $P = 12$ at $t = 0.0025$, (d) at $t = 0.00125$.
11. $E = 8 \sin 250t - 12 \cos 250t$.   **13.** (a) $i = 2.0 \sin 4000t$,
(b) $E = 2\sqrt{61} \sin (4000t - 50.2°)$.   **15.** 1.5 at $\theta = \pi/6$.   **17.** 11 at $\theta = \pi/2$.
19. (a) Maximum $y = v_0^2 \sin^2 \theta/2g$, (b) $\theta = 45°$.   **21.** (a) $q = 0.080 \cos 80t$,
(b) $i_C = -6.4 \sin 80t$, (c) $i_R = 4.0 \cos 80t$,
(d) $i = 4.0 \cos 80t - 6.4 \sin 80t = 7.55 \cos (80t + 58°)$, (e) $P = 8.0 \cos^2 80t$.
23. (a) $5.65 \cos 377t$, (b) $0.0113 \cos 377t$, (c) $-4.26 \sin 377t$, (d) $1.74 \sin 377t$,
(e) $6.96 \sin 377t + 5.65 \cos 377t$, (f) $12.1 \sin^2 377t$.
25. (a) $4.18 \sin (400t + 103.2°)$ (b) $14.4 \cos^2 400t$.   **27.** $0.5I_0 E_0 \pi(2 + m^2)$.
29. Isosceles triangle with two 45° angles.

Page 199, Sec. 10.5

1. A constant current.   **3.** (a) $2/\pi$, (b) $5.46/\pi$, (c) $4/\pi$, (d) $5.46/\pi$.   **5.** 0.637.
7. (a) 0.424, (b) 0.176, (c) 0.520.   **9.** (a) 5.2 amp, (b) 48.6 watts.   **11.** 4.7.   **13.** 4.12.
15. (a) 5330 volts, (b) 5540 volts, (c) 5440 volts.   **17.** (a) 5.46 watts, (b) 328 joules.

Page 205, Sec. 11.1

1. 1.49.   **3.** 0.549.   **5.** $3e^{3x}$.   **7.** $-\frac{2}{3}e^{-x/3}$.   **9.** $-kAe^{-kt}$.   **11.** $2(6x - 1)e^{3x^2 - x - 5}$.
13. $\frac{1}{2}(e^t + e^{-t})$.   **15.** 1.   **17.** 241.   **19.** $1/x - 1$.   **21.** $e^{-5t}(A - 5At - 5B)$.
23. $[e^{-3y}(5y + 1) - 2y - 1]/y^2 e^{2y}$.   **25.** $(3e^{-3x} \sin 5y + y)/(5e^{-3x} \cos 5y - x)$.
27. $-(5e^{-5x} + 3 \sin 3x)/2\sqrt{e^{-5x} + \cos 3x}$.   **29.** $-e^{-5x}(5 \cot 2x + 2 \operatorname{cosec}^2 2x)$.
31. $e^{-3x}[3x \cos 3x - \sin 3x(3x + 2)]/x^3$.   **33.** (a) $A = 4$, $m = 2.5$; (b) $A = 10$,
$m = -5$; (c) $A = 6$, $m = -0.347$.   **35.** (a) 1.2, (b) $5e^{-0.05t}$, (c) $0.25e^{-0.05t}$,
(d) 5, (e) 100.   **37.** 0.625 at $t = 0.0139$.   **39.** 0.806 at $t = 0.0039$.   **41.** E.
45. (a) $6.0 - 7.5e^{-40t} + 1.5e^{-200t}$, (b) $9.0(e^{-40t} - e^{-200t})$,
(c) $-1.5e^{-40t} + 7.5e^{-200t}$, (d) 6.0 volts.

Page 213, Sec. 11.3

1. -0.75.　**3.** (a) $i = Ae^{-40t}$, (b) $i = Ae^{-t/RC}$, (c) $v = Ae^{-t/RC}$,
(d) $y = Ae^{(d-b)x(a-c)}$, (e) $y = -20x^2 + A$.　**5.** $y = Ae^x$, Ae^{-x}, $A\cos x$ or $A\sin x$.
7. $i = 4.45e^{-80t}$.　**9.** $i = (E/R)e^{-(R+R_1)t/L}$.　**11.** $i = 2.5e^{-50t}$.　**13.** $i = (E/R)e^{-R_Tt/L}$.
15. $q = CEe^{-t/R_TC}$.　**17.** (a) $T = 325e^{-0.04778t}$, (b) $123°$F, (c) 87.3 min.
19. (a) $v_R = Ee^{-t/RC}$, (b) $v_C = E(1 - e^{-t/RC})$, (c) $q = EC(1 - e^{-t/RC})$,
(d) $0.865EC$.　**21.** (b) $v = 12 - 12e^{-50t}$.

Page 218, Sec. 11.4

1. (a) $1/(x - 2)$, (b) $3/(3x - 2)$, (c) $-3/(2 - 3x)$, (d) $-x^{-2}$, (e) $6x/(3x^2 - 2)$,
(f) 2, (g) $(\cos\theta + \sin\theta)/(\sin\theta - \cos\theta)$, (h) $-1/t$, (i) $4/(2x + 5)$,
(j) $[4\ln(2x + 5)]/(2x + 5)$.　**5.** 0.
7. $2[(x - 3)\ln(x - 3) - x\ln x]/x(x - 3)[\ln(x - 3)]^2$.
9. $(1 - x^2)/(x^4 + x^2 + 1)$.　**11.** $2\sec^2 2x\ln 2x + (\tan 2x)/x$.
13. $2e^{2x}[2(\ln x + x) + (1 + x)/x]$.　**15.** (a) $100t = \ln\sin 50t - \ln v$,
(b) $v(50\cot 50t - 100)$.　**17.** $1/e$.　**19.** -1.　**21.** -0.64.　**23.** $k^x\ln k$.
25. $x^x(1 + \ln x)$.　**27.** $-200\cos^9 20t\sin 20t$.　**29.** $1/(x + 2)\ln 10$.
31. $x^{x^2+1}(1 + 2\ln x)$.　**33.** 0.0096.

Page 222, Sec. 11.5

1. (a) $e^{x-2} + C$, (b) $-0.5e^{-2x} + C$, (c) $e^{mt+6}/m + C$,
(d) $e^{(k+m)x}/(k + m) + C$, (e) $-3e^{-x/3} + C$, (f) $e^{-3t} + C$, (g) $10^{2x}/2\ln 10 + C$,
(h) $2^{10}x^{11}/11 + C$, (i) $-2e^{-x/2} + C$, (j) $-\frac{1}{3}e^{-3x+2} + C$, (k) $\frac{1}{2}t^2 + C$.
3. (a) $x \geq 0$, (b) $x \geq 3$, (c) $x \geq \frac{3}{4}$, (d) $x \leq \frac{3}{4}$.　**5.** $\frac{1}{2}$.　**7.** $\frac{1}{2}$.　**9.** $2L/R$.　**11.** $(1/R)\ln E$.
13. $E^2L/2R^2$.　**15.** $v = E(1 - e^{-t/RC})$; $v \leq E$.　**17.** 0.618 or -1.618.　**19.** (b) 0.15
coulomb, (c) $\frac{9}{14}$ joule.　**21.** $e^{5x^2} + C$.　**23.** $-e^{-x^3}/3 + C$.　**25.** $e^{4x}/4 + C$.
27. $e^{2x^2}/4 + C$.　**29.** $a^{2x^3-5}/6\ln a + C$.　**31.** $-2/(2 + 3x) + C$.
33. $-1/(2 + 3x^2) + C$.　**35.** $\ln A(t + 2)$.　**37.** (a) $v_R = 14(e^{-200t} - e^{-500t})$;
$v_C = 4e^{-500t} - 10e^{-200t} + 6$, (b) 6 volts, (c) $v_R = 0$; $v_L = 6$; $v_C = 0$,
(d) $v_R = 0$; $v_L = 0$; $v_C = 6$.

Page 227, Sec. 12.1

1. $3\cosh 3t$.　**3.** $12\tanh^2 4x\,\text{sech}^2 4x$.　**5.** $(1 - 8x)\,\text{cosech}^2(4x^2 - x)$.
7. $3\cosh 3x$.　**9.** $\frac{1}{2}\sinh(2x + 3) + C$.　**11.** $\cosh 3x^2 + C$.
13. $\frac{1}{2}x + \frac{1}{4}\sinh 2x + C$.　**15.** $\sinh x/2 - \frac{1}{2}x + C$.　**17.** $0.2\sinh 5u - 0.4\tanh 5u + C$.　**19.** 1.086 square units.　**21.** 0.49 in.^{2}.
23. $x - \tanh x + C$.　**25.** $2\cosh 2x\cos 3x - 3\sinh 2x\sin 3x$.　**27.** 0.
29. $e^x(\coth x + \ln\sinh x)$.　**31.** $5(\cosh 5x)^{\cosh 5x}\sinh 5x(1 + \ln\cosh 5x)$.
33. (a) $y = Ae^x$, Ae^{-x}, $A\cosh x$ or $A\sinh x$; (b) $y = Ae^{3x}$, Ae^{-3x},
$A\cosh 3x$ or $A\sinh 3x$; (c) $y = Ae^{\sqrt{k}x}$, $Ae^{-\sqrt{k}x}$, $A\cosh\sqrt{k}x$ or $A\sinh\sqrt{k}x$.
35. 3.0.

Page 231, Sec. 12.2

1. (a) $2/\sqrt{1-4x^2}$, (b) $2x/(1+x^4)$, (c) $4/\sqrt{2-x^2}$, (d) $1/\sqrt{4+x^2}$,
(e) $3/(9-x^2)$, (f) $\sqrt{3}/\sqrt{3x^2-4}$.   **7.** $-1/x\sqrt{x^2-1}$.   **9.** $2/65$.   **11.** $\frac{4}{3}$.
13. -0.408.   **15.** 0.571.   **17.** $2\cos 2x \arcsin 2x + (2\sin 2x)/\sqrt{1-4x^2}$.
19. $1/(1-x^2) + x(\arcsin x)/(1-x^2)^{3/2}$.   **21.** $-x(\arccos x/a)/\sqrt{a^2-x^2} - 1$.
23. $\sqrt{4+x^2}$.

Page 235, Sec. 12.3

1. $\frac{1}{2}\arctan\frac{x}{2} + C$.   **3.** $\sinh^{-1}\frac{x}{2} + C$.   **5.** $\frac{1}{2}\arcsin x$.

7. $(1/2\sqrt{3})\tanh^{-1}(2x/\sqrt{3}) + C$.   **9.** $-\frac{1}{2}\ln(1-2x) + C$.   **11.** $\pi/6$.   **13.** $\frac{1}{2}\ln 2$.
15. 0.267.   **17.** -0.110.   **19.** (a) 1.32 amp, (b) 1.33 amp.

21. $\arctan 2x - \frac{1}{2}\ln(4x^2+1) + C$.   **23.** $\frac{1}{2}\ln(x^2+4) - \arctan\frac{x}{2} + C$.

25. $\arcsin\frac{x}{2} + 2\sqrt{4-x^2} + C$.   **27.** $\frac{1}{2}\arctan[(x-3)/2] + C$.

29. $\frac{1}{2}\arctan[(x-2)/2] + C$.   **31.** $\cosh^{-1}(x-2) + C$.
33. (a) $-1 \le x \le 5$, (b) yes.

Page 240, Sec. 13.1

1. $\ln A(x^3+5)$.   **3.** $\frac{1}{2}x^2 - 5/x + C$.   **5.** $1/(3-x) + C$.

7. $\frac{1}{2}\ln(x^2+4) - 2\arctan\frac{x}{2} + C$.   **9.** $\ln A(x-2)$.   **11.** $\frac{1}{4}\ln A(x^4+4x)$.

13. $-\frac{1}{3}(3-\sin 2x)^{3/2} + C$.   **15.** $\ln A(e^x-1)$.   **17.** $-\frac{1}{2}\arctan(e^{-x}/2) + C$.
19. $2\arctan e^x + C$.   **21.** $\sinh^{-1}e^x + \sqrt{1+e^{2x}} + C$.   **23.** $-1/x + C$.
25. $\arctan(\ln x) + C$.   **27.** $(x^2-1)/2x + C$, or $\sinh(\ln x) + C$.

29. $\frac{1}{4}\arcsin\frac{x^4}{3} + C$.   **31.** $\sinh x + C$.   **33.** $\frac{1}{2}[\sinh^{-1}(x/2)]^2 + C$.

35. $\frac{1}{8}(2\theta - \sin 2\theta)^2 + C$.   **37.** $3\arctan x^{1/3} + C$.   **39.** $\ln(e^x+2) + C$.
41. $\sinh^{-1}(x-1) + C$.   **43.** $\frac{1}{2}[\ln\sin(3x^2-2e^{-5x})]^2 + C$.   **45.** $61/3$.   **47.** ∞.
49. 0.66.

Page 246, Sec. 13.3

1. $x + 2\ln(x-1) + C$.   **3.** $-2x - 1.5\ln(1-2x) + C$, if $x \le \frac{1}{2}$.
5. $\frac{1}{2}[x^2 - 7\ln(x^2+4)] + C$.   **7.** $x^2/2 - 2x - 3\ln(x^2+9) + \frac{14}{3}\arctan\frac{x}{3} + C$.
9. $-ax^2/2k - acx/k^2 - [(ac^2+bk^2)\ln(c-kx)]/k^3 + C$.
11. $e^x - \ln(e^x+1) + C$.   **13.** $\ln[A(x^2-4)^{1/2}(x-2)^{3/4}/(x+2)^{3/4}]$.

15. $\frac{1}{2}\ln A(x^2-4x)$.   **17.** $\frac{x^2}{2} + \ln(x^2-5) + C$.   **19.** $\ln[A(2x+3)/(x+2)]$.

21. $-\sqrt{1.5}\arctan(\sqrt{2}x/\sqrt{3}) + \sqrt{2}\arctan(x/\sqrt{2}) + C$.

23. $\dfrac{1}{2a}\ln[k(a+x)/(a-x)]$.   **25.** $\dfrac{1}{2a}\ln[k(x+a)/(x-a)]$.   **27.** -0.275.

29. 0.288.   **31.** 0.1076.
33. $i = 2.5[15t + 2\ln(t+1) - 107\ln(t+6) + 107\ln 6]$.

Page 251, Sec. 13.3, second section

1. $\ln [Ax/\sqrt{x^2 + 3}]$.   **3.** $1/x + 4 \ln [A(x - 1)/x]$.

5. $\ln [Ax/(x - 1)] - 4/(x - 1)$.   **7.** $\ln Ax(x - 1)^5 - 4/(x - 1) + 3x$.

9. $\frac{1}{4} \ln [Ax(x^2 + 3)]$.   **11.** -0.059.   **13.** $\ln [A(x^2 - 2x + 2)]$.

15. $2 \ln A(x^2 - 2x + 2) + 2 \arctan (x - 1)$.

17. $-4\sqrt{4x - x^2} + 8 \arcsin [(x - 2)/2] + C$.

19. $\ln [A(x^2 - 2x + 4)/(x + 2)^2]$.   **21.** $\frac{1}{3} \ln [A(x + 3)/x] - 3/(x + 3)$.

23. $\ln [Ax^2/(1 - x^2)^{\frac{1}{2}}(3 - x^2)^{\frac{5}{2}}]$.   **25.** $\frac{1}{4} \arctan (e^x/2) - \frac{1}{2}e^{-x} + C$.

27. 0.117 cubic unit.

Page 254, Sec. 13.4

1. $(2x - 1)^{\frac{3}{2}}(3x + 1)/15 + C$.   **3.** $-\frac{2}{3}(4 + x)\sqrt{2 - x} + C$.

5. $2(4 - x)/\sqrt{2 - x} + C$.   **7.** $-0.8(10 + x)(5 - 2x)^{\frac{1}{4}} + C$.

9. $\frac{1}{15}(3x^2 - 4)(x^2 + 2)^{\frac{3}{2}} + C$.   **11.** $\sqrt{9 - x^2} - 3 \tanh^{-1} (\sqrt{9 - x^2}/3) + C$.

13. $\frac{1}{3}(x^2 - 32)\sqrt{x^2 - 8} + 8\sqrt{8} \arctan \sqrt{(x^2 - 8)/8} + C$.   **15.** 15.73.

17. $2(\sqrt{3} - \pi/3)$.   **19.** 0.62.   **21.** 1.67.

Page 257, Sec. 13.5

1. $\frac{1}{2} \tan^2 \theta + C$.   **3.** $\frac{1}{5} \cosh^5 x + C$.   **5.** $\theta/8 - (\sin 8\theta)/64 + C$.

7. $\frac{1}{3}(3 - \sin^2 x) \sin x + C$.   **9.** $\frac{1}{8}[5x/2 - 2 \sin 2x + \frac{3}{8} \sin 4x + \frac{1}{6} \sin^3 2x] + C$.

11. $-\frac{1}{3} \cos^3 x + \frac{2}{5} \cos^5 x - \frac{1}{7} \cos^7 x + C$.

13. $(12x - 3 \sin 4x - 4 \sin^3 2x)/192 + C$.   **15.** $\frac{1}{3} \tan^3 \theta - \tan \theta + \theta + C$.

17. $\frac{1}{3}\tan^3 \theta + \frac{1}{5} \tan^5 \theta + C$.   **19.** $\frac{1}{2} \tan^2 x + \ln \cos x + C$.

21. $\tan x + \frac{2}{3} \tan^3 x + \frac{1}{5} \tan^5 x + C$.   **23.** $-\frac{1}{8} \cos 4x + C$.

25. $\frac{1}{4}x - \frac{1}{32} \sin 4x - \frac{1}{24} \sin 6x + \frac{1}{40} \sin 10x - (1/128) \sin 16x + C$.

27. $\frac{1}{2}\theta - \frac{1}{4} \cos \left(2\theta + \dfrac{\pi}{6}\right) + C$.   **29.** $\frac{1}{3} \sin^3 \theta + C$.   **31.** 0.403.   **33.** $\dfrac{\pi}{2}$.

35. 243,000 joules.

Page 261, Sec. 13.6

1. $4.5 \arcsin (x/3) + 0.5x\sqrt{9 - x^2} + C$.   **3.** $-(9 - x^2)^{3/2}/3 + C$.

5. $\frac{1}{16}[\arctan (x/2) + 2x/(4 + x^2)] + C$.   **7.** $-\frac{1}{15}[(3x^2 + 8)(4 - x^2)^{\frac{3}{2}}] + C$.

9. $-[(2x^2 + 3)(9 - 4x^2)^{\frac{3}{2}}]/40 + C$.   **11.** $[(3x^2 + 2)(x^2 - 1)^{\frac{3}{2}}]/15 + C$.

13. $\ln A(9 - 4x^2)^{-\frac{1}{8}}$.

15. $\frac{1}{6}(2x^2 - x - 9)\sqrt{4 - (x - 1)^2} + 2 \arcsin [(x - 1)/2] + C$.

17. $0.5[(x - 3)\sqrt{x^2 - 6x} - 9 \cosh^{-1} (\frac{1}{3}x - 1) + C$.

19. $\sqrt{x + x^2} - \sinh^{-1} \sqrt{x} + C$.   **21** 1.   **23.** 21.5.   **25.** πab.

27. $4\pi In/\sqrt{4r^2 + k^2}$.

Page 266, Sec. 13.7

1. $x^3/3 + C$.   **3.** $\frac{1}{2} \sin^2 x + C$.   **5.** $x^3(3 \ln x - 1)/9 + C$.

7. $x \sin x + \cos x + C$.   **9.** $\frac{1}{4}[2(x^2 - 1) \ln (1 + x) - x^2 + 2x] + C$.

11. 0.285.   **13.** 1.296.   **15.** $\frac{1}{9}[3x^3 \ln (1 + x^2) - 2x^3 + 6x - 6 \arctan x] + C$.

17. $\frac{1}{8}e^{4x} - \frac{1}{2}x + C$.   **19.** $\frac{1}{4} \cosh 2x + C$.   **21.** $x[(\ln x)^2 - 2 \ln x + 2] + C$.

23. $(2x - 1) \sin x + 2 \cos x + C.$

25. $-\frac{2}{9}[3(1 - x)^{3/2} \arc\sin \sqrt{x} - 3\sqrt{x} + x\sqrt{x}] + C.$ **27.** $\frac{1}{2} \tan^2 x + C.$

29. $0.002e^{-50t}(2 \sin 100t - 5 - \cos 100t) + C.$ **35.** $\frac{1}{2}x - \frac{1}{12} \sin 6x + C.$

37. $\frac{1}{32}[\sin 4x(2 \cos^2 2x + 3) + 12x] + C.$

39. $q = [4 - e^{-50t}(4 \cos 20t + 10 \sin 20t)]/290.$

41. $q = [2e^{-50t}(10 \sin 500t - \cos 500t) - 101 \cos 200t + 103]/5050.$

Page 269, Miscellaneous Exercises

1. $\frac{1}{2}(\arc\tan x)^2 + C.$ **3.** $\ln [A(x^2 - 4)^{1/8}/x^{1/4}].$

5. $-(6x^2 + 1)(1 - 4x^2)^{3/2}/120 + C.$ **7.** $0.275.$ **9.** $0.05.$

11. $\frac{1}{3}(x^3 + 3x) \ln x - \frac{1}{9}x^3 - x + C.$ **13.** $\frac{1}{2}x^2 - 9x + 84 \ln (x + 9) + C.$

15. $\frac{1}{5}(3 + 2x)^{5/2} + C.$ **17.** $0.333.$ **19.** $x/4\sqrt{4 + x^2} + C.$ **21.** $0.044.$

23. $0.193.$ **25.** $1.0\dot{3}.$

Page 276, Sec. 14.2

1. (a) $\partial z/\partial x = 6x$, (b) $\partial z/\partial y = -4y$, (c) 12, (d) -4, (c) is slope in zx plane, and (d) is slope in zy plane.

3. (a) $2xy + e^{x-y} - y \cos xy$, (b) $x^2 - e^{x-y} - x \cos xy.$ **7.** (b) $-Z \sin \phi$; in (a), X is constant and in (b), Z is constant.   **9.** $\partial z/\partial x = y(1 + \ln x - \ln y)$; $\partial z/\partial y = x(\ln x - \ln y - 1).$ **11.** $\partial z/\partial x = 2 \ln y/[x(\ln xy)^2]$; $\partial z/\partial y = -2 \ln x/[y(\ln xy)^2].$ **13.** $\partial z/\partial x = -y \cos xy(1 + \cosec^2 xy)$; $\partial z/\partial y = -x \cos xy(1 + \cosec^2 xy).$ **15.** $\partial Z/\partial R = R/Z; \partial Z/\partial C = -1/4\pi^2 f^2 C^3 Z.$ **17.** $-29.6.$ **19.** (a) $\partial z/\partial x = yz/x$, (b) $\partial z/\partial y = z(1 + \ln xy).$

Page 279, Sec. 14.3

1. $\partial^2 z/\partial x^2 = -1/(x + y)^2 - 4e^{-2x} \sin 3y$; $\partial^2 z/\partial y^2 = -1/(x + y)^2 + 9e^{-2x} \sin 3y$; $\partial^2 z/(\partial y \, \partial x) = \partial^2 z/(\partial x \, \partial y) = -1/(x + y)^2 + 6e^{-2x} \cos 3y.$ **3.** (a) $z + 4x = 3$, (b) 16.5, (c) $-2.$ **7.** $\partial^2 z/\partial u^2 = \partial^2 z/\partial v^2 = \partial^2 z/\partial w^2 = -1/(u + v + w)^2$; $\partial^2 z/(\partial v \, \partial u) = \partial^2 z/(\partial u \, \partial v) = w - 1/(u + v + w)^2$; $\partial^2 z/(\partial w \, \partial u) = \partial^2 z/(\partial u \, \partial w) = v - 1/(u + v + w)^2$; $\partial^2 z/(\partial w \, \partial v) = \partial^2 z/(\partial v \, \partial w) = u - 1/(u + v + w)^2.$

9. (a) $\partial^2 z/\partial x^2 = 4y^3/(x - y)^3$; $\partial^2 z/\partial y^2 = 4x^3/(x - y)^3$; $\partial^2 z/(\partial y \, \partial x) = \partial^2 z/(\partial x \, \partial y) = (x^3 - 3x^2y - 3xy^2 + y^3)/(x - y)^3$, (b) where $x = 0$ and $y = 0.$

Page 283, Sec. 14.4

1. $dz = y \, dx + x \, dy.$ **3.** (a) $z = 270$, (b) 11.52, (c) 11.4.   **5.** Increasing at 0.031 ohm per min.   **7.** 1% increase.   **9.** (a) 0.023 ohm increase, (b) 3.5% decrease, (c) 4% decrease.   **11.** 7% too high.   **13.** $\pm 1.7.$

Page 289, Sec. 14.6

1. No maximum or minimum.   **3.** No maximum or minimum.   **5.** 4, 4, and 4.   **7.** $\sqrt{2}.$ **9.** Two points dividing the length into three equal parts.   **11.** Length $=$ width $= = 2.29$ ft; height $= 3.43$ ft.   **13.** (a) v_p varies linearly with v_g, (b) inverse variation.   **15.** Each side $= 5$ in., bent upwards at 60°.   **17.** $y = 3x + 35.$

Page 294, Sec. 15.1

1. $1 - x + x^2 - x^3 \cdots$. **3.** (a) 1.01, (b) 0.975, (c) 0.92, (d) 5.2.

9. $\dfrac{E}{X}(1 - R^2/2X^2)$. **11.** 2.080. **13.** 0.4518. **15.** 1.105. **17.** 0.7869. **19.** Et/R.

Page 301, Sec. 15.3

3. 0.2079. **5.** 0.2231. **7.** 0.2526. **9.** -0.2013. **11.** 0.0677.
13. $-2x + 8x^3/3! - 32x^5/5! \cdots$. **15.** $2(x + x^3/3 + x^5/5 \cdots)$.
17. $-1 - 5x - x^2/2 - 11x^3/3 - 13x^4/4 - 35x^5/5 \cdots$.
19. $2x^2/2! - 8x^4/4! + 32x^6/6! \cdots$.
21. $q \simeq Et/R$. **23.** $x \simeq \frac{1}{3}$. **25.** $x = 0$ or $x \simeq 0.2$.
27. (a) $y = 2 - 2x + x^2 - \frac{2}{3}x^3 \cdots$, (b) close to $x = 0$, (c) 1.55.

Page 304, Sec. 15.4

1. $x + x^2 + x^3/3 \cdots$. **3.** $1 - x^4/3! + 2x^8/7! \cdots$.
5. $10t - 50t^2 - 250t^3/3! \cdots$. **7.** $x + 5x^3/3! + 101x^5/5! \cdots$.
9. $x - x^3/3 + 2x^5/15 \cdots$. **11.** $1 - x^2/3 + x^4/18 \cdots$. **13.** 0.794.
15. $1 + x^2 + 2x^4/3 + 17x^6/45 \cdots$. **17.** $x + x^3/3 + x^5/5 + x^7/7 \cdots$.
19. $2(x^2/2 - x^3/2 + 11x^4/24 - 5x^5/12 \cdots)$. **21.** 0.588. **23.** 0.842.

Page 306, Sec. 15.5

1. 0.3818 **3.** 0.008124. **5.** 0.06185. **7.** 0.2214. **9.** 0.1988. **11.** 0.132.
13. 3.90.

Page 311, Sec. 15.6

1. 0.6691. **3.** 2.197. **5.** 4.457. **7.** 0.7953 radian. **9.** 0.5469 radian.
11. -0.244. **13.** $x \simeq 1.85$. **15.** $x \simeq 2.96$. **17.** $x = 1$ or $x \simeq -0.49$. **19.** 1.84
or 4.54. **21.** 0.80 and an infinite number of higher values.

Page 314, Sec. 15.7

1. (a) $1\underline{/60°}$; $0.5 + j0.866$, (b) $2\underline{/0}$, $2 + j0$, (c) $1\underline{/-45°}$; $0.707 - j0.707$,
(d) $3\underline{/57.3°}$, $1.62 + j2.52$. **11.** $e^{-\pi}$. **13.** $\pm \pi$. **15.** -0.5. **17.** -1. **19.** e^{-10t}.

21. $-4 \sin \dfrac{\pi}{4} = -2.83$. **23.** $2e^{-4t}(2 \cos 2t + 3 \sin 2t)$.

25. $-e^{-5t}(3 \cos 3t + 5 \sin 3t)/34 + C$.

Page 319, Sec. 16.3

1. (a) 2, (b) 1, (c) 3. **3.** $y = -0.5 \cos 10t + A$.
5. $y = 3 \ln x + 0.5 \sin 4x + A$. **7.** $y = -0.5e^{-x^2-5} + A$.
9. $y = \arcsin(2x/\sqrt{6}) + A$. **11.** $y = \ln[A(x + 2)/(x + 4)^2]$.
13. $y = [\arctan(t/\sqrt{3})]/2\sqrt{3} - 2$. **15.** $y = (6 \cos^3 x - 3 \cos^2 x + 1)/3 \cos^3 x$.
17. $y = A(\cos 20t - j \sin 20t)$. **19.** $y = Ae^{3x/8}$. **21.** $y = 3 + Ae^{-5x}$.
23. $y = 2e^{5x}$. **25.** $i = (E/R)(1 - e^{-Rt/L})$.

Page 325, Sec. 16.4

1. (a) 6, (b) 0, (c) $-3/7$, (d) 60, (e) no steady state occurs, (f) E/R.
3. $y = -5 + Ae^{3x}$. **5.** $q = CE + Ae^{-t/RC}$. **7.** $y = 2x + 1 + Ae^{2x}$.
9. $y = -\frac{3}{4}x + \frac{3}{8} + Ae^{-2x}$. **11.** $y = -0.4e^{-10x} + Ae^{-5x}$.
13. $y = -0.6e^{0.5t} + Ae^{1.75t}$. **15.** $i = 0.5(\cos 200t + \sin 200t) + Ae^{-200t}$.
17. $v = 10^{-3}(7\cos 1000t + 6\sin 1000t)/17 + Ae^{-4000t}$.
19. (a) $dT/dt = -k(T - 70)$, (b) $T = 70 + Ae^{-kt}$, (c) $T = 70 + 230e^{-0.030t}$,
(d) $108°$. **21.** $i = (E/R)(1 - e^{-Rt/L})$. **23.** $v = -0.1e^{-20t} + 1.1e^{-2t}$.
25. $q = 0.05(\sin 500t - \cos 500t + e^{-500t})$.

Page 328, Sec. 16.5

1. $i = (E/R)(1 - e^{-Rt/L})$.
3. $v = A(R\omega L \sin \omega t - \omega^2 L^2 \cos \omega t - R^2 e^{-Rt/L})/(R^2 + \omega^2 L^2)$.
5. $v = A(\cos \omega t + RC\omega \sin \omega t - e^{-t/RC})/(1 + R^2 C^2 \omega^2)$.

Page 332, Sec. 16.6

1. (a) $2y\,dy = e^x\,dx$, (b) $2y^{-1}\,dy = 5\,dt$, (c) $2\,dy/(5y + 8) = dt$, (d) impossible, (e) $2\,dy = (8t + 5)\,dt$, (f) $y^{-\frac{1}{2}}\,dy = (x - 3)^{\frac{1}{2}}\,dx$,
(g) $dy/\sqrt{1 - 4y^2} = \sqrt{x}\,dx$, (h) impossible, (i) $3y^{-1}e^{y^2}\,dy = 2e^{-x}\,dx$. **3.** $y = Ax$.
5. $y = Ae^{-3x}$. **7.** $y = 3 + Ae^{-x}$. **9.** $y = 5/3 + Ae^{-3x}$. **11.** $i = \sin(\theta + A)$.
13. $y = \sqrt{Ae^x - 4}$. **15.** $y = \sqrt{4 - Ae^{-2x}}$. **17.** $y = \frac{1}{2}\arcsin(4\tan 0.5x + A)$.
19. $y = \frac{1}{3}\ln(A - 3xe^{-x} - 3e^{-x})$. **21.** $y = 2 + A(2 + x)/(2 - x)$.
23. $y = \sqrt{4\arcsin 0.5x + x\sqrt{4 - x^2} + A}$.
25. $\ln(Ax/y) = 1/x + 1/y$; y cannot be expressed as a function of x.
27. $i = 2(1 - e^{-1000t})$. **29.** $v = 2\sin(\theta + \pi/3)$. **31.** $y = 3\sinh(x - 1)$.
33. $y = -\sin^2 2\theta$. **35.** $y = -\frac{1}{9}(x^3 + 8)$. **37.** $y = 3 - 3\sin(1 - t^2)$.

Page 339, Sec. 16.8

1. (a) Exact, (b) not exact, (c) exact, (d) exact, (e) exact, (f) not exact, (g) not exact, (h) not exact. **3.** (a) $[F(y)f'(x) + \phi'(x)]\,dx + f(x)F'(y)\,dy$, (b) each side equals $f'(x)F'(y)$. **5.** $y = x^2(\frac{1}{5}\sin 5x + A)$. **7.** $y = 0.25(x^3 - 2x + A/x)$.
9. $y = e^{-3x}(\sin x - x\cos x + A)$. **11.** $y = t\ln A\sqrt{1 + t^2}$.
13. $y = A/\sqrt{1 + \ln x}$. **15.** $y = [(x^2 + A)/\arctan x - 1]^{\frac{1}{2}}$.
17. $y = (5\sin 5t + 2\cos 5t)/29 + Ae^{-2t}$. **19.** $y = (2 + x)/x^3$.
21. $y = 0.5(1 - e^{-2.5x})$. **23.** $y = 0.5(\sin x - \cos x + e^{-x})$.
25. $i = \sin 100t - \cos 100t + e^{-100t}$. **27.** $y = \sqrt{(2x^2 + 6x + 19)/(x + 2)}$.
29. $y = \cos x(2e^{-2x} + 1)$. **31.** $y = \sqrt{2}\,e^{-x}\sqrt{x + 2}$.

Page 342, Sec. 16.9

1. (a) Nonhomogeneous, (b) nonhomogeneous, (c) homogeneous, (d) homogeneous, (e) homogeneous. **3.** $y = x(\ln x + A)$. **5.** $y = x\ln Ay$.
7. $y = x(\ln Ax)^2$. **9.** $y = \sqrt{x^2 - x}$. **11.** $y = \frac{1}{2}(x^2 - 1)$.

13. $x\sin\dfrac{y}{x} - y\cos\dfrac{y}{x} = x\ln x$.

Page 345, Miscellaneous Exercises, Sec. 16.10

1. $y = 2x(1 + \ln x)$.   **3.** $y = 2 \sin [\arc \sin 0.5x - 2\sqrt{4 - x^2}]$.
5. $y = A \cos x$.   **7.** $y = 2/(2 - \sqrt{x^2 - 3})$.   **9.** $i = \frac{10}{3}(e^{-200t} - e^{-500t})$.
11. $i = A\omega C(\cos \omega t + \omega RC \sin \omega t - e^{-t/RC})/(1 + R^2\omega^2C^2)$.
13. $y = [A - x - 3 \ln (3 - x)]/(3 + x)$.   **15.** 75.6°.   **17.** $s = 96$ ft; $v = 64$
ft/sec. Distance $= 2/3$ distance of free fall.   **19.** 17.1 min.   **21.** 1.3 sec after opening.
23. 56 lb.   **25.** (a) 3.25 sec, (b) 121 ft/sec, (c) 300 ft/sec at $t = \infty$.

Page 350, Sec. 17.1

1. (a) $y = At + B$, (b) $y = \frac{1}{2}kt^2 + At + B$, (c) $y = 2e^x - \frac{1}{4}x^4 + Ax + B$,
(d) $y = \frac{8}{15}t^{5/2} + At^2 + Bt + C$, (e) $y = (1 + 2x)^{5/2}/15 + Ax + B$,
(f) $y = (1 - 2x)^{3/2}/6 + Ax + B$, (g) $y = -\ln (1 + x) + Ax + B$.
3. $y = 3 \cos 2x$.   **5.** $y = \arc \sin x - 3x$.   **7.** $y = -8 \ln (1 - 0.5x) - x$.
9. $y = 0.5(1 + 2x) \ln (1 + 2x) - x + 1$.   **11.** $i = \cos 6t + \sin 6t$.
13. $v = 125t^3/6 - 4t$.   **15.** $y = 0.5(1 - \cos 4x + \sin 4x)$.
17. $y = 3.13e^{3.16x} - 1.93e^{-3.16x} - 1.2$.   **19.** $q = 1.2 \times 10^{-2}(1 - \cos 250t)$.
21. $q = \frac{1}{3}(0.0003 \sin 100t - 0.0001 \sin 300t)$.   **23.** $y = -\frac{5}{2}x - \frac{1}{4} \cos 4x + 2e^{-2x}$.
25. $q = 1.2 \times 10^{-3} (1 - \cos 500t)$.

Page 359, Sec. 17.2

1. (a) $m^2 - 5m + 4 = 0$, (b) $Lm^2 + Rm + 1/C = 0$, (c) $Lm^2 = -1/C$,
(d) $Lm^2 + Rm = 0$, (e) $m = -50$, (f) $m^3 + 4m^2 - 3 = 0$.   **3.** $y = A + Be^{25x}$.
5. $y = A + Be^{3x} + Ce^{-3x}$.   **7.** $y = Ae^{2.5x} + Be^{-2x}$.   **9.** $i = (At + B)e^{-100t}$.
11. $y = e^{-2t}(A \cos 2t + B \sin 2t)$.   **13.** $y = e^{1.5x}(A \cos 1.12x + B \sin 1.12x)$.
15. $i = (At + B)e^{-t/\sqrt{LC}}$.   **17.** $y = 2e^{40x} - 2e^{20x}$.
19. $y = 2e^{-7.07x} - 2e^{7.07x}$.   **21.** $y = 0$.   **23.** $i = 0.8e^{-1000t} \sin 1000t$.
25. $s = 0.5e^{-0.25t} \cos 2.25t$.   **27.** $i = E\sqrt{C/L} \sin (t/\sqrt{LC})$.   **29.** $i = Ete^{-t/\sqrt{LC}}/L$.
31. $i = 2Ee^{-Rt/2L} \sin (\sqrt{4L/C - R^2}/2L)/\sqrt{4L/C - R^2}$.

Page 363, Sec. 17.3

1. (a) $y = 1.25$, (b) $q = CE$, (c) $y = -0.5e^{-x}$, (d) $y = \sin t$.
3. $q = 1.25 \times 10^{-5} + Ae^{-4000t} + Be^{-16000t}$.
5. $q = 4 \times 10^{-5}e^{-5000t} + (At + B)e^{-10000t}$.
7. $y = -2.4e^{-5x} + Ae^{-1.55x} + Be^{-6.45x}$.   **9.** (a) $2L$, $\frac{1}{2}C$, $2E$, where L, C, and E
are original values, (b) frequency is doubled.
11. $y = [15 - e^{-5x}(66 \cos 3x + 76 \sin 3x)]/51$.   **13.** $y = 2x - \frac{4}{3} + \frac{7}{3}e^{-3x}$.
15. $i = 0.2 \cos 1000t + 1.6 \sin 1000t - e^{-1000t}(0.2 \cos 500t + 3.6 \sin 500t)$.
17. $y = -5xe^{2x} + e^{2x} + e^{3x}$.   **19.** $y = -2 + 2e^x - (x + 2)e^{-x}$.

Page 368, Sec. 17.4

1. (a) $R = 4$, (b) $R = 2\sqrt{10}$, (c) $R = 0.84$.   **3.** (a) $10^3/18\pi$ cps, (b) $1.2/\pi C$ cps
or $0.075/\pi L$ cps, (c) 702 cps, (d) 118 cps.
5. (a) $q = 2.5 \times 10^{-3}e^{-500t} - 6.25 \times 10^{-4}e^{-2000t}$, (b) $i = 1.25(e^{-2000t} - e^{-500t})$.
7. (a) $q = 10^{-3}e^{-120t}(6 \cos 160t + 4.5 \sin 160t)$, (b) $i = -1.5e^{-120t} \sin 160t$.

9. (a) $i = e^{-200t} \sin 400t$, (b) $q = 2 \times 10^{-3} - 10^{-3}e^{-200t}(2 \cos 400t + \sin 400t)$.
11. (a) $i = 2 \cos 200t + \frac{2}{3}(e^{-100t} - 4e^{-400t})$,
(b) $q = 0.01 \sin 200t + (e^{-400t} - e^{-100t})/150$.
13. (a) $i = 2 \sin 5000t - 5 \cos 5000t + e^{-500t}(5 \cos 3500t - 2.14 \sin 3500t)$,
(b) $q = -4 \times 10^{-4} \cos 5000t - 10^{-3} \sin 5000t + e^{-500t}(4 \times 10^{-4} \cos 3500t + 1.48 \times 10^{-3} \sin 3500t)$.
15. $i = 3 \cos 200t + 2 \sin 200t - e^{-100t}(3 \cos 300t + \frac{7}{3} \sin 300t)$.
17. $i = 2.5 - e^{-333t}(2.5 \cos 471t + 1.77 \sin 471t)$.

Page 373, Sec. 17.5

1. $y = Ae^{-x} + B$.   **3.** $y = \pm \cos (x + A) + B$ or $\pm \sin (x + A) + B$.
5. $y = \ln \sec (x + A) + B$.   **7.** $y = \frac{1}{2}e^x + Ae^{-x} + B$.   **9.** $y = e^{2x}$.
11. $y = x^2 - x + Ae^{-2x} + B$.   **13.** $y = (3x + 1)/(x + 1)$.
15. $y = x^3 - \frac{3}{2}x^2 + \frac{1}{4}$.   **17.** $y = A/(B - x)$.   **19.** $x = y(\ln y - 1)$.
21. $y = e^{x^2/4}$.

Page 374, Miscellaneous Exercises

1. $y = 4x^2 + Ax + B$.   **3.** $y = Ae^{1.46x} + Be^{-5.46x}$.
5. $y = 1 + e^{-2x}(A \cos 2x + B \sin 2x)$.   **7.** $y = 1/(1 + x)$.
9. $y = \frac{1}{2}x^2 - \frac{1}{4}x + Ae^{-4x} + B$.   **11.** $y = \frac{1}{3}e^{3x} + A \cos 2x + B \sin 2x$.
13. $x = \tan y$.   **15.** $i = 0.1 \sin 1000t - 0.125e^{-600t} \sin 800t$.
17. (a) 236 ft per sec, (b) 1310 ft.   **19.** 0.9°.   **21.** (a) When it is completely unsubmerged or completely submerged, (b) 32 ft per sec per sec downwards or upwards, (c) $s = \frac{1}{2}h[1 - \cos (8t/\sqrt{h})]$.

Page 379, Sec. 18.1

1. (a) $\bar{y} = 3/(s + 8)$, (b) $y = 3e^{-8t}$, (c) $\bar{y} = 1/(s + 8)$.   **3.** $5\omega/(s^2 + \omega^2)$.
5. $-LE/(Ls + R)$.   **7.** $CV(Ls + R)/(Ls^2 + Rs + 1/C)$.

Page 383, Sec. 18.2

3. $800/s(s + 300)$.   **5.** $(2s^5 + 24)/s^5(s - 20)$.
7. $(3\omega + 4s)/(s^2 + \omega^2)(Ls^2 + Rs + 1/C)$.

Page 386, Sec. 18.3

1. $y = \frac{1}{4}t^2$.   **3.** $i = \frac{1}{2}(\cos \sqrt{5}\, t + 1.2\sqrt{5} \sin \sqrt{5}\, t)$.
5. $v = 2 - 2 \cos \sqrt{2}\, t + 2.12 \sin \sqrt{2}\, t$.   **7.** $y = 5/7 - 4e^{-t}/5 + 3e^{-3.5t}/35$.
9. $i = e^{-3t}(2 \cos 4t - 1.5 \sin 4t)$.   **11.** $v = 4e^{-2t} - 3e^{-t} + te^{-t}$.
13. $y = (39t - 12 + 12e^{-2t} \cos 3t - 5e^{-2t} \sin 3t)/169$.
15. $y = 2(2 \sin 5t - \sin 10t)/375$.   **17.** $q = 2(\cos 5t - \cos 10t)/75$.

Page 388, Sec. 18.4

1. (a), (b), and (c) cannot be solved directly, (d) can be solved by Laplace transforms.   **3.** $i = Ee^{-Rt/L}/R$.   **5.** $y = (31e^{10x} - 3e^{-4x})/14$.
7. $i = -e^{-150t} + \cos 50t + 2 \sin 50t$.   **9.** $y = 2.2e^{-5x} - 3xe^{-20x} - 0.2e^{-20x}$.

11. $y = 0$.   **13.** $q = (1 + 2e^{-30t} - 3e^{-20t})/30$.   **15.** $i = 50e^{-40t} \sin 20t$.

17. $\theta = 0.05e^{-\pi t}(2 \cos 2\pi t + \sin 2\pi t)$.

19. $i = (e^{-250t} - 13e^{-1000t} + 12 \cos 2000t + 18 \sin 2000t)/65$.

21. $v = 0.004(2 - 2e^{-100t} \cos 200t - e^{-100t} \sin 200t)$.

Page 400, Sec. 19.3

1. No. It has an infinite discontinuity.

3. $v = (12/\pi)(\cos \theta - \frac{1}{3} \cos 3\theta + \frac{1}{5} \cos 5\theta \cdots)$.

5. $i = 1 + (4/\pi)(\cos \theta - \frac{1}{3} \cos 3\theta + \frac{1}{5} \cos 5\theta \cdots)$.

7. $i = 0.75A + (A/\pi)(-\cos \theta + \frac{1}{3} \cos 3\theta - \frac{1}{5} \cos 5\theta \cdots) +$
$(A/\pi)(\sin \theta + \sin 2\theta + \frac{1}{3} \sin 3\theta + \frac{1}{5} \sin 5\theta + \frac{1}{3} \sin 6\theta \cdots)$.

9. $v = 0.75 + \dfrac{1}{\pi}(\cos \theta - \frac{1}{3} \cos 3\theta + \frac{1}{5} \cos 5\theta + \cdots + 5 \sin \theta - \sin 2\theta +$
$\frac{5}{3} \sin 3\theta + \frac{5}{5} \sin 5\theta - \frac{1}{3} \sin 6\theta \cdots)$.

Page 405, Sec. 19.4

1. $i = \dfrac{4}{\pi}(\sin \theta - \frac{1}{2} \sin 2\theta + \frac{1}{3} \sin 3\theta \cdots)$.

3. $v = \pi + 2(\sin \theta + \frac{1}{2} \sin 2\theta + \frac{1}{3} \sin 3\theta \cdots)$.

5. $i = 3 - \dfrac{8}{\pi^2}\left(\cos \theta + \dfrac{1}{3^2} \cos 3\theta + \dfrac{1}{5^2} \cos 5\theta \cdots\right)$
$- \dfrac{4}{\pi}(\sin \theta + \frac{1}{2} \sin 2\theta + \frac{1}{3} \sin 3\theta \cdots)$.

7. $i = \dfrac{\pi^2}{6} - 2\left(\cos \theta - \dfrac{1}{2^2} \cos 2\theta + \dfrac{1}{3^2} \cos 3\theta \cdots\right)$.   **9.** $a_0 = 7/3$; $a_m = -6/m^2\pi^2$
if m is odd; $a_m = -2/m^2\pi^2$ if m is even; $b_m = (-3m^2\pi^2 + 4)/m^3\pi^3$ if m is odd;
$b_m = -3/m\pi$ if m is even.

Page 408, Sec. 19.5

1. $i = 1.91 + 3 \cos \theta +$
$\dfrac{12}{\pi}[(\cos 2\theta)/3 \times 1 - (\cos 4\theta)/5 \times 3 + (\cos 6\theta)/7 \times 5 \cdots]$.

3. $v = 2/\pi - \dfrac{4}{\pi}[(\cos 2\theta)/3 \times 1 + (\cos 4\theta)/5 \times 3 + (\cos 6\theta)/7 \times 5 \cdots]$.

5. $a_0 = e$; $a_m = -2(e + 1)/(m^2\pi^2 + 1)$ if m is odd, or
$\qquad = 2(e - 1)/(m^2\pi^2 + 1)$ if m is even.
$\qquad b_m = 2\pi m(e + 1)/(m^2\pi^2 + 1) - 4/m\pi$ if m is odd, or
$\qquad = -2\pi m(e - 1)/(m^2\pi^2 + 1)$ if m is even.

Index